A Guide to Insecticide Toxicology

NIPA® GENX ELECTRONIC RESOURCES & SOLUTIONS P. LTD.
New Delhi-110 034

About the Authors

D.S. Reddy - Author of book "Applied Entomology"

Dr. L.R. Chowdary, presently working as Scientist (Entomology) in AICRP (Cotton) at Regional Agricultural Research Station, Lam, Guntur, ANGRAU. He completed his B.Sc. (Agriculture) from College of Agriculture, Dharwad, (UAS, Dharwad) (ICAR-AIEEA-UG) during 2008, M.Sc. (Agril. Entomology) and Ph.D. (Entomology) (Gold Medal) for securing highest O.G.P.A from UAS, Raichur. As a researcher he worked on the aspects of ecology, pest management and toxicology. He has to his credit with several awards from the professional societies and also received best oral and poster presentations in conferences, symposiums and conferences. He has published 78 research articles in reputed international and national journals, 34 abstracts in international/national seminars, conferences and symposium, 4 review articles, 15 book chapters, 03 books, 05 extension bulletins and 42 popular articles to his credit.

Dr. Thammali Hemadri, Presently working as Senior Research Fellow (Entomology) in Pesticide Residue and Food Quality Analysis Laboratory (NABL accredited), University of Agricultural Sciences, Raichur, Karnataka. He completed his B.Sc. (Hons.) Horticulture from College of Horticulture, Anantharajupeta, Dr. YSR Horticultural University, A.P.; M.Sc (Entomology) COA, Mandya, UAS, Bangalore, Karnataka and Ph.D. (Entomology) from COA, Raichur, UAS, Raichur, Karnataka. During M.Sc and Ph.D he got gold medals for securing highest OGPA. He is expertise in insecticide toxicology, pest management, pesticide formulation analysis and pesticide residue analysis.

A Guide to Insecticide Toxicology

D.S. Reddy
Author of book "Applied Entomology"

L.R. Chowdary
Scientist (Entomology)
AICRP (Cotton) at Regional Agricultural Research
Station, Lam, Guntur, Acharya N.G. Ranga Agricultural University

Thammali Hemadri
Senior Research Fellow (Entomology)
Pesticide Residue and Food Quality Analysis Laboratory (NABL Accredited)
University of Agricultural Sciences

NIPA® GENX ELECTRONIC RESOURCES & SOLUTIONS P. LTD.
New Delhi-110 034

NIPA GENX ELECTRONIC RESOURCES & SOLUTIONS P. LTD.
101,103, Vikas Surya Plaza, CU Block
L.S.C. Market, Pitam Pura, New Delhi-110 034
Ph : +91-11-43860225, Mob.: +91 9717133558, 9540816132
E-mail: newindiapublishingagency@gmail.com
Website: www.nipaersources.com

Print ISBN: 978-93-58877-70-0

ebook ISBN: 978-93-58876-96-3

Composed and Designed by NIPA®.

All diagrams in this book were drawn by professionals, and credit goes to Dr. Saraswati Mahato, Ph.D (Agril. Ento), PRFQAL, UAS, Raichur; Vinay, M.Sc. (Agril. Economics), UAS, Raichur; Krishna R, M.Sc. (Entomology), UAS, Raichur; and Lakshmikant, M.Sc. (Entomology), UAS, Raichur; B. Havila, Polytechnic of Agriculture, Darsi, ANGRAU.

Preface

Ten to thirty per cent of the nation's crop yield is lost to pests. In order to maintain good yield output and limit the population of pests below the economic threshold, pesticides are essential. India consumes 381 g.a.i of pesticide per hectare, far less than the global average of 500 g.a.i per hectare.

Throughout history, the application of pesticides, including insecticides, has emerged as a crucial and mandated aspect of agriculture to ensure crop productivity. The challenge of attaining long-term development without harming the environment has never been higher, given the continuously growing population and decreasing environmental conditions. Many insecticide classes, such as neonicotinoids, pyrethroids, carbamates, and organophosphates, have been produced over time; each has a distinct mechanism of action, physiological target, and level of efficacy. With the discovery of highly selective pesticides that target processes other than brain function and the identification of new and targeted target locations, the arsenal of pest management techniques has grown to include more treatments with distinct and targeted modes of action.

A wide range of stakeholders are involved in the toxicology of pesticides, including consumers, agriculturalists and scientists. The emphasis in this textbook for researchers studying insecticide toxicology as well as undergraduate and graduate students is on the fundamental ideas and experimental methods that form the basis of the field's progress and future directions. Nonetheless, the various stakeholder groups influence how pesticide toxicity is perceived, researched, and, where practical, integrated from several points of view. The first section of the book provides an overview of the necessity for pesticides, their usage patterns, and the significance of pest insects for agricultural productivity as well as serving as a foundational text for an introduction to insecticide toxicology.

The book includes descriptions and explanations of the basic introduction of evolution of insecticides and their use, Insecticides classification (conventional and novel), banned insecticides, proinsecticides, mode of action, pesticides formulation, pesticide residue analysis, Insects bioassay, Resistance and resurgence in insects, synergism, insecticide poisoning and therapy, impact of pesticides on honey bees and novel technologies developed by agrochemical industries etc.

Authors

Contents

1

Introduction

History of Chemical Control, Insecticide Toxicology and Properties

History of pest control probably started when the first person swatted a mosquito or pulled off a tick, the battle for the control of our planet didn't start until organized agriculture, when pests attacked the plants, and we grew to eat and threatened our very survival.

A pesticide is any substance or mixture of substances used to stop, kill, avoid, or lessen the effects of a pest (insects, mites, nematodes, weeds, rats, etc.). This includes insecticide, herbicide, fungicide, and many other substances used to control pests. Pesticide was defined in different ways at different times and in different places. But pesticide is still generally the same: it is a (mixed) substance that is poisonous and effective on target organisms but safe for non-target organisms and environments.

Chemical control, commonly known as pest control, is the use of chemicals to reduce insect populations. Herbicides, fungicides, insecticides, rodenticides, acaricides, nematodes, and fungicides are all examples of pesticides. Insecticides are a class of chemicals used to combat insect pests. A substance or combination of compounds used to kill, repel, or otherwise prevent insects is called an insecticide. When it comes to controlling pests, insecticides are most powerful tools available for management. They have a wide range of positive effects, including high efficiency, speed of healing, general applicability, adaptability to different settings, adaptability to shifting agronomic and ecological conditions, and low cost. When insect pest numbers approach or surpass the economic threshold, insecticides are the sole viable instrument for emergency action in pest management. The usage of pesticides, a widely used method, can be central to such a system. The use of chemical pesticides will remain pivotal in these initiatives. In many cases, chemical control of pests is the best option available. Some people's view of pesticide use for pest management as an ecological sin is incorrect. Chemical pesticides are a reliable and useful tool for the biologist, so long as they are used in accordance with good ecological principles.

The Indian economy is mostly based on agriculture, and roughly 18 percent of food grains are wasted due to plant infections, pests, weeds, rodents, and other pests, and many pesticides are employed to reduce these losses. Pesticides are agrochemicals that are used to prevent, repel, mitigate, or eradicate pests. Pesticides include insecticides, fungicides, rodenticides, herbicides, and others.

Insecticides are one type of pest control method; they may be chemical or biological in nature and work by either killing the insect or stopping it from engaging in damaging behavior. Natural and synthetic insecticides are both used to kill insects and other pests, and they come in a wide variety of forms and methods of application, such as sprays, baits, slow-release diffusion, dust, and so on. Integrated Pest Management (IPM), the use of bio-pesticides, and other environmentally friendly methods for insect pest control are gaining popularity, and in recent years the bacterial genes coding for insecticidal proteins have been incorporated into various crops to deal with the mortality of the pests feeding on them. As more people become aware of the importance of environmentally responsible farming methods, the use of bio-pesticides and IPM will likely increase.

There are three main phases in the history of pesticides. In the first phase, natural pesticides, like sulphur in ancient Greece, were used to get rid of bugs. In the second phase, from the 1870s to 1945, inorganic synthetic pesticides were used. During this time, most pesticides were made from natural materials or inorganic chemicals. Since 1945, the third phase has been the time of organic synthetic pesticides. Since 1945, man-made organic pesticides like DDT, 2,4-D, and later HCH and dieldrin put an end to the use of natural and artificial pesticides. Since then, most pesticides have been made by people, and they are called chemical pesticides. The use of chemical pesticides, especially those made from organic compounds, is a major sign of human society. These pesticides protect crops and make them easier to grow.

Insecticides may have existed as early as 2500 BC, when the Sumerians utilized sulphur as both an insecticide and an acaricide. The earliest accounts of cultural control, particularly the manipulation of planting dates, date to 1500 BC. Around 1200 BC, plant extracts were used to cure grain storage in China, and arsenic sulphide was also used to treat human lice. Chalk and wood ash were employed to keep insects out of enclosed spaces. The employment of sulphur, fumigants, oil sprays, oil and bitumen sticky bands, oil and ash, and other preparations by the ancient Greeks and Romans to control insects is generally acknowledged. Pliny and Elder recommended using arsenic as an insecticide and mentioned using soda and olive oil to treat legumes. In 77 AD, Pliny also mentioned using sulphur. Predatory ants (*Oecophylla smaragdina*) were first observed using bamboo bridges to migrate between trees in citrus groves about 300 A.D. to eradicate pest beetles and caterpillars. Around the 16th century, the use of plant insecticides such as pyrethrum, derris, quassia, and tobacco leaf infusion expanded. While nicotine may have been the first organic insecticide when it was used to naturally control aphids in crushed tobacco leaves as early as 1763, the most effective natural insecticide has been pyrethrum, a mixture of natural esters extracted from Chrysanthemum flowers primarily grown in Kenya.

Arsenic, which was used as ant bait and combined with honey in the middle of the 1600s, was likely the first stomach poison. Dusting of Paris green (a copper salt of arsenic) on plant foliage was employed in late 1810 to suppress the Potato Colorado Beetle, the codling moth, and other leaf-eating insects. Lead arsenate, one of the most efficient inorganic insecticides, was first introduced in 1892 to control pests including the gypsy moth, apple maggot, and numerous soil insects. During this time, other inorganic compounds containing mercury, tin, or copper were also utilized as stomach poisons. The high toxicity of these chemicals to mammals was the one significant limitation in the application of the majority of them. On the advice of C.V. Riley of California, the Vedalia beetle, *Rodolia cardinalis*, was imported into California from Australia for the control of cottony-cushion scale, *Icerya purchasi*, on citrus plantations, and this resulted in the first notable success in biological control in 1888. Sulphur, arsenicals, fluorides, soaps, kerosene, and other botanicals—among which nicotine, rotenone, pyrethrum, sabadilla, and quassia appear to have been the most commonly employed would be included in a list of insecticides used in the 19th century. Additionally, the use of biological control agents had started.

Significant advancements in the synthesis of pesticides were made over the first four decades of the twentieth century. Following the discovery of dichlorophenyl trichloroethane's (DDT's) insecticidal capabilities, chemical insect control underwent a significant transformation. In 1939, Paul Muller of the J.R. Geigy Company made this discovery. During World War II, the Western Allies used it primarily to combat malaria and typhoid. It was at this time that hexachlorocyclohexane (HCH) and the cyclodiene chemicals were first synthesized, ushering in the era of chlorinated hydrocarbon insecticides. These chlorinated hydrocarbons were initially well received, but their stability and hydrophobicity led to environmental contamination and bioconcentration in the bodies of numerous species, prompting restrictions on their usage or outright bans. Even though their application has been restricted or outright prohibited in developed nations, the original generation of insecticides, also known as chlorinated hydrocarbons, are still in use. However, India has just recently outlawed the use of DDT and HCH in agriculture.

After that, scientists all over the world (begun in Germany) began synthesizing organophosphorous insecticides. Gerhard Schrader is credited with developing three of these chemicals, and they saw widespread use: HETP, Parathion, and Schradan. However, hundreds of additional OP insecticides have been synthesized in an effort to reduce mammalian toxicity and boost efficacy, making organophosphorous chemicals the most widely used class of insecticides despite their lack of persistence. In 1940, Swiss researchers predicted the presence of a second class of insecticides called carbamates, but it wasn't until 1950, when the American insecticide carbaryl was introduced, that the class really took off, leading to the synthesis of many more carbamates. All of these substances (organophosphates) are neurotoxins and, more precisely, inhibitors of acetylcholinesterase.

The high expense of using pyrethrum prompted the development of various analogues of the substance in search of a less expensive alternative. Several synthetic pyrethrin analogs were developed between 1949 and the early 1970s; the first synthetic pyrethroid was launched in 1976; and since then, this class of insecticide has become the second most widely used type. At low concentrations, they are effective, and they are not harmful to animals.

Until this point, insecticide development was mostly led by chemorational design, which was gradually replaced by biorational design based on a basic understanding of the physiology and ecology of insects and crops. IGRs include chitin synthesis inhibitors, juvenile hormone mimics, ecdysone agonists, pymetrozine, and other new agents like pheromones, *Bacillus thuringiensis*, avermectins, and formamidines.

The global market for pesticides has expanded phenomenally in spite of intense environmental pressures. New insecticides, herbicides, and fungicides as well as their formulations are being introduced with greater frequency, but with conscious efforts to minimize the risks to humans and the environment. India is the world's second-biggest producer of technical grade pesticides, producing 46,000 metric tons of them annually. Asia's pesticide sector is the largest in the world overall. In terms of value, it utilizes around 3% of all pesticides used worldwide. However, per hectare consumption of pesticides in India is very low compared to other developed countries. The majority of pesticides are insecticides, and the two most significant crop segments that use them most heavily are cotton and rice. With over 50% of all pesticide consumption, cotton excludes a wide range of other crops from the advantages of chemical crop protection. Organophosphates (50%) and pyrethroids (19%), organochlorines (18%), carbamates (4%) and bio-pesticides (1%), respectively, are the top five insecticides sold in India.

Historical Timeline in the development of pesticides

Year	Pesticide development	Remarks
----BC	Red squill was said to have been used to kill rats on early stone tablets.	-
12000BC	Initial observations of insects by humans and their cultures	-
8000BC	The beginning of farming	Cereals provide a main source of food, a way to store food between harvests in the established villages

Year	Pesticide development	Remarks
2500 BC	Mites and insects were controlled with sulphur by ancient Sumerians	-
1200 BC	Biblical armies sowed conquered fields with salt and ashes to make land unproductive	Probably the first herbicide that kills weeds before they grow
---- BC	The ancient Romans used hellebore to control rodents and insects	One of the first poisons made
1000 BC	Homer talks about how sulphur chemicals are used.	-
324 BC	Chinese people use ants to get rid of caterpillars on citrus trees.	Early biocontrol/IPM use
70 AD	Pliny the Elder mentions green lizard gall protecting apples from worms and decay.	Early organic chemical usage
900 AD	Use of Arsenic in control of garden insects in China.	Early pesticides were made from artificial stomach poison.
1300 AD	Marco Polo says that mineral oil can be used to treat camels with mange.	
1300 AD	Pyrethrum is a combination that is said to have come from a secret place. It is said that Marco Polo brought it to Europe.	The biological extract of pyrethrum is still used, and it has given rise to current synthetic pyrethroids.
Several centuries	South American locals used powders made from Sabadilla plants to get rid of lice.	-
1669	The first time arsenic was used to kill insects in the Western world	Honey ant bait
18th century	Petroleum, kerosene, creosote, and turpentine were used as pesticides.	-
As early as 1763	In France aphids were killed with ground tobacco.	
1787	Soap mentioned as insecticide and turpentine emulsion recommended to kill/repel insects	-
1809	France found that nicotine could kill bugs.	-
1825	Michael Faraday created BHC.	However, the insecticidal properties are unknown.
As early as 1848	Rotenone was used as insecticide.	Not widely used until the 1920s, then experiencing explosive growth in the 1930s
1860	As inorganic chemicals gain popularity as pesticides, Paris Green, an arsenical compound, was employed to control the Colorado potato beetle in the Rocky Mountain Region.	-
1867	Unknown inventor discovers that the dye Paris Green killed insects	For insect-chewing organisms

Year	Pesticide development	Remarks
1873	DDT was initially created in a laboratory (Othmar Ziedler).	But insecticidal properties not detected until 1939
1877/78	To destroy sucking insects, kerosene emulsified in soap was designed to kill sucking insects.	Prof. John Cook, Mich. Ag. College.
1882	France discovered Bordeaux mixture to combat plant diseases.	Copper sulphate was the main stay for many years.
1883	John Bean invents the pressure sprayer to apply insecticides, which inspires FMC to start making fire engines.	Key innovation that makes it easier to apply chemicals to crop surfaces
1886	To get rid of scale insects in California, inorganic lime sulphur washes and also hydrogen cyanide spray were used.	One of the first occurrences of insect resistance to a poison was caused by hydrogen cyanide.
1892	In Massachusetts, lead arsenate was discovered to be an effective against gypsy moth.	F.C. Moulton, MA State Bd. of Ag.
1893/1906	Lead arsenate has been discovered to be efficient against a wide range of insects, and the use of homemade preparations expanded.	Widely accepted by home gardeners
1894/1900	The development of steam/mechanical/horse-driven spray apparatus	-
1921/22	First airplane application of insecticides to a field (cotton, Louisiana, 1922)	-
1932	First use of methyl bromide as a fumigant (France)	-
1932/39	The Swiss company Geigy (Dr. Paul Hermann Muller, also known as Pauly Mueller) discovers DDT while searching for insecticides and seed disinfectants.	Compound had extraordinary lethal power and weather resistance when exposed to the elements; Mueller was awarded the Nobel Prize.
1940	France and England have identified BHC's insecticidal properties.	-
1941/42	DDT was used on crops and to control human lice in Switzerland.	DDT is made available to other countries by Geigy.
1942	Liquefied gases are utilized as a propellant in insecticide aerosols.	-
1942/45	DDT was made available for use in the U.S., first by the military and then by civilians and farmers by July 1945. It stopped the spread of typhus in war-torn Europe.	USDA and the War Production Board were in charge of putting the drug on the market.
1940's	Value of D-D mixture as a nematicide identified.	Much less expensive than other compounds, resulting in increased consumption
1949	Captan was the first dicarboximide fungicide to be introduced.	-

Year	Pesticide development	Remarks
1972	*Bacillus thuringiensis* (Berlinger) (Bt), a biological, registered as an insecticide	This opened the door for more Bt applications and biologicals in general.
1979	First registered synthetic pyrethroid insecticides (fenvalerate and permethrin).	Reduced application rates and substitution of older chemicals with regulatory and resistance issues.
1994	Imidacloprid is the first neonicotinoid insecticide to be registered.	Nicotine-based insecticides are extremely promising
1997	Fipronil is registered as a Type fiprole systemic insecticide.	-

Brief history of toxicology

Definitions

The term toxicology is typically defined as "the science of poisons." Toxicology is defined as "the study of the adverse effects of chemicals or physical agents on living organisms" in order to better reflect our understanding of how various agents can harm humans and other organisms.

Toxicology is the study of the harmful effects of chemicals on living creatures, including humans. The term comes from the Greek words "toxicos" (meaning "poisonous") and "logos" (meaning "study"). Multiple meanings can be attributed to it. In its broadest sense, toxicology is "the study of the harmful, deleterious, and/or poisonous effects of chemicals on living organisms," or "the study of the signs and symptoms of poisoning, as well as its causes and treatments, and methods for detecting poisoning."

Toxicology is the medical discipline concerned with poison itself as well as its causes, consequences, and detection. Therefore, it is the study of poisons as a scientific discipline" (Du Bois and Geiling, 1959). Toxicology is described by Fogleman (1963) as "the study of limits of the biological effects of a chemical or mixtures of chemicals." Insect toxicology is further defined by Sun (1968) as "the study of economic poisons, their effects, mechanism of action, and metabolism of toxicant in insects."

Toxicology is the study of the harmful effects of chemicals on living systems, and insecticide is a toxic substance. The toxicity of chemicals depends on the chemical's constitution, the route of exposure (oral, dermal, and inhalation), the dose, and the organism. Insecticide toxicity is typically expressed as LD_{50} or LC_{50}. Its values are measured in milligrams per kilogram of body weight or parts per million.

History of toxicology

According to a common Chinese myth, Shennong, also known as "the divine farmer," brought farming to China around 2696 BC (Wu, 1982). He's also called the "father of Chinese medicine" because he wrote a book called "On Herbal Medical Experiment Poisons." He was known for trying 365 different kinds of herbs, most of which probably killed him in the end.

The Athenian philosopher Socrates was found guilty on two counts, both of which involved Greek goddesses and gods, in the year 399 BC. For spreading radical views among young Athenians, he was condemned to death and put to death by swallowing a beverage containing hemlock, a toxic alkaloid from the plant *Cornium maculatum* (Apiaceae) (Stone 1988).

Later, between the years 50 and 400, AD, Romans frequently utilized poisons in public executions and assassinations.

The Swiss-born Renaissance physician, alchemist, astrologer, and botanist Phillip von Hohenheim, better known as "Paracelsus" (1493-1541), practiced his craft in Austria. His famous German quote "*Alle Ding' sind Gift, und nichts ohn' Gift; allein die Dosis macht, daß ein Ding kein Gift ist*" (translated as "All things are poison and nothing is without poison, only the dose permits something not to be poisonous") made him famous. He was the first to describe how the toxicity of a poison could be quantified by its "lethal dose" (LD). Because of this, he is often referred to as the "father of toxicology."

Mathieu Orfila (1787-1853) was born in Spain and worked as a French chemist and toxicologist. He played a major role in forensic toxicology, and was credited with being the founder of toxicology as a distinct scientific discipline, which he established in 1815.

DDT (Di-chloro Diphenyl Trichloroethane) was first produced in 1874 by Swiss chemist Paul Hermann Müller (1899-1965). For his work in developing and introducing DDT, he was given the Nobel Prize in Physiology or Medicine in 1948 (Grandin, 1948). Unfortunately, as chronicled by Rachel Carson (1962) in her book Silent Spring, the indiscriminate spraying of DDT has created numerous undesired environmental repercussions, and as a result of much unfavourable publicity, DDT was banned in the United States in 1972 and in many other parts of the world.

Sub disciplines of toxicology

The two primary branches of toxicology are toxicokinetics and toxicodynamics. The former refers to the processes involved in (1) taking in a harmful chemical, (2) distributing it throughout the body, bio transforming it, or metabolically eliminating it, and (3) excreting it. The field of toxicology known as toxicodynamics studies how poisons influence living things in various ways, including as a (1) irritant, (2) corrosive, (3) teratogenic or sterilizing agent, (4) asphyxiant, (5) carcinogen, (6) mutagen, and (7) anesthetic or narcotic.

Toxicology is separated into numerous sub disciplines. Almost 20 separate sub disciplines are commonly acknowledged, with eight of them being strongly established:

(1) Aquatic toxicology (2) Chemical toxicology (3) Ecotoxicology (4) Entomotoxicology (insect toxicology) (5) Environmental toxicology (6) Forensic toxicology (7) Medical toxicology and (8) Toxicogenomics

Aquatic toxicology

The field of study known as "aquatic toxicology" examines the negative consequences of human and non-human actions and substances on aquatic creatures at all levels of organization, from the molecular and cellular to the population and ecological.

Ecotoxicology

Ecotoxicology focuses on how harmful substances affect populations, communities, and entire ecosystems of living things.

Entomotoxicology or insect toxicology

Insect toxicology is mostly about how poisons slow or stop insects from developing, growing, changing into new forms, reproducing, or dying. It also talks about how pesticides work, how they kill bugs, and how bugs can become resistant to them. It is multidisciplinary and includes: (1) entomology (anatomy, morphology, and taxonomy); (2) chemistry (of inorganic and organic insecticides); (3) insect biochemistry; (4) insect ecology (chemical ecology, behaviour, and population dynamics); (5) genetics (related to insecticide resistance); (6) insect physiology; (7) statistics; and (8) techniques (related to application and bioassay). As a result, to fully understand how an insecticide works and how tolerance builds up, you need to know the basic biochemical, genetic, and physiological processes that happen when an insect's biological systems are poisoned.

Environmental toxicology

It is an area of science that looks at the harmful effects of different chemical, biological, and physical agents on living things. It is also known as "entox."

Forensic toxicology

Toxicology and other fields such as analytical chemistry, pharmacology, and clinical chemistry are used to assist in the medical or legal examination of death, poisoning, and drug usage.

Medical Toxicology

Medical Toxicology is the branch of medicine concerned with the study of poisoning and other undesirable health effects caused by drugs, environmental toxins, and even microorganisms.

Toxicogenomics

Toxicogenomics is the study of how a given drug affects a certain cell or tissue in an organism, and how that information is collected, interpreted, and stored.

General Properties of Insecticides

1. Except for ready-to-use dust and granules, concentrated pesticides must be diluted before application.
2. They are highly toxic and available in different formulations.

Properties of an ideal insecticide or pesticide

1. It should be easy to get on the market and come in different forms.
2. It needs to be poisonous and kill the pest that needs to be stopped.
3. It should not be phytotoxic to the crops on which it is used.
4. It should not be toxic to non-target species like animals, natural enemies etc.
5. It should be less harmful to human beings and other animals.
6. It shouldn't leave residues on crops like vegetables
7. It should have wide range of compatibility.
8. It shouldn't be harmful to bees, fish, or other beneficial organisms.
9. It should have higher tolerance limits.
10. It should have quick knock down effect.
11. It should be stable when applied.
12. It shouldn't have any tainting affects and should be free from bad odour.
13. Should cost less

2

Bioassay of Insects

A bioassay, also known as a biological assay, is a scientific experiment used to quantify the impact of a substance on a living organism. According to Finney (1952), bioassay is defined as measurement of potency of any stimulus (physical, chemical, biological, physiological) by means of reactions which it produces in living matter. It is an essential procedure in insecticide development and the assessment of environmental pollutants, as well as the evaluation of the effects of interventions *viz.* pesticides and herbicides on the environment. In the field of biostatistics, a bioassay refers to a set of techniques used to determine the quantity or intensity of an agent or stimulus based on the response of the subject. Typically, many methods are employed to assess the strength or effectiveness of an insecticide by examining its impact on living organisms.

An insecticide is a pesticide used to kill or eliminate insect pests in agriculture, households, and industries. Judicious use of insecticides may be a factor in the increase of agricultural productivity, but by their nature of having high toxicity to non target organisms and capability to develop resistance through widespread use, most insecticides have high potential to significantly affect and alter ecosystems. Many are toxic to humans and animals (both domestic and wildlife), and can accumulate as concentrates in the food chain and water resources, giving rise to serious environmental contamination and pollution.

Chemicals are usually described by their relative toxicity, and any chemical, even one that is normally thought to be safe, can become harmful to living things depending of the dosages exposed. So, even something as common as water has an LD_{50} of just over 80 g/kg, while sugar (sucrose) has an LD_{50} of 30 g/kg and alcohol (ethanol) has an LD_{50} of 13.7 g/kg. These substances can be harmful in large amounts. So, most insecticides, like other toxic chemicals, show varied levels of toxicity. Chemicals with a relative toxicity of 50 mg/kg or less are usually thought to be highly toxic, while chemicals between 50-500 mg/kg are usually thought to be moderately toxic.

A good technique to compare the toxicity of different compounds or the susceptibility of different insect species or the same species from different habitats in insecticide research is to find doses of equivalent toxicity. The three ways to bioassay compounds to determine critical doses is direct assaying, which gradually increase doses up to the critical point to kill individual animals. This method is impractical for insects. The other two methods involve indirect assaying, which involves introducing batches of insects to normal doses and monitoring their responses that may include death, knockdown, deformation, or discoloration, depending on the compound's expected effects. Bioassays can be based on quantitative responses like survival time, but measuring survival times is challenging, making them unsuitable for insecticide testing. Quantal response bioassays, the simplest and most popular insecticide research bioassay test, is the third approach. In dose-response or concentration-response bioassays, the explanatory variable is a range of dosages or concentrations and the response is dead or alive, knocked down or standing, deformed or not, and discoloured or not.

Quantal response experiments require data on the proportions of each batch responding to the insecticide in a specific way. The goal is to determine the dose level that is just enough to cause death (or a certain response) in the specified proportion of insects and then use that estimate to make comparisons. It is often easier to estimate the population's median (50%) response rate.

The median lethal dosage is a quantitative measure of a species tolerance under specific conditions or locations. It is a decisive biological property that is determined by other physiological and physical variables such as age, sex, rearing conditions, and temperature. In older literature, it is frequently abbreviated as MLD, but this can be confused with "minimum lethal dose LD_{50} is typically used to refer to a 50% deadly dose, whereas LD_{90} and LD_{95} refer to 90% and 95% lethal doses, respectively. Other dosage variables are abbreviated as LC_{50} (concentrations), LT_{50} (lethal time exposures), KD_{50} (knockdown dosages), and ED_{50} (effective doses). LD_{50} and other parameters indicate the toxicity of the pesticide employed and reflect the insect's tolerance. The toxicity decreases as the LD_{50} value increases.

Laboratory bioassay approaches should accurately imitate field circumstances to predict field control efficacy from lab-measured resistance. However, due to practical considerations, resistance detection and monitoring methods are designed to make bioassays reliable, replicable, consistent, and robust enough to withstand variations in operator skills, materials, extraneous factors, and handling procedures. For instance, direct bioassays on simulated field circumstances utilising larvae of different resistance levels were planned to measure resistance, but space and operational constraints limit sample size. Over the past two decades, leaf-dip, larval dip, topical application, and vial-residue bioassay procedures have offered alternatives to mimic field conditions. Bioassays collect quantitative response data from each unit that gets therapy. The unit in tests where each insect is individually treated is the insect, while the unit in spray or diet experiments is the group. Experimental precision requires constant units, such as insects from the same region, age, stage, sex, nutrition, and rearing conditions. The key bioassay characteristics and variables are

Properties of a good bioassay

1. Reproducibility
2. Results easily observed and measured
3. Relatively low cost
4. Preferably of short duration (less opportunity for confounding factors); more replication
5. Linear dose-response

Principle of Bioassay

The active principle to be tested should give the same measured response across all animal species. The chosen approach should be reliable, sensitive, and reproducible, with minimal mistakes related to biological variation and methodology.

Types of Bioassays

There are three main types of bioassays

1. Direct assays
2. Indirect assays based upon quantal responses (all or none)

1. Direct Assay

It is possible to directly measure the amounts of the standard and test preparations that are needed to get a certain reaction.

2. Indirect Assay

The initial stage in indirect bioassays involves determining the effect of each preparations concentration on the reaction. The required dose for a given reaction is then determined by applying the relationship to each preparation.

i. Quantal Assay

This response is "all or none," which signifies either no response or the utmost possible response. Bio samples of these substances can be performed using the endpoint method. The measurement of a predetermined response generated by the threshold effect. Quantal responses refer to population responses that are binary in nature, where the outcome is either present (1) or absent (0), such as death.

Concentration of Unknown= Dose of the Standard Dose of the Test Concentration of Standard

1. Important factors that influence bioassays

A) Stage of the insect for bioassay

The toxicity of a pesticide might vary significantly depending on the age and developmental stage of the organism. Insecticides display varying levels of toxicity towards distinct developmental stages of insects. For instance, lepidopteran adults and early instars are more vulnerable to insecticides compared to older stage larvae. Insects that have acquired resistance to a specific insecticide can be identified during the later stage of their development, i.e. third instar. These resistant insects can be easily differentiated from susceptible ones using diagnostic tests. Therefore, the standard method of applying the insecticide topically is performed on the third instar insects.

B) Selection of insecticide

The selection of an insecticide cannot be generalised to the entire group of insecticides, as certain molecules within a class of insecticides may exhibit greater resistance than others. Additionally, comparisons are made between different insecticides within the same class to assess the heterogeneity among the tested insects. For instance, multiple strains of *H. armigera* that are resistant to pyrethroids exhibited greater resistance to deltamethrin compared to cypermethrin or fenvalerate. Therefore, tracking resistance to a particular pesticide molecule may not accurately reflect the resistance to the entire class of chemical group. Nevertheless, laboratories typically opt for a chemical that exhibits a median resistance response relative to other compounds in the same class, in order to facilitate convenience. The choice of pesticide dosage is determined by the lethal dose of the target insect. For instance, dosages that elicit a response ranging from 25% to 75% are particularly valuable for determining the LD_{50}.

C) Selection of doses

This parameter is crucial in the field of bioassay. The bracketing approach is typically employed for dose selection of pesticides. When choosing the doses or concentrations of the insecticide for the experiment, it is advisable to distribute them uniformly across the entire range of mortality. Given that toxicity is logarithmically linked to dosage, it is preferable to use a dose range that follows an arithmetic series, such as 10, 20, 30, 40, 50, 60, 70, 80, 90, and 100. The control batches undergo identical treatments, with the exception of the poison. Consequently, the control insects must be treated with the solvent employed for diluting the solutions. Replications are ideally performed on different days within a short period of time, provided that day-to-day variability is not a source of error, and the treatment doses employed within each replicate should be ordered from lowest to highest. For any bioassay, we will initially check whether selected insecticides have 100% mortality. Then, we will use a bracketing method to get an accurate lab dose.

Bracketing Method – It is best to do replications of different doses within a short time frame. The treatment doses used in each test should be arranged from lowest to highest. Generally, the test concentration is in ppm and their response is in per cent mortality. Here we aim to identify the correct concentration of insecticide, which yields 50 %. If not get it, then we do bracketing again and check their responses. This way finally we will get the concentration, which gives 50 % mortality. Close bracketing gives more accurate results.

Eg: Determine the LC_{50} of a pesticide for a specific insect species using the bracketing method.

i). Selection of Test Concentrations

- Lower Bracketing Concentration (LBC) = 10 ppm
- Upper Bracketing Concentration (UBC) = 100 ppm

ii). Test Concentrations

- 10 ppm, 20 ppm, 30 ppm, 40 ppm, 50 ppm, 60 ppm, 70 ppm, 80 ppm, 90 ppm, 100 ppm

iii). Exposure and Observation

- Expose the insects to each concentration for 24 hours and calculate mortality (refer 2.3)
- Record the mortality rates:
- 10 ppm: 0% mortality
- 20 ppm: 10% mortality
- 30 ppm: 20% mortality
- **40 ppm: 30% mortality**
- **50 ppm: 40% mortality**
- **60 ppm: 60% mortality**
- 70 ppm: 80% mortality
- 80 ppm: 90% mortality
- 90 ppm: 95% mortality
- 100 ppm: 100% mortality

iv). Interpolation and Close Bracketing

- Based on the mortality data, estimate the LC_{50} using a probit analysis.
- If the estimated LC_{50} is not yielding 50% mortality, perform another round of bracketing using closer concentrations to the estimated LC_{50}, e.g., **45 ppm, 55 ppm, 65 ppm.**

By utilizing replications and close bracketing, the bracketing method allows for the accurate determination of the LC_{50} with a relatively small number of tests. This systematic approach ensures the reliability and precision of the toxicity assessment, making it a practical and effective method in pesticide toxicology studies.

D) Bioassay response

In most bioassays, the test insects death is used to show that it was poisoned by the insecticide being tested. However, some chemicals have strong effects on susceptible strains that are overcome by resistant strains. In these situations, EC_{50}, MIC_{50}, or IC_{50} are estimated instead of LD_{50} to find the middle growth-regulating response. These kinds of studies with cryotoxins are becoming more popular.

E) Application Method

Toxins (pesticides) are applied to organisms in bioassays through Potter's tower, microapplicators, residual bioassays employing glass, paper, or leaves for contact poisons, diet incorporation, and diet surface coating for oral toxins. Topical application of conventional pesticides is preferred due to its simplicity of handling and precision in administering known amounts of toxin (pesticide) on test insect. An insecticide for example, indoxacarb is an oral toxicant, topical bioassay does not accurately depict its toxicity, while some chemicals are water-soluble and not suitable for topical use. The application of diet inclusion and surface coating is used for cry toxins.

F) Optimum conditions for Bioassay

After the bioassay, test insects should be kept at the optimal temperature and humidity for growth and development. For *H. armigera*, maintain 25 ± 1^{0}C and 70 ± 5% relative humidity, while for *Pectinophora gossypiella*, maintain 27 ± 1^{0}C, 60 ± 5% RH, 9h light, 15h dark. This change in test insect optimal conditions may affect bioassay response.

G) Diet

In the lab, insects are raised on natural or artificial diets for various purposes. Developing artificial diets can be challenging and time-consuming, but once optimised, they are straightforward to prepare and use because they are made from commercial ingredients, available year-round, and cost-effective. Only by following proper quality control procedures, high-quality insects can be raised on artificial diets that are comparable or even better than their natural food. Many artificial diets have been developed and different species are being successfully raised on them. Ideally, bioassays should evaluate pesticide bioefficacy, all else being equal. An inappropriate diet can vitiate or amplify the harmful reaction of organisms to toxicants. The larvae must be fed with fresh food often throughout the bioassay. Semisynthetic diets can be utilised worldwide with little dietary effect on organisms since they are constant. Natural diets *viz*. leaves or other plant parts (pods/fruits) vary by age, stage, and diversity.

i. Types of insect artificial diets

A. **Holidic diets**: Chemically specified or holidic diets have chemical formulas for components. Only known pure compounds are in this diet.

B. **Meridic diets:** Meridic or semi-synthetic diets contain unprocessed plant or animal substances such plant tissue, liver powder or extracts, yeast, or wheat germ. Meridic diets contain largely specified compounds and one or more unknown ingredients.

C. **Oligidic diets:** Diets composed of unrefined resources. They are intended to resemble real food and are thought to include all of the necessary nutrients along with indigestible inert substances. Oligidic diets are utilised for large-scale insect rearing.

H) Test insect health

Prior to being exposed to the specified pesticide and procedure, it is essential to thoroughly analyse health of the test insects obtained from the field and if not good health will result in inaccurate outcomes. Larvae obtained from the field may contain parasitoids and illnesses, rendering them susceptible to the pesticides easily. Additionally, unclean conditions in insectaries can greatly contribute to inaccuracies in bioassays. When dealing with populations that consist of both healthy and unhealthy individuals, it can be difficult to differentiate between insect genotypes that are resistant or susceptible, as a result, one of the recommended strategies for monitoring resistance is to directly collect eggs of the target insect species from fields, raise them to the appropriate stage, and then perform bioassays. In order to determine the initial vulnerability of the insect, larvae of the same developmental stage can be subjected to testing or alternatively, they can be raised in a laboratory until the they reach F_1 generation and then can be used for testing.

I) Sampling

In determining the efficacy of dose-mortality analyses, sample size and dose selection are the two most important factors. Insect resistance monitoring samples should reflect the resistance distribution observed in field populations. The precision of analyses is also affected by sample size; for instance, samples obtained from a small section of the crop may have originated from a single pair of moths, or at most a few pairs, wherein a small sample size may not accurately reflect the resistance profile of the typical field populations. In a similar vein, the F_1 offspring produced by mass mating of moths from a limited number of larvae (30-40) may not serve as a representative sample due to the possibility that only a few moths (occasionally only one or two pairings) contributed disproportionately to the offspring. For example, in the case of *H. armigera*, a sample size of 1000-2000 eggs collected from widely separated fields to represent 4-5 sq. km of potential hosts would be adequate as the moth's nocturnal dispersal range is 1-2 km.

J) Sample size

Bioassay interpretations can be wrong if the sample numbers are too small. The sample size is based on how common resistant insects are likely to be in the groups that are being studied. Based on their research in 1984, Robertson et al. found that 240 insects were needed for a normal bioassay to give a reliable result, though 120 insects were usually enough. Roush and Miller (1986) says that a sample size of 100 insects is needed for the discriminating dose fixation based on the LD_{99} of susceptible strains and this is done when resistance levels are expected to be higher than 10%, and it is assumed that only resistant insect genotypes will survive the dose.

In the experiment, groups of insects are given different doses of pesticide. The size of each group is usually based on how useful it is, with higher number of test insects per group usually produces more accurate results, but using over 30 to 50 per batch doesn't help more unless the population is very different (Busvine, 1971).

If resistance is 1%, at least 1500 larvae are needed to detect resistant genotypes with 95% confidence with the same discriminating dosage. A precise discriminating dose that kills 99.9% of susceptible and 0.1% of resistant population holding a sample size of 300 would discover resistant alleles at 95% probability if the resistant genotype frequency is 1%. A minimum sample size of 100 numbers in discriminating dose assays and 250 for log dosage probit assays with 50 insects per dose at five concentrations is advised for conventional pesticides with resistance estimated to be >1%. Repeating dosage-mortality studies and utilising a "Common-line Model" to compare bioassay findings is the most significant aspect in dose response experiments (Finney, 1971). Time-dose response tests vary time while keeping dose constant, unlike dose-mortality experiments, which vary dose but maintain mortality.

K) Operator skills

If insects are not handled properly with forceps or fine brushes, they may be harmed and bioassay effects exaggerated. Topical pesticide application must be done carefully to avoid squirting and use a blunt end needle to gently apply a single drop on a test insect like *H. armigera* larvae's pro-thoracic region. Careless technical work can harm bioassays. Bioassay results can also be misleading if the syringe is not adequately cleaned between insecticide applications or doses of the same insecticide. For log dose probit tests, start topical administration with a low dose and move to higher doses. A good bioassay requires treating the insect at the right stage, identifying it, changing the diet periodically, and maintaining hygiene. Naturally, a tiny error makes results unreliable.

L) Safety Statement

Safety is key while using insecticides. Consult the Material Safety Data Sheet (MSDS) for proper Personal Protective Equipment (PPE) before handling any pesticides. Before doing bioassays, lab staff should get pesticide management training.

M) Analysis Method

Tabulation and analysis are crucial to bioassays and vary by approach. Time course analysis determines lethal time, usually 50% of test insects perish. Period-mortality data can only be analysed using probit or logit-regression models for various insect groups and each period. Survivorship curves can be fitted to a Weibull function (Pinder *et al.,* 1978) to estimate median fatal times with upper and lower 95% confidence limits using SAS. Analysing time and dose-mortality data is often reliable. Log-dose probit regression examines dose-mortality relationships (Finney, 1971). Commercial software can estimate LD_{50} and other LD values and their 95% confidence limits. For toxicological evaluations, insect populations are serially diluted with pesticides to determine a dose-mortality regression response. At least a minimum of five toxicant concentrations must be evaluated on each population and the regression equation gives the dose response as LD_{10}, LD_{50}, LD_{90}, LD_{99}, etc. LD_{50} is the dose that kills 50% of the test population. Abbott's formula is to be adopted in treated samples if control mortality is recorded between 5% to 20%.

N) Presenting results

Each probit analysis includes LD_{50} or LD_{90} estimates. A different statistical approaches provide parameters for fitting quantal response data to regression models. Once assay results show graded response based on toxicant concentration, log-dose probit analysis is performed. Log dose probit analysis is used to estimate a lethal concentration (LC_{50}) that kills 50% of target populations and confidence limits (CLs) for data variability. If controls exceed 5%, Abbott's formula is used to adjust response.

$$\text{Corrected response} = \frac{(\%\text{ response in treatment} - \%\text{ response in control})}{(100 - \%\text{ response in control})} \times 100$$

O) Bioassay Acceptance criteria

1. Response bracketing median (50%) response
2. Mortality response in untreated control should be <10%
3. Contamination should be <10%
4. There should be a tight fit between the lower and upper fiducial limits
5. The calculated Chi-square values should be less than the tabulated value at probability value of 0.01 and any analysed data with high probability values are rejected.

6. Assays should be discarded with probability value of > 0.01
7. Analyzed data should not have the negative slope values and also should not have high fiducial limits.

2. Preparation of test solutions for bioassay

The median lethal dose (LD_{50}) of insecticides accurately compares their toxicity. The lower the insecticide estimated LD_{50} value, the higher its toxicity or potency and the insect population's tolerance to insecticide treatment. Dilution of pesticide active components must be standardised to reliably evaluate and compare toxicities.

2.1 Preparation of stock solutions

Make a stock solution of insecticide active ingredient (a.i.) from a source and use it to make all bioassay treatments. A stock solution of 100 µg/ml will be developed for this study. However, any concentration can be diluted from this initial concentration. For laboratory tests, use technical-grade insecticides with 95-99% purity and balance the active component accordingly. Insecticide active ingredients (a.i.) vary; hence, a 100% stock solution (SS) is created using the correction factor (CF):

$$CF = \frac{100\ \%}{\text{Percent a.i of the insecticides}}$$

For a technical insecticide with 99.5% a.i., $CF = \frac{100\ \%}{99.5\ \%} = 1.005$

The weight of the technical insecticide required can be calculated based on the CF, and the formula for preparing the desired volume and concentration can be utilised.

Concentration of insecticides × volume × CF

To prepare 2.5 ml of 10,000 µg/ml stock solution, the weight of insecticide needed will be 10,000 µg/ml × 2.5 ml × 1.005 = 25,125 µg = 25.125 mg = 0.025 g.

To acquire the 100% stock solution, 0.025 g of technical-grade insecticide is weighed into a 6-ml screw cap vial using an analytical weighing balance. Subsequently, 2.5 ml of technical-grade acetone is added as a solvent.

2.2 Preparation of insecticide concentrations for tests

Based on a preliminary test, at least five insecticide concentrations with a 15–85% insect mortality range are prepared from the stock solution (SS), and serial dilution is performed from higher to lower concentration.

From the stock solution, serial dilutions are prepared using the equation $C_1V_1 = C_2V_2$, where C_1 = initial concentration, V_1 = initial volume, C_2 = final concentration, and V_2 = final volume.

To prepare 2 ml of 5,000 µg/ml from 10,000 µg/ml stock solution, the volume needed using the formula above will be (10,000 µg/ml) (x) = (5,000 µg/ml) (2 ml) = 10,000x = 10,000, where x = 1 ml SS + 1 ml acetone.

Serial dilution is continued using the above equation, or a 1:1 dilution of the following 10-12 concentrations is performed sequentially from highest to lowest concentration. The vial cover is sealed with parafilm to prevent evaporation, and the created pesticide dilutions are stored in a refrigerator (4°C) or freezer (ideally -20°C). After preparing an insecticide, the pipettor tips are replaced and appropriately discarded.

2.3 Abbott formula- Correction for control mortality

In bioassays, it is usual for a proportion of the insects in the control batches to die during the experiment owing to natural reasons or the solvent treatment, and the Abbott formula is frequently used. The formula credited to Abbott (1925) had already been used by Tattersfield and Morris (1924) and is commonly in the form

$$P = \frac{1}{\sqrt{(2\pi)}}$$

Where P is the corrected mortality, Po is the observed mortality, and Pc is the control mortality, all stated as percentages.

2.4 Probit analysis-a statistical method in bioassays

Chester Ittner Bliss introduced probit analysis in 1934 in Science, his main goal, as an entomologist at the Connecticut Agricultural Experiment Station was to find a pesticide that controlled grape leaf-eating insects (Greenberg, 1980). By graphing the insects responses to varied pesticide concentrations, he could tell that one was more effective than the other was, but he did not have a statistically reliable way to compare them. The most logical way is to fit a regression of response *vs* concentration or dose and compare pesticides. The dose response was sigmoid, and regression was only employed on linear data at the time. Thus, Bliss proposed a straightening of the sigmoid dose-response curve. In 1952, University of Edinburgh statistics professor David Finney wrote Probit Analysis, based on Bliss theory. Today, probit analysis is the preferred statistical tool for dose-response correlations.

Probit analysis is used to examine a wide range of dose-response or binomial response tests. Probit analysis is a technique commonly used in toxicology to determine the relative toxicity of chemicals to living organisms. It works by testing an organisms response to various concentrations of each of the chemicals in question and then comparing the concentrations at which a response occurs. As previously stated, the response is always binomial (e.g., death/no death), and the connection between the response and the different concentrations is always sigmoid. Probit analysis transforms sigmoid to linear and then does regression on the relationship.

The output of the probit analysis can be used by the researcher to compare the amounts of different chemicals needed to produce the same response after a regression has been performed. The most commonly used results of contemporary dose-response experiments are the LC_{50} (liquids) or LD_{50} (solids), although there are numerous other endpoints that can be used to compare the various toxicities of substances. The concentration (LC_{50}) or dose (LD_{50}) at which 50% of the population reacts is represented by the LC_{50}/LD_{50}.

Take into consideration, for instance, contrasting the toxicity of insecticides A and B to aphids. In the event where pesticide A has an LC_{50} of 50 ug/L and pesticide B has an LC_{50} of 10 ug/L, pesticide B is more toxic than pesticide A since it requires only 10 ug/L in comparison to 50 ug/L required to kill 50% of aphid population.

2.5 How does probit analysis work and how to get from dose-response curve to LC_{50}

The simplest technique is to utilise a statistical package such as SPSS, SAS, XLSTAT, and so on, but it is important to grasp the methodology's history in order to fully comprehend the subject.

Step 1: Convert % mortality to probits (short for probability unit)

Method A: Determine probits by looking up those matching to the percentage of responses in Finney's table (Finney 1952):

Table 3.2: Transformation of percentages to probits

%	0	1	2	3	4	5	0	7	8	9
0	—	2.67	2.95	3.12	3.26	3.30	3.46	3.62	3.59	3.66
10	3.72	3.77	3.82	3.87	3.92	3.00	4.01	4.05	4.08	4.12
20	4.16	4.19	4.23	4.20	4.29	4.33	4.30	4.39	4.42	4.45
30	4.48	4.50	4.53	4.50	4.69	4.01	4.64	4.07	4.69	4.72
40	4.76	4.77	4.80	4.82	4.85	4.87	4.00	4.92	4.95	4.97
50	5.00	5.03	5.05	5.08	6.10	5.13	5.16	5.18	5.20	6.23
60	5.25	5.28	5.31	5.33	5.30	5.30	5.41	5.44	6.47	6.50
70	5.52	5.55	5.58	6.01	5.04	6.07	6.71	6.74	6.77	5.81
80	5.84	5.88	5.92	5.95	6.99	6.04	6.08	6.13	6.18	6.23
00	6.28	0.34	0.41	0.48	0.56	6.64	6.76	0.88	7.05	7.33
—	0.0'	0.1	0.2	0.3	0.4	0.6	0.0	0.7	0.8	0.9
90	7.33	7.37	7.41	7.40	7.51	7.58	7.06	7.76	7.88	8.09

For example, a 17% answer would result in a probit of 4.05. Furthermore, for a 50% answer (LC_{50}), the appropriate probit will be 5.00.

Method B: Hand calculations (Finney and Stevens 1948):

The probit Y, of the proportion P is defined by:

$$P = \frac{1}{\sqrt{(2\pi)}} \int_{-\infty}^{Y-5} e^{-iu^2} du.$$

The maximum and minimum working probits are used in the conventional analysis method:

$$Y_{max.} = Y + \frac{Q}{Z}$$

$$Y_{min.} = Y + \frac{P}{Z^2}$$

And the range 1/Z where

$$Z = \frac{}{\sqrt{(2\ )}} e - 1(Y - 5)$$

Method C: The percentage of responses to probits is automatically converted by computer programmes like XLSTAT, SAS, and SPSS.

Step 2: Take the log of the concentrations.

If performing computations by hand, this can be done manually. Alternatively, the computer programme of choice can describe this step.

Step 3: Graph the probits versus the log of the concentrations and fit a line of regression.

Method A

The least squares method is a manual method of fitting the line by eye that minimises the distance between the line and the data. The accuracy of this method can be surprising, although it is undoubtedly more accurate to calculate a regression by hand or using a computer programme. Confidence intervals can also be obtained using computer programmes and manual calculations.

Method B

According to Finney and Stevens (1948), the following approach is used by hand to compute the linear regression:

$Y = 5 + (x - \mu)/\sigma$

To begin, set p = r/n for the proportion of respondents and q = 1-p for their complement. A line fitted by eye may be utilised to provide a comparable set of predicted probits, Y. The probits of a given value of p should be roughly linearly related to x, the stimulus measure. Next, using one of the following formulas, the working probit associated with each proportion is determined:

$y = Y + Q/Z - q/Z.$

$y = Y - P/Z + p/Z,$

Subsequently, a series of anticipated probits is calculated based on the weighted linear regression equation of working probits on x. Each y value is assigned a weight, nw, with the weighting coefficient, w, defined as:

$W = Z^2/PQ.$

Method C

Make use of a computer programme. Maximum Likelihood is used by SPSS to estimate linear regression. The easy steps below will walk you through running the probit analysis in SPSS.

To use the Data Editor, enter at least three columns: 1. Number of respondents per container 2. Total number of individuals per container 3. Concentrations

In the shown screen, "almost" represents the number of individuals that reacted per container, "total" represents the total number of individuals per container, and "a_conc" represents the concentrations. The data in row 7 of the given example represents the results of the control group, where none out of ten participants replied at a concentration of zero.

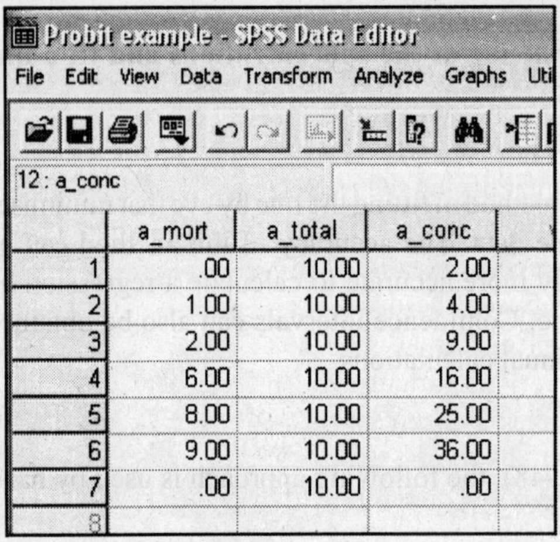

Probit example - SPSS Data Editor

File Edit View Data Transform Analyze Graphs Uti

12 : a_conc

	a_mort	a_total	a_conc
1	.00	10.00	2.00
2	1.00	10.00	4.00
3	2.00	10.00	9.00
4	6.00	10.00	16.00
5	8.00	10.00	25.00
6	9.00	10.00	36.00
7	.00	10.00	.00
8			

Once you have set up your columns, proceed to the "analyse" tab and select "regression" followed by "probit".

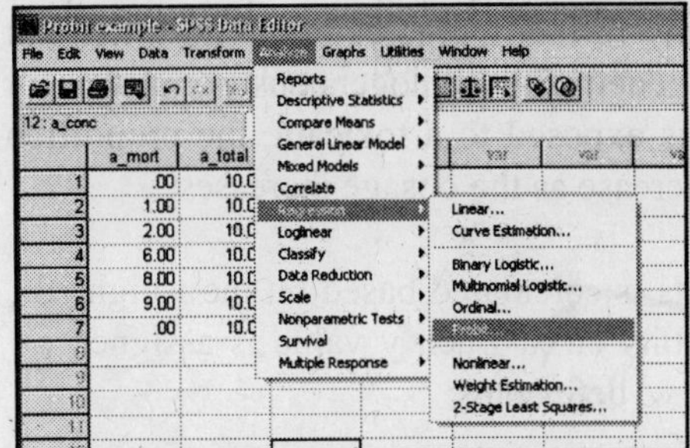

Assign the column containing the numbers of responses as the "Response Frequency", the total number per container as the "Total Observed", and the concentrations as the "Covariates". Remember to choose the logarithm with a base of 10 in order to convert your concentrations.

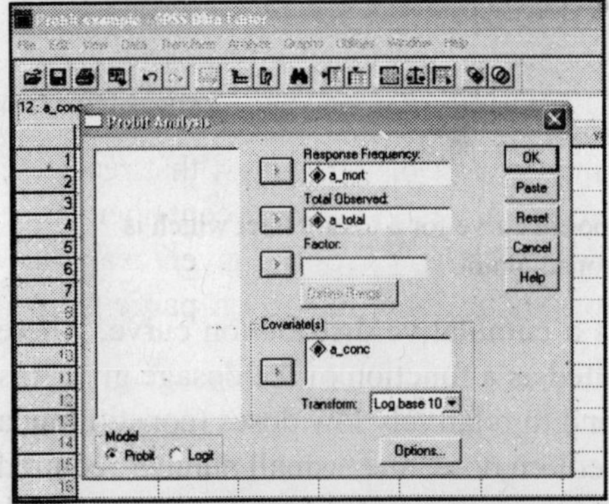

Probit transformed response

Toxicity of chemical is determined by quantifying the response on test animal to a series of increasing dose. The relationship between animal and administered dose can be graphically presented as dose-response relationship curve. Dose response relationships means the relationship between the dose of a chemical substances administered or received and the incidence of an adverse health effect in exposed population. In toxicology, the dose is very important which determine its impact on organism. It is the dose which makes substances a poison. The right dose differentiates a poison and a remedy. At high doses, all the chemicals are toxic, at judicious doses they are useful and at very low doses they do not have a detectable toxic effect.

A dose-response relationship is based on the following important assumptions:

There is always a threshold dose below which no effect occurs. Once effect occurs, response increases as dose increases.

Once a maximum response is reached, any further increases in the dose will not result in any increased effect. When a genetically homogeneous population of animals of the same species and strain is exposed to a toxicant, the proportion exhibiting a particular toxic effect will increase as the dosage increases.

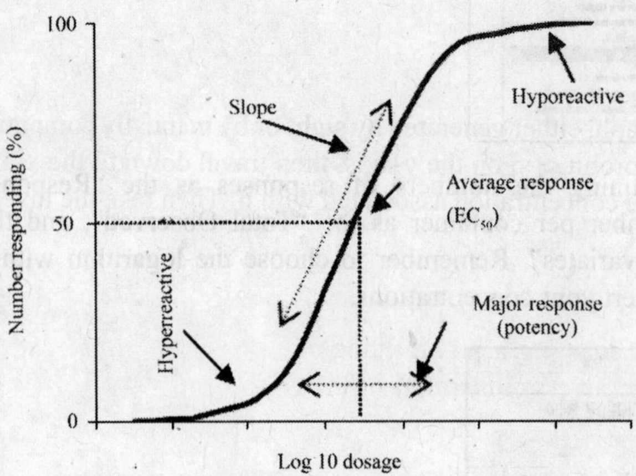

Fig. 1: Typical sigmoid cumulative dosage-response curve for a toxic effect which is symmetrical about the average (50 percent response) point.

This is shown schematically (Fig. 1.) as a cumulative distribution curve, where the number of animals responding is plotted as a function of the dosage given (as a $\log_{10}$ function). A few individuals respond to relatively low doses (constituting a hyperreactive group), most respond to medium dose, and a small number required a relatively high dose (constituting a hyporeactive group) before they are affected. The lowest dose at which an effect is discernable is referred to as the no observable effect level (NOEL).

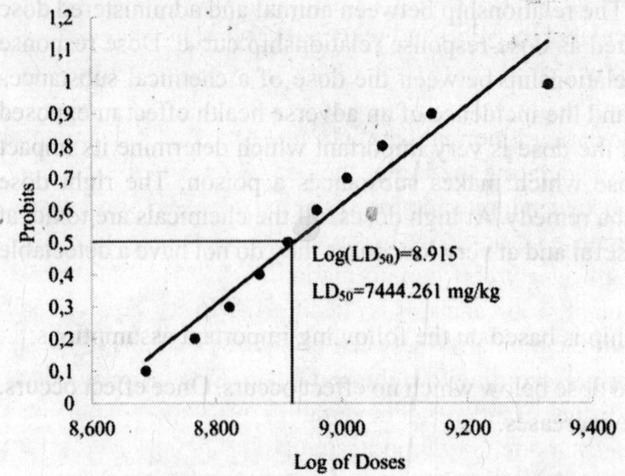

In an insect-bioassay, the data transformation is needed in the dose/concentration-mortality. Mortality data is converted to probit values in y-axix. Dose/Concentration in x-axis is transformed to log-dose/concentration.

Response curve, to obtain a straight line

Step 4: Find the LC_{50}

Method A

Using your hand drawn graph, either generated by sight or by manually computing the regression, locate the probit of 5 on the y-axis, then travel down to the x-axis and calculate the log of the concentration associated with it. Then take the inverse of the log. You have the LC_{50}.

Method B

The LC_{50} is calculated by locating a probit value of 5.00 in the probit list and subsequently obtaining the inverse logarithm of the concentration with which it is correlated.

Step 5: Determine the 95% confidence intervals:

Method A

The following equation is used for hand calculations:

The approximate value of the standard error is $\pm 1/b \sqrt{(Snw)}$.

b = estimate of the slope of the line

Snw = summation of nw

w = weighted coefficient from table III = Z^2/PQ (Finney, 1952)

Method B

This is automatically calculated by SPSS and other software applications.

Logit *vs.* Probit

Logit is a method for translating binomial data into linearity that is quite similar to probit. Logit works by calculating the log of the odds: logit (P) = log P/(1-P). However, the connection between logit and probit is nearly identical: logit = $(\pi/\sqrt{3})$ × probit. In general, if response *vs.* dosage data is not normally distributed, Finne[illegible] proposes applying the logit over probit transformation (Finney 1952). Althou[illegible] the multivariate application of probit analysis is outside the scope of this webpa[illegible] it is worth mentioning that the similarity between probit and logit does not [illegible] true in a multivariate setting. According to Hahn and Soyer, logit provides a [illegible] fit when there are several values of independent variables, whereas probit pr[illegible] a better fit for random effects models with moderate data.

3. Bioassay methods

1. Topical application
2. Potter spray tower
3. Surface coating/ residual bioassays (Residual deposit or film)
   - → A. Leaf surface coating
   - → B. Filter paper surface test
   - → C. Adult glass vial test (Residual deposit or film)
4. Immersion method (Dipping method)
   - → 4.1. Direct insect dipping method
   - → 4.2. Plant parts dipping method method
     - → A. Leaf dipping method
     - → B. Square dipping bioassay for *H. armigera*
5. Diet incorporation
   - → 5.1. Diet incorporation for filter
   - → 5.2. Surface coating of semi-synthetic diet
6. Diet covering bioassay
7. Leaf spray bioassay
8. Sticky card assay
9. Injection method
10. Aqueous solution method
11. Sand witch method
12. Fumigation method

3.1 Topical application

Concerning contact poisons, this technique is extremely practical. The technique proves to be appropriate for administering technical insecticides that are only available in restricted quantities and possess adequate contact activity. The hand-held Hamilton repeating dispenser has supplanted methods that were once conventional, including the utilisation of Burkhard's microapplicators and Potter's towers. By precisely dispensing a known quantity of toxins onto the insect body, this method has become one of the most practical. For example, dorsal surface of the prothoracic region of third instar *H. armigera* larvae was treated with a pre-calibrated 1µl solution of technical grade insecticide dissolved in acetone and a 50µl Hamilton repetitive manual dispenser was utilised to apply the solution.

The topical application method is most effectively utilised when treating organisms ith a minimum surface area capable of absorbing 1µl of insecticide in acetone. ical application bioassays on small insects, including the first instar larvae erous lepidopteran species. This bioassay technique is also applicable to

f this method are

el of precision and roducibility can be achieved.

er of tests can be run in a very short period of time.

of insects (10-20) are required for each replication.

sive equipment is required.

chemicals and solvents are used.

6. The fact that LD_{50} values for any species are reasonably stable and reproducible from laboratory to laboratory as long as identical testing protocols are followed.

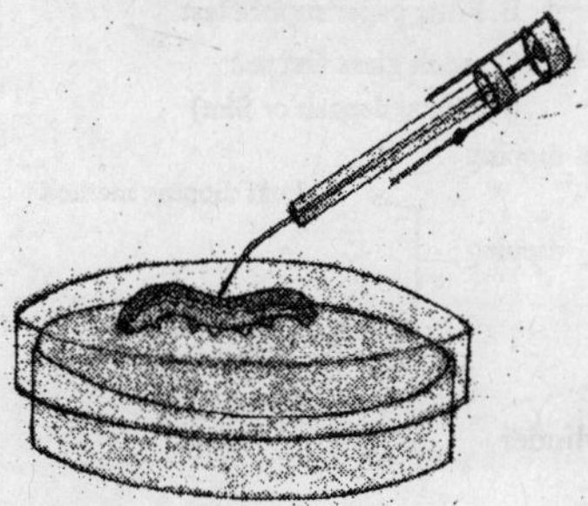

3.2 Potter spray tower

The Potter Spray Tower was named after C. Potter, who created the spray apparatus at Rothamsted Experimental Station. The Potter Tower is recognised worldwide as the reference instrument for laboratory chemical spraying procedures and the device is essential for research into the biological effects of contact poisons on organisms.

High-grade stainless steel with an appealing polish is used in the tower's construction. The Potter Spray Tower has height of 120 cm with standard sample reservoir volume of 20 cc, and working pressure is 15 1b/sq.in. Pneumatically controlled spray table and fast removable atomizers are features of the tower, which has all controls conveniently located at the front. This air-operated spraying device covers a 9 cm diameter circular area with a uniform spray deposit. Potter's spray tower was created to shield operators from pesticides and other harmful substances.

Principle

It operates at constant atmospheric pressure, 15 1b/sq.in. The input connection provides a steady supply of 15 1b/sq.in and is controlled by an on/off switch and an exhaust valve. A sensitive needle valve on the instrument's left side allows for direct readings on the pressure gauge or manometer (an added feature). This pressure operates both the air jack and the nozzle head.

Uses

It is suitable for studying the biological effects of chemicals applied directly to organisms or as a residual film. The automatic safety load and unload system enables operators to set petri dishes on the spray table away from the spray tube, lowering the risk of hazardous material contamination.

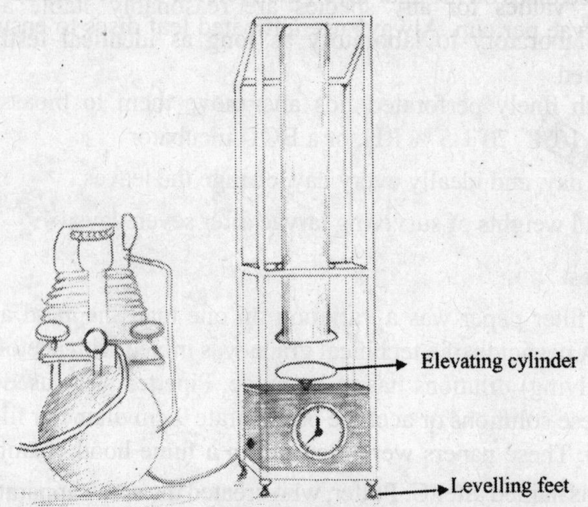

3.3 Surface coating/ residual bioassays (Residual deposit or film)

The technique, known as residual tests, includes immersion in a thin film of diluted solutions of formulated insecticides on leaf and paper surfaces. Glass vials are covered with a thin film of insecticide solution in acetone by evaporating the solvent by continuous rolling of the vials, and insects are released onto the treated surface, where they are exposed to the insecticide. The leaf residue assays closely mimic field exposure conditions and have been used to track insecticide resistance in *H. armigera*, whiteflies, aphids, and mites.

A. Leaf surface coating

Bioassays with leaf-feeding insects should be conducted with the leaves of the most commonly growing host plant type.

1. Cotton leaf disc residue bioassays are commonly conducted using plastic cups with inside dimensions of 6.8 × 5 (d x h).
2. Properly disperse one gram of agar in 99 millilitres of water by means of continuous shaking. The solution should be heated to a simmer before being cooled to 65°C and 0.3% anti-mould solution is to be added.
3. To create the anti-mould solution, combine 53.5 ml water with 4.5 ml phosphoric acid and 42 ml propionic acid, then pour the solution into the plastic containers until a 0.5 cm thick layer forms.
4. Wash tender cotton leaves under tap water and gently sandwich them in blotting paper to remove water. The leaves are punched into 6.5 cm discs with a metal cap. The discs are coated with 100 µl of diluted poison on each side and air-dried. Use the bottom of a 0.5 × 5 cm test tube to carefully disperse the toxin over the leaf.

5. Place the toxin-coated discs on the agar layer, then release one second instar or ten first instar larvae per cup. Always use untreated leaf discs to ensure optimal control.
6. Close the cups with finely perforated lids and move them to bioassay chambers with 25 ± 10°C, 70 ± 5 % RH, or a BOD incubator
7. At least every other day, and ideally every day, change the leaves.
8. Record mortality and weights of surviving larvae after seven days.

B. Filter paper surface test

The test method used on filter paper was a variation on one that Sheppard and Hinkle (1987) described. A pesticide of a technical grade was mixed with acetone, and six to eight serial (halving) dilutions has to be made. Pipettes were used to put one-mL portions of these solutions or acetone on separate 9 cm diameter filter papers (Whatmans No. 2). These papers were then put in a fume hood chamber to dry for 24 hours. After that, they were covered in metal foil and frozen (-25 °C) until they were needed. The filter paper was put in a new, throwaway 9-cm-diameter plastic petri dish right before it was used. After that, known insects will be put into it, and the death rate will be calculated.

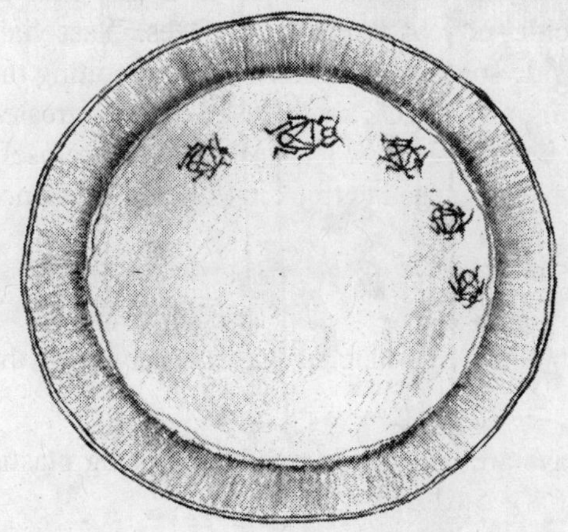

C. Adult glass vial test (Residual deposit or film)

Insecticide resistance in *H. virescens* or *H. armigera* has been closely monitored using the adult vial test, which is particularly helpful for determining sex-allele-linked inheritance. What follows is the procedure.

1. For the test, glass vails are used. Take acetone and rinse the vails. Then, dry them in a 120°C oven.
2. Label the vials and start giving out the insecticides diluted in acetone one at a time, starting with the lesser amounts.

3. Pipette 500 µl of toxin solution into each 25 mL glass vial. Lay the vials carefully on their sides and roll them on a motorised roller or a bench-top surface until the solvent has evaporated completely. Ensure that the solution is evenly and completely dispersed throughout the inner surface of the vial.
4. Coat control vials with acetone.
5. This procedure is well-suited for testing moths or flies. Provide a diet of 10% sugar to moths that are one day old for a duration of 2-3 hours. Afterwards, release each moth individually into separate vials. Close the vials with stoppers made of cotton or glasswool, 3 hours after feeding.
6. Move the vials to the insectary, maintaining a temperature of 25 ± 1°C and a relative humidity of 70 ± 5%, or alternatively, place them in a BOD incubator.
7. Replace the diet consisting of cotton swabs soaked in a solution containing 5% sucrose and 5% honey in water on a daily basis. Keep track of the number of deaths each day for a period of three days.

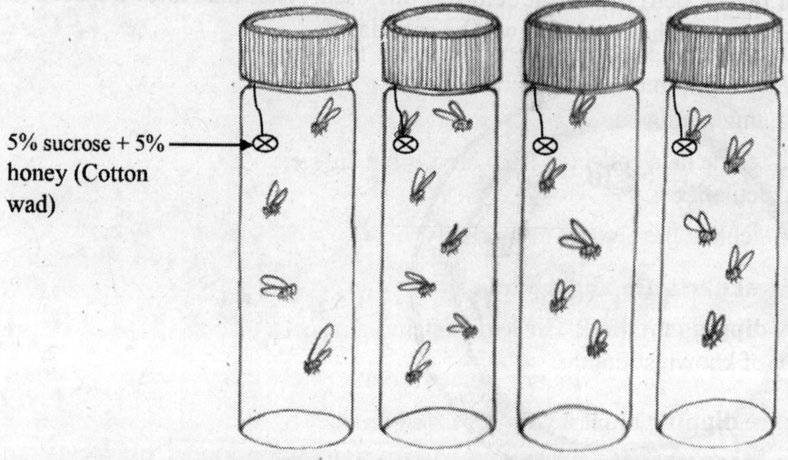

3.4 Immersion method (Dipping method)

This technique is utilized in situations when the use of topical application or injection methods is not feasible, such as small plant-feeding insects, stored-product insects, housefly larvae, insect eggs etc. The immersion approach is further divided into two types. 1) Direct insect dipping method, 2) Plant parts dipping method.

3.4.1 Direct insect dipping method/Immersion

The insects are captured using a pair of forceps and briefly immersed in an aqueous solution containing the chemical.

The method is described as below,

1. Collect at least 150 third-instar larvae directly from fields and place them in separate cups.
2. Sort larvae weighing 30-40 mg by eliminating those that are underweight or overweight.
3. The suggested field application dose should be set as a thumb rule for the test. Using the stepwise dilution method, different concentrations of insecticide have to be prepared as desired.
4. Dip the larvae individually and place them on blotting paper for a few seconds. Treat at least 100 larvae with the recommended dose and maintain 20-30 untreated larvae as control.
5. Place the larvae individually in separate wells of multi-cell trays or in preferred plastic vials, depending on the test insect being investigated. Position these trays/vials in BOD insect growing chambers, duly ensure that the temperature and relative humidity (RH) are set at appropriate levels to facilitate optimal growth and development of test insect.
6. Data is collected at 24-hour intervals for a duration of three to four days, depending on the insect species being studied and the type of insecticide being employed.
7. Exclude data collected from larvae that are sick or parasitized from the final calculations.
8. Calculate the percent of mortality.

3.4.2 Plant parts dipping method

A. Leaf dipping method: The leaf containing insects is dipped in the insecticidal solution of known strength.

B. Square dipping bioassay for *H. armigera*

A sufficient number of cotton squares has to collected from the unsprayed cotton field. The collected squares has be carefully cleaned and then dipped in test solutions for 10 seconds, and transferred to paper-padded trays for air drying. After 60 minutes of air drying, squares were placed in plastic cups, and larvae were inserted into the cups to feed on the treated squares.

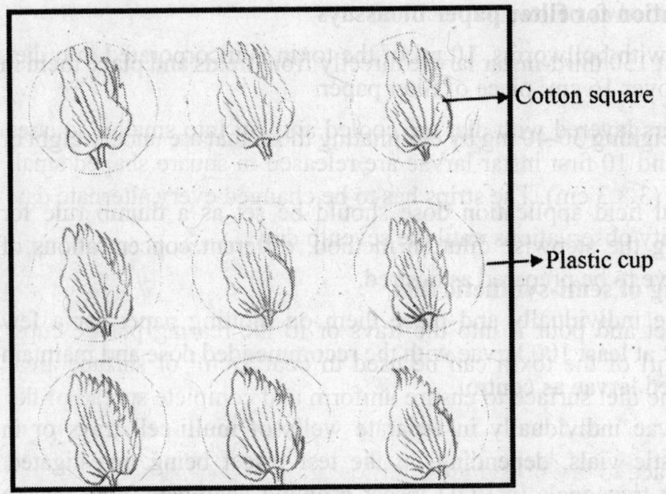

3.5 Diet incorporation

In order to conduct oral toxicant bioassays, researchers created diet incorporation or surface-coating studies. The availability of large doses of the toxin, its thermal stability, and a consistent bioactivity under bioassay conditions are three factors that these tests depend on.

To determine the toxicity of oral toxicants to specific insects, the following approach is followed for testing and their inclusion in a diet. The below flow chart depicts how to conduct a Cry I Ac bioassay on *H. armigera* using first-instar larvae.

1. Pipette 3 ml of toxicant solution into a 40 ml plastic cup
2. Add lukewarm diet (60°C) to make a total volume of 30 ml
3. Shake the cup vigorously for 1 minute to mix
4. Pour the diet into a 24-cell insect-rearing tray to 0.5 cm height
5. Allow the diet to cool under UV lamps for 1 hr
6. Release first instars into the diet trays (1 per well)
7. Cover trays with semi-permeable wrap and close the lid
8. Secure lid with rubber bands to prevent escape
9. Keep control with released larvae on untreated diet
10. Store unused trays with diet in a refrigerator for a week
11. Change the diet every 2-3 days
12. Record mortality at 8-hour intervals for 7 days for LT_{50} calculations
13. Record mortality at alternate days for 7 days for LC_{50} calculations
14. Record weights of surviving larvae at the end of 7 days for EC_{50} calculations

3.5.1 Diet incorporation for filter paper bioassays

1. For bioassays with bollworms, 10 ml of the toxin is incorporated into diet and is poured over 16 cm^2 piece of filter paper.
2. The filter papers layered with diet are cooled and cut into smaller squares of 2 × 2 cm, and 10 first instar larvae are released in square shaped small plastic cups of (3 × 3 cm). The strips has to be changed every alternate day.
3. Record mortality observations until the seventh day.

3.5.2 Surface coating of semi-synthetic diet

1. Prepare the diet and pour it into the trays or to the rearing plastic cups. Generally, 10 µl of the toxin can be used to coat 1 cm^2 of surface area. Gently swirl the diet surface to ensure uniform and complete spread of the solution.
2. Allow the surface to dry in a laminar airflow under UV light for 2-3 hours for surface sterilization.
3. Release first instar larva @ one per well and also maintain control using untreated diet.
4. Change the diet on alternate day and record mortality until the seventh day. The body weight of surviving larvae has to be recorded on the final day of bioassay.
5. This method has the advantage of obtaining constantly reliable results because the toxin is unlikely to be affected by either improper mixing or heat as it can occur in the diet-incorporation method and moreover, less quantity of the toxin is required for the assay, compared to the diet-incorporation method.

3.6 Diet covering bioassay

Using a Potter's tower (Burkard, England) with a pressure of 1 bar, two millilitres of poison is placed on a petridish that has 10 cubes of the desired insect's synthetic food. The control possesses two millilitres of clear water. When the pesticide is applied, it should touch the bottom of the diet cubes. The cubes should then be moved to the sprayed surface of the Petri dish using forceps. The diet cubes are placed at room temperature for 30 minutes before being moved to different cells in the 16 well or 25 well polystyrene bioassay box. Each cell contains one third-instar larvae, and the tray should be closed and kept at 25 ± 2 °C, 60–65% RH, and a 16:8 h light: dark photoperiod for 72 hours, after which mortality rate is measured. This method is used to do bioassays on lepidopterans.

3.7 Leaf spray bioassay

Lemon leaves are sprayed for 3 seconds with a hand sprayer (2.1 kg/cm^2 pressure; 75-cm spray distance) containing six to eight dilutions of commercial insecticides suspended or dissolved in distilled water, with TritonX100 added as described above (Morse and Brawner, 1986). Control leaves were sprayed with distilled water and Triton XI00. The bioassay approach is appropriate for thrips.

3.8 Sticky card assay

Insecticide resistance in whiteflies has been widely studied using the sticky card test (Prabhaker et al., 1988). The test has benefits of being user-friendly for field workers.

1. Spray on yellow cards (7.5 x 12.5 cm) with a thin layer of sticky adhesive
2. Spray serial dilutions of formulated insecticide on the card with help of potter's tower
3. Spray control cards with water
4. Carry treated cards to the field in cool boxes and expose to whiteflies for 1-2 minutes
5. Place cards on styrofoam slabs at room temperature in humid conditions
6. Record mortality observations after 24 hours

The test offers certain advantages over the vial approach. Insecticides used in vial testing were shown to deteriorate faster than those used on sticky cards, and unlike vial tests, the sticky card test does not impose a fumigation effect or give untreated places for insects to seek refuge.

Adult age for bioassay: Preliminary observations should be conducted to determine whether older or younger flies from the same population would die faster on sticky cards with or without pesticides. If this is the case, bioassays on field populations with different age distributions may produce inconsistent results with unsatisfactory control mortality rates.

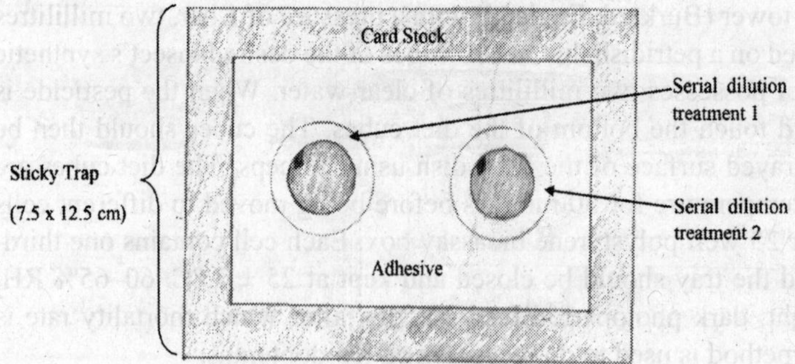

3.9 Injection method

The toxin is injected directly into the organism's body system using a hypodermic fine stainless-steel needle (27 or 30 gauge thickness; 0.41 or 0.30 mm in diameter), but sometimes small glass needles with diameters ranging from 0.1-0.16 mm are used to inject toxicant into minute insects. A desired quantity of toxicant is measured using a micrometer. The pesticide is often diluted in propylene glycol or peanut oil and injected intraperitoneally (into the bodily cavity).

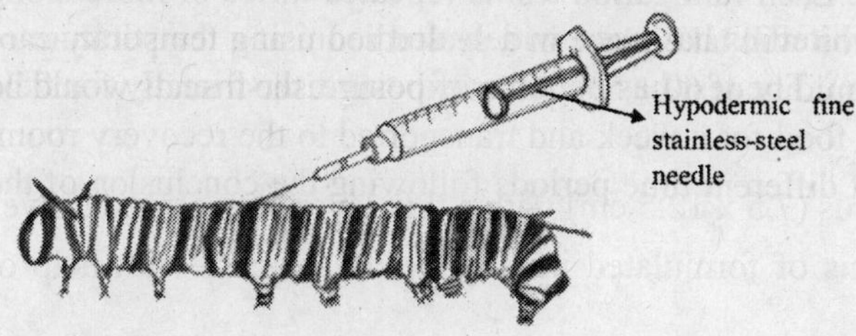

3.10 Aqueous solution method

This approach involves dispersing the insecticide residue in a small quantity of a solvent that is capable of mixing with water, such as acetone or alcohol, to form a solution or suspension. Next, a limited population of delicate aquatic species, such as mosquito larvae, micro crustaceans, or fish, is subjected to the solution. The quantity of residue in a treated sample is determined by comparing its toxicity to that of the standard. The quantity of solvent in water is minimized to conduct control tests to identify any potential toxicity. This method offers the benefit of maintaining continuous contact between the entire organism and the media. There have been assertions of consistent distribution of the poison, although it is possible for the suspended substance to settle. The high sensitivity is believed to stem from the toxicant's circulation and absorption through the gills or similar organs.

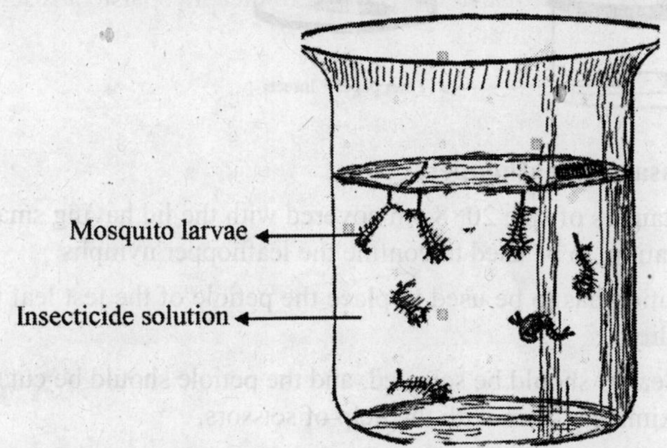

3.11 Sandwich method - The pesticide concentration is placed between two leaves, and the test insect is allowed to feed on it. This is usually done for caterpillars that feed on leaves.

3.12 Fumigation method

Insecticides possessing fumigant property are tested using fumigation method which is an appropriate technique for testing fumigants against pests damaging stored products. The insecticide is added to a hermetically sealed container along with the release of insects. Each fumigation test is repeated thrice or more along with control. The fumigation will take place in a sealed room at a temperature of 30 ± 2 °C and a relative humidity of 60 ± 5%. After exposure, the insects would be fed with small quantity of food for a week and transported to the recovery room. Adult mortality is noted at different time periods following the conclusion of the exposure period.

Advantages

1. Determination of potency of the insecticide (LD_{50})
2. To find the relative toxicity of different insecticides
3. To find the status of insect resistance to different insecticides
4. Screening of new chemistry molecules
5. To find the effect of synergism and potentiation of a pesticide molecule

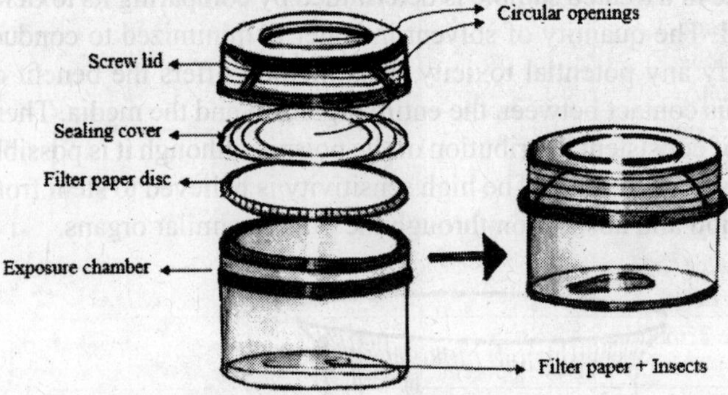

Chimney method bioassay for leafhoppers

1. Small plastic containers of size 20×8 cm covered with the lid having small holes on it for aeration, to be used to confine the leafhopper nymphs.
2. 10 % sucrose solution has to be used to place the petiole of the test leaf to maintain its turgidity.
3. Uncontaminated leaves should be selected, and the petiole should be cut to a length of approximately 4 cm with the help of scissors.
4. The leaves should be dipped into the test solution (insecticide) for 5 seconds holding the leaf by the petiole.
5. After treatment, the leaves should be dried in the open air for approximately 5 minutes.

6. The petiole of a test leaf has to be placed into a small glass vial containing 10% sucrose solution to maintain its turgidity up to 4 days.
7. Glass vial along with the test leaf should be placed into the small plastic containers covered with plastic lid (with small holes) to confine the nymphs of leafhopper
8. Immediately after the test leaves are dried, second to third instar nymphs should be released on to each leaf by using a camel hair brush and cover with its lid.
9. One set of untreated control treated with distilled water has to be maintained simultaneously.

IRAC Bioassay Methods

IRAC No.	Insect species	Stage	Bioassay method	Insecticide group
001	Green peach aphid (*Myzus persicae*)	All Stages of Development	Dipping	Carbamates (1A) Organophosphates (1B)
002	Psyllids (*Psylla* spp.)	All Stages of Development	Dipping	Organophosphates (1B)
003	Two spotted Spider Mite (*Tetranychus urticae*) European Red Mite (*Panonychus ulmi*)	Eggs	Dipping	Clofentezine, Hexythiazox (10A)
004	Citrus red spider mite (*Panonychus citri*) European Red Mite (*Panonychus ulmi*) Twospotted Spider Mite (*Tetranychus urticae*)	Adults	Dipping	Carbamates (1A) Propargite (12C)
006	Red flour beetle Tribolium castaneum	All Stages of Development	Filter-paper	Organophosphates (1B)
008	Tobacco Whitefly Bemisia tabaci	Adults	Dippping	Amitraz (19)
009	Pear Leaf Blister Moth (*Leucoptera scitella*) Apple leaf miner moth (*Lithocolletis blancaedella*)	Larvae Eggs	Dippping	Benzoylureas (15)
010	Western flower thrips (*Frankliniella occidentalis*)	Adults	Dipping	Organophosphates (1B) Cyclodiene organochlorines (2A) Cyclodiene organochlorines (2A) Pyrethroids-Pyrethrins (3A)

IRAC No.	Insect species	Stage	Bioassay method	Insecticide group
				Juvenile hormone analogues (7A) Benzoylureas (15) Diacylhydrazines (18)
012	European Red Mite (*Panonychus ulmi*)	Adults	Petri-dish	METI acaricides and insecticides (21A)
013	European Red Mite (*Panonychus ulmi*)	Adults	Dipping	METI acaricides and insecticides (21A)
014	Western flower thrips (*Frankliniella occidentalis*)	Larvae	Dipping	Benzoylureas (15)
015	Tobacco Whitefly (*Bemisia tabaci*) Glasshouse whitefly (*Trialeurodes vaporariorum*)	Adults	Dipping	Organophosphates (1B) Pyrethroids-Pyrethrins (3A) Neonicotinoids (4A) Pyridine azinomethine derivatives (9B) Flonicamid (29)
016	Tobacco Whitefly (*Bemisia tabaci*) Glasshouse whitefly (*Trialeurodes vaporariorum*)	Nymphs	Dipping	Organophosphates (1B) Pyrethroids-Pyrethrins (3A) Neonicotinoids (4A) Pyriproxyfen (7C) Pyridine azinomethine derivatives (9B) Buprofezin (16) Tetronic and Tetramic acid derivatives (23)
017	Codling Moth (*Cydia pomonella*)	Larvae	Diet	Compounds of Unknown or Uncertain MoA (UN) Organophosphates (1B) Pyrethroids-Pyrethrins (3A) Neonicotinoids (4A) Spinosyns (5) Avermectins-Milbemycins (6) Fenoxycarb (7B) Juvenile hormone analogues (7A) Benzoylureas (15) Diacylhydrazines (18) Indoxacarb (22A) metaflumizone (22B) Diamides (28)

IRAC No.	Insect species	Stage	Bioassay method	Insecticide group
018	Diamondback Moth (*Plutella xylostella*)	Larvae	Dip	Compounds of Unknown or Uncertain MoA (UN) Carbamates (1A) Organophosphates (1B) Cyclodiene organochlorines (2A) Phenylpyrazoles (Fiproles) (2B) Pyrethroids-Pyrethrins (3A) Spinosyns (5) Avermectins-Milbemycins (6) Benzoylureas (15) Diacylhydrazines (18) Indoxacarb (22A) metaflumizone (22B) Diamides (28)
019	Pea aphid (*Acyrthosiphon pisum*) Black bean aphid (*Aphis fabae*) Soybean aphid (*Aphis glycines*) Melon & Cotton Aphid (*Aphis gossypii*) Potato Aphid (*Macrosiphum euphorbiae*) Green peach aphid (*Myzus persicae*)	Adults Nymphs	Dip	Carbamates (1A) Organophosphates (1B) Pyrethroids-Pyrethrins (3A) Neonicotinoids (4A) Pyridine azinomethine derivatives (9B) Diafenthiuron (12A) Tetronic and Tetramic acid derivatives (23) Flonicamid (29)
020	Beet armyworm (*Spodoptera exigua*) Fall armyworm (*Spodoptera frugiperda*) Cotton leafworm (*Spodoptera litura*)	Larvae	Diet	Diamides (28)
022	Tomato leaf miner (*Tuta absoluta*)	Larvae	Dip	Spinosyns (5) Indoxacarb (22A) Diamides (28)
023	Green peach aphid (*Myzus persicae*)	Nymphs	Feeding	Diamides (28)
024	Melon & Cotton Aphid (*Aphis gossypii*)	Adults, Nymphs	Feeding	Diamides (28)

IRAC No.	Insect species	Stage	Bioassay method	Insecticide group
026	Housefly (*Musca domestica*)	Adults	Feeding	Neonicotinoids (4A)
031	Cabbage Flea Beetles (*Phyllotreta* spp)	Adults	Vial	Pyrethroids-Pyrethrins (3A)
032	Asian Citrus Psyllid (*Diaphorina citri*)	Nymphs	Dipping	Carbamates (1A) Organophosphates (1B) Pyrethroids-Pyrethrins (3A) Neonicotinoids (4A) Spinosyns (5) Avermectins (6) Diamides (28)
034	Rice Stem Borer (*Scirpophaga incertulas*) Pink Borer (*Sesamia inferens*)	Larvae	Dipping	Indoxacarb (22A) Diamides (28)
035	Cotton leafhopper (*Amrasca* spp.)	Nymphs	Dipping, Chimney method	Organophosphates (1B) Pyropenes (9D) Flonicamid (29)

3

Insecticide Formulations

Formulation is a word that can be used in a lot of different ways, but at its core, it means putting things together in the right way, usually by following a formula. Insecticides come in different forms, which are called formulations. A formulation is just the way a product is used. The chemical or biological makeup of a pesticide and its active ingredient (also called the active substance) determine how it works on living things. Most of the time, the active ingredient is mixed with other things, and the product as it is sold will need to be weakened even more before it can be used. Formulation makes a drug easier to handle, store, and use. It can also have a big effect on how well it works and how safe it is.

Insecticides come in many different forms because the active ingredients vary in how easily they dissolve, how well they kill pests, and how easy they are to handle and move. The pesticide is made up of both active and non-active (used to be called "inert") chemicals. The manufacturer will need to know the identity of the pest, as well as details about its feeding behavior, reproduction, and life cycle, in order to create an effective formulation. Manufacturers also take into account factors including surface and equipment type, runoff and drift rates, pest habits, and safety when formulating insecticides. Inactive materials (adjuvants/additives) are often added to insecticides to make them stick to the application surface or spread out over leaves. There are solvents (liquids) that dissolve the active ingredient, carriers (liquids or solids) that aid in the delivery of the active ingredient, etc.. Boosting the efficacy of active ingredients requires additives, such as spreaders, stickers, wetting agents, compatibility agents, foaming agents, etc.

Compound inputs encompass a combination of physical, chemical, and biological properties.

- The application input includes factors such as pests, plants, equipment, climatic conditions, and public health.
- The marketing input includes factors such as user-friendliness, attractiveness, safety, durability, and economy.
- The manufacturing input includes production equipment and quality control facilities.

Steps for development (formulation)

- Basic Research Including Preparation for the Lab, Physical, and Chemical Examinations
- Investigational Stage: The concepts of bio-efficacy, phytotoxicity, shelf life, and analytical method are important in the field of formulation.
- Research and Development: Toxicology; Field Trials on a Small Scale
- Commercial: The formulation process, tank mix compatibility, and packaging development are key aspects of product development in various industries. These processes involve the careful formulation of ingredients, ensuring compatibility between different components, and designing appropriate packaging for the final product.

How to choose a desirable formulation

It is common for a single active ingredient to be available in multiple formulations. As an example, deltamethrin, an active ingredient, is available in granules, suspension concentrates, dusts, and aerosols.This suggests that different formulations of the same active ingredient exhibit distinct behaviors. Hence, the utilization of deltamethrin dust is highly suitable for the application within wall voids, as it effectively adheres to the inner wall surface and effectively manages pests that crawl, whereas deltamethrin aerosol is better suited for the control of pests that fly. Understanding the distinctive attributes of a particular formulation facilitates the selection of an appropriate product to meet specific requirements and enhances the efficacy of its utilization.

Formulations process

The toxicant will undergo dilution with additional substances in order to achieve the desired formulation, at which point it will be referred to as the active ingredient (a.i). The additional components incorporated into the formulation are commonly referred to as adjuvants. An adjuvant refers to a substance that augments the physical characteristics of an active ingredient. Alternatively, it may lack inherent pesticidal properties. Several examples of substances include xylene, talc, flour, and bran.

The process of formulating insecticides necessitates the incorporation of diverse auxiliary substances, including dust, diluents, solvents, emulsifiers, wetting and dispersing agents, stickers, deodorants, and masking agents. Accessory agents are assigned names that indicate their particular functions or their ability to improve the formulation. For instance, a substance referred to as a 'spreader' would facilitate the even distribution of the pesticide across a given surface. The carrier is a crucial element within a formulation as it serves the purpose of transporting the active ingredient to the desired surface. The carriers utilized in formulations can be classified as either dry or liquid.

Irrespective of their origins or methods of production, pesticide active ingredients exhibit significant variations in their physical and chemical properties. One notable variable among these properties is solubility. Certain substances have the ability to dissolve in water, while others lack this property and instead exhibit solubility in oils or organic solvents. Nevertheless, numerous solvents of this nature are inaccessible to individuals who apply them or pose potential safety hazards. Several active ingredients exhibit limited solubility in various solvents. Active ingredients are formulated into end-use products based on two key factors: solubility and application. Typically, liquid pesticide products can be categorized into one of the following types.

Solution

The formation of a solution occurs through the process of dissolving a solute in a solvent. A veritable solution is a composite substance that is impervious to separation through filtration or other mechanical methodologies. Once prepared, a genuine solution exhibits no tendency to separate, and thus does not require any form of agitation, such as shaking or stirring, to maintain the homogeneity of its constituent components within the solution. Transparent solutions possess the characteristic of light transmission, enabling the passage of light through them.

Dissolution: is the movement of mass from the solid surface into the liquid phase, also known as solubilization. Dissolution is illustrated by dissolving sugar into water by stirring.

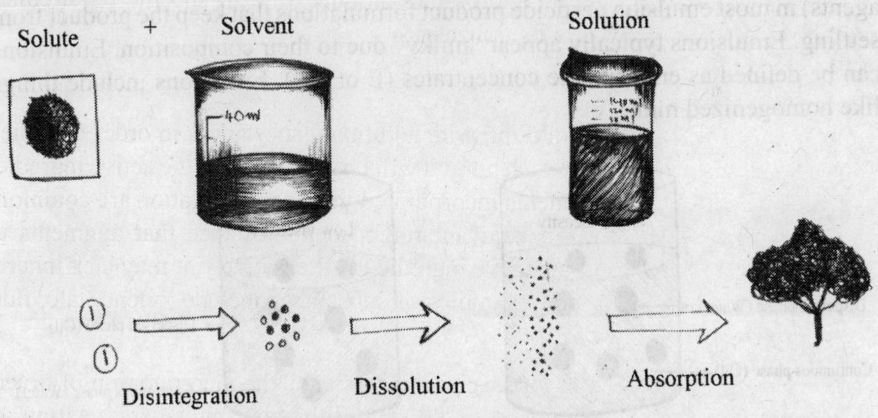

Suspension

One type of liquid mixing is a suspension. However, a suspension is created when tiny solid particles are evenly distributed throughout a liquid. To keep particles dispersed evenly in a suspension, shaking is required. If not, the components of a suspension mixture that have not dissolved will rise to the surface. The majority of suspensions are opaque or foggy and will not allow any light to flow through. When combined with water to create a spray, pesticide suspensions form a more diluted form of the original solution that is not water soluble. Shake suspension formulation thoroughly before using, as directed on the label. The label will also instruct the user to only use spray equipment with sufficient agitation to spread the final mixture throughout the spray tank.

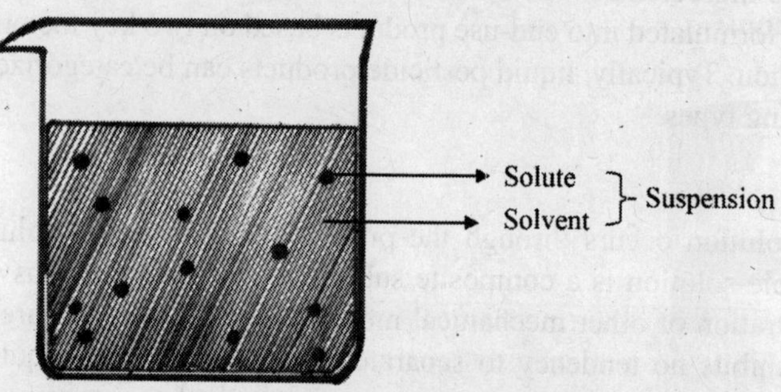

Emulsion

Emulsions are a type of suspension in which one liquid is suspended within another. The individual qualities and identities of each component are preserved. An active ingredient is first dissolved in an oil-based solvent, such as vegetable oil, before being further diluted with water to create an emulsion. An emulsion may require agitation to prevent separation. There are additives (emulsifiers or emulsifying agents) in most emulsion pesticide product formulations that keep the product from settling. Emulsions typically appear "milky" due to their composition. Emulsions can be defined as emulsifiable concentrates (E or EC). Emulsions include things like homogenized milk.

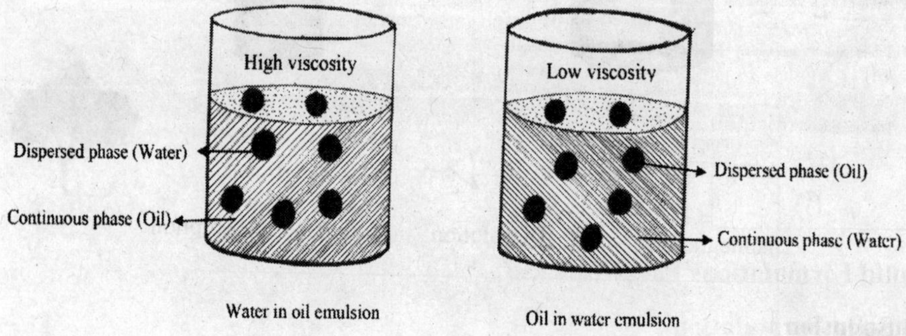

Dispersion: A mixture refers to the phenomenon where one phase of a substance becomes separated from another phase of a different substance. Phase separation occurs due to the lack of solubility between the components, preventing them from dissolving into each other.

The term "dispersed phase" refers to the phase that exists in the form of colloidal particles and is subject to scattering. The dispersion medium refers to the medium in which the colloidal particles are dispersed. In the context of a starch solution, it can be observed that starch functions as the dispersed phase, whereas water serves as the dispersion medium.

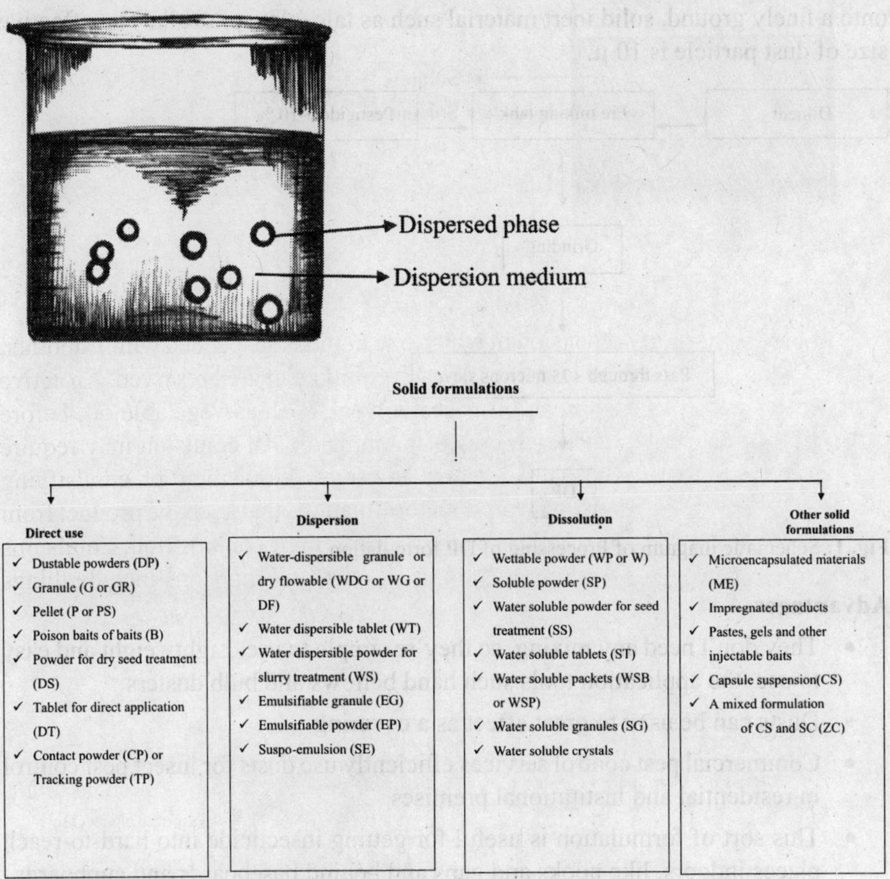

Solid Formulations For Direct use

I) Solid Formulations

There are two broad categories of solid formulations: those that require no additional water for usage, and those that must be diluted before being sprayed. Dusts, granules, and pellets are all ready-to-use solid forms, while wettable powders, dry flowables, and soluble powders require the addition of water before use.

1. Dustable Powder (DP)

Dusts, denoted as D, are prepared as insecticide formulations that are readily available for use without the need for dilution. The active substance exists in two forms: a crystalline solid with a particle size ranging from 1 to 10 micrometers, or a liquid or waxy compound that is absorbed onto an inert mineral carrier. The active ingredient is commonly found in concentrations below 10% by weight, while the remaining portion consists of a finely ground mineral diluent. Dusts are produced through the process of sorption, wherein an active ingredient is adsorbed onto a finely ground, solid inert material such as talc, clay, or chalk. The effective size of dust particle is 10 μ.

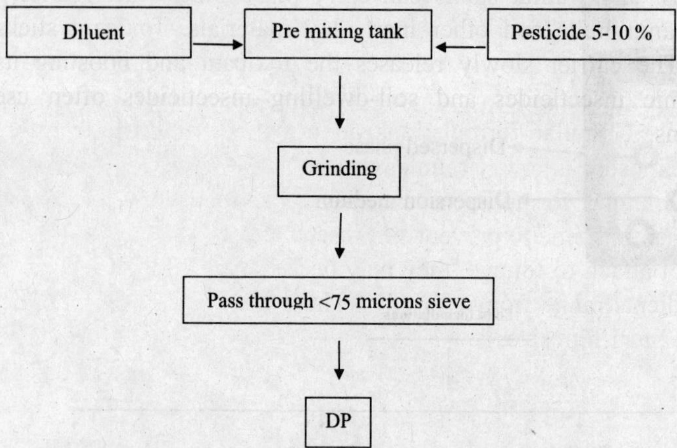

Fig. 1: Schematic diagram of Processing of DP formulation

Advantages

- They don't need any mixing, so they're simple to use. Lightweight and easy to use, the application tools such hand bellows and bulb dusters
- Dusts can be used to great effect as a covering
- Commercial pest control services efficiently use dusts for insect pest control in residential and institutional premises
- This sort of formulation is useful for getting insecticide into hard-to-reach places indoors, like nooks and gaps and behind baseboards and cupboards.

Disadvantages

Outdoor applications rarely use dust formulations due to their high drift potential.

Eg : Quinalphos 1.5 % DP

Consitutents role	Name of the constituent	Contents (% w/w)
Active ingredient	Quinalphos Tech (based on 70% w/w a.i.)	2.14
Inert carrier and adhesive agent	China Clay	25.00
Inert carrier	Soap Stone	72.86
Total		100.00

2. Granule (G or GR)

Granules are pesticide pellets larger than dust. A dry, ready-to-use granular formulation contains a small quantity of pesticide and an inert carrier. Clay, pulverized corn cobs, and walnut shells can carry insecticides (inert carrier). Diluents include peanut hulls and other inert clay materials. Toxicant sticks to these particles. The carrier slowly releases the toxicant and boosting its permanence. Systemic insecticides and soil-dwelling insecticides often use granular formulations. Granular formulations do not blow or drift, therefore inadvertent human exposure is low. Granules are dry, big granules with 2–20% active component. Granular formulations are easy to apply and wander lessA granule is given at planting time to prevent soil insects from damaging the roots. Although they do not attach to foliage, they may become stuck in the whorls of plants. Active ingredients range from 1 to 15%. The most widely used granular formulation are of 0.2 to 1 mm size.

Advantages

- Readily usable
- Simple to apply,
- Can pass through dense foliage,
- There is a limited risk of drift, and particles will settle fast.
- The weight of a formulation facilitates its movement from foliage to a non target location, such as soil or water. Simple application equipment.
- A slow-release coating may lead it to degrade at a slower rate than WPs or ECs.

Disadvantages

- Calibration is more difficult on application equipment than on spray equipment. (Weight is used to measure the quantity of particles released instead of volume)
- Active ingredients are expensive per unit weight
- In dry conditions, the pesticide may not work because it requires moisture to work
- There are hazardous conditions on steep slopes, frozen soil, and near non-target plants

- In addition to targeting organisms, it may also be attractive to non-target organisms, such as birds, squirrels.
- The distribution of certain devices is difficult to achieve uniformly.

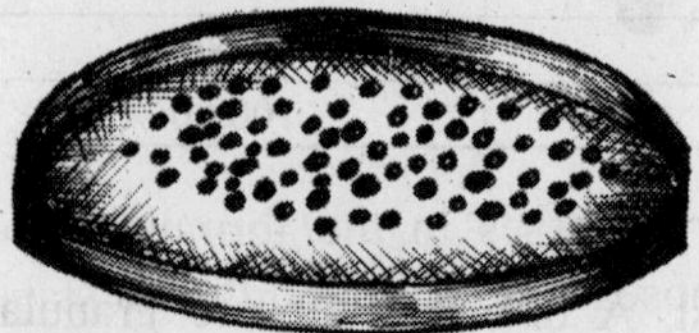
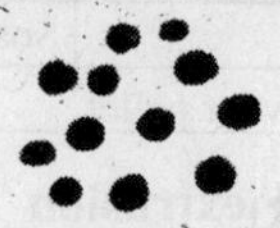

Eg: Chlorpyriphos 10 % GR

Consitutents role	Name of the constituent	Contents (% w/w)
Active ingredient	Chlorpyriphos Technical	10.00
Solvent (Evaporated during manufacturing process)	Aromax	2.00
Inert granules	Bentonite (850/420 microns)	Q.S.
Total		100.00

3. Pellet (P or PS)

Pellets are similar to granules, however they are made differently. Combining an active substance with an inert ingredient produces a slurry (a thick liquid combination). Using a die and pressure, the slurry is extruded into a homogeneous in size and shape but is significantly larger than a granule. The pellets, like granules, are ready to use, apply dry, and contain a small quantity of active component (typically 10 to 20% by weight), as well as an inert carrier. Drift is not an issue with this formulation. Use on steep slopes or near the root systems of non-target plants poses additional risks. Pellets offer a high level of applicator safety.

Advantages

- The pellets are ready to use
- Simple to use, even by hand
- The hazard associated with the applicator can be minimized
- There is no drift
- It is an efficient approach for spot treatment

Disadvantages

- The active ingredient is costly
- Hazardous conditions can arise on steep slopes, in close proximity to desired vegetation, and on frozen soil
- Shipping large numbers can be difficult
- Hard to disperse evenly around obstructions

Eg: Arka neem seed powder pellet

Spheronization and **extrusion** are both key processes in the formulation of granules or pellets, but they serve different purposes and occur at different stages in the manufacturing process. Extrusion is about shaping the material into a consistent form, while spheronization is about transforming that form into smooth, round granules or pellets, enhancing the final product's performance and handling characteristics.

Aspect	Extrusion	Spheronization
Purpose	To shape material into cylindrical or rod-like forms	To convert cylindrical extrudates into spherical granules or pellets
Process	Material is forced through a die to create elongated shapes	Extrudates are rolled into spheres using a rotating frictional surface
Output Shape	Cylindrical rods or strands	Spherical granules or pellets
Stage in Manufacturing	Typically the first step before spheronization	Follows extrusion in the granulation process
Function in Formulation	Provides a consistent initial shape for further processing	Refines the shape for better flow, uniformity, and handling
Material State	Usually involves a wet mass	Uses the extrudates produced from the extrusion process
Application	Suitable for initial shaping of material	Suitable for achieving final spherical shape and uniformity
Advantages	Enables controlled and consistent shape formation	Enhances packing density, reduces dust, and improves controlled release properties

4. Poison baits of Baits (B)

Poisonous baits consist of pesticide-infused food-like substances that are intentionally formulated to entice and be consumed by insects or other pests, ultimately leading to their demise through poisoning. The poison baits are composed of a base or carrier substance that is appealing to the targeted pest species, along with a chemical toxicant present in relatively small amounts. Baits typically possess a relatively modest concentration of active ingredient, typically ranging from 1/4 to 5%. The toxic substance is combined with a variety of carriers or attractants, including bran, orange pulp, corn cobs, and sugars. Baits are frequently employed for the purpose of managing rodent populations, encompassing both mice and rats, and are additionally utilized for the control of various other insect species. Poison baits are employed for the purpose of managing fruit flies, chewing insects, wireworms, white grubs in the soil, household pests such as roaches, and slugs. Baits are frequently employed for the control of subterranean soil pests, including ants, mole crickets, and cutworms. Bait formulations have the potential to be utilized in both indoor and outdoor settings, strategically positioned or dispersed in areas where they can be ingested by the intended pest population.

Advantages

- This formulation is for immediate utilization
- It is not necessary to cover the entire area as pests are attracted to the bait
- Implement measures to manage pests that exhibit mobility within a given geographical region

Disadvantages

- The given stimulus has the potential to elicit interest and appeal among both children and domesticated animals
- The potential consequences of outdoor usage include the mortality of domestic animals and unintended harm to non-target wildlife
- The potential consequences of outdoor usage include the mortality of domestic animals and unintended harm to non-target wildlife
- Dead pests might generate an odor problem
- Other animals may become poisoned as a result of eating poisoned pests

4.1. Grain Bait (AB)-Bromodialone 0.005%

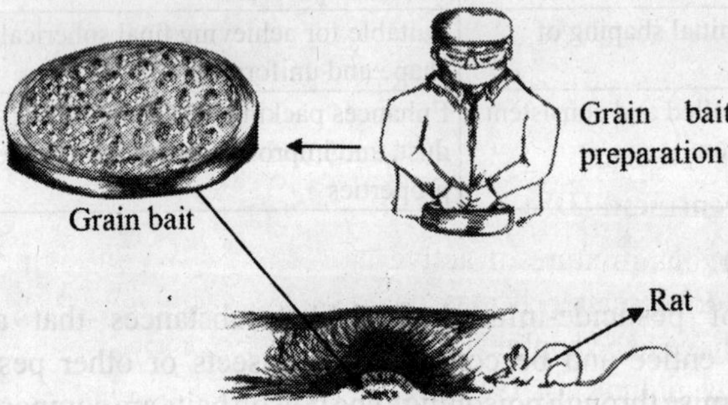

4.2.Granular Bait (GB)

4.3. Block Bait (BB)- Bromodialone 0.005%

4.4. Paste Bait (PB)- Brodifacoum 0.005%

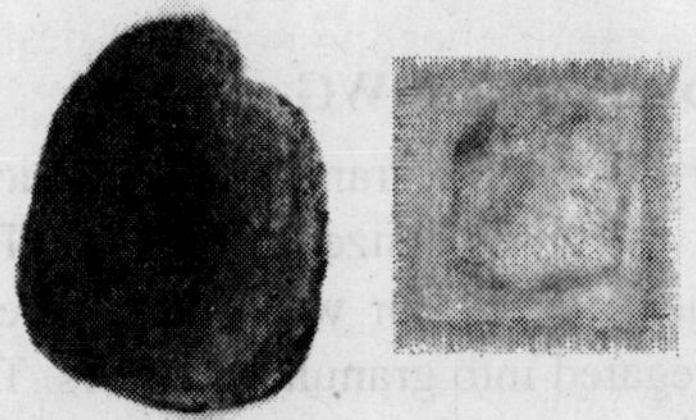

5. Powder for dry seed treatment (DS)

DS consists of a homogeneous mixture of active ingredient, filler and requisite formulants, including colouring matter. It may be coated on seeds in dry form or the surface on which these have to be applied may be pre-wetted before application. Two important requirements include non interference with plantability of seed and no undesirable effect on seed viability.

6. Tablet for direct application (DT)

These are designed for immediate field application without prior dispersal or dissolving in water. Tablets are pre-formed solids with regular shape and dimensions, typically circular, and with either flat or convex faces, with a distance between faces smaller than the diameter.

7. Contact powder (CP) or Tracking powder (TP)

Tracking powder is a specialized dust used for observing and managing pests like insects and rodents.

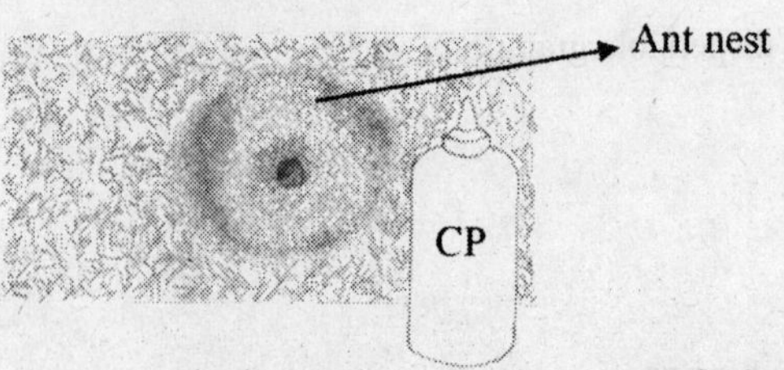

Solid Formulations for Dispersion

1. Water Dispersible Granule or Dry Flowable (WDG or WG or DF)

Similar to wettable powder formulations, water-dispersible granular formulations consist of the active component produced as granule-sized particles. The production process for dry flowables is identical to that for wettable powders, with the exception that the powder is first aggregated into granular particles. The look of dry flowables is similar to that of granules, but their behavior is more like to that of wettable powders. Just like a wettable powder, they become a spray suspension after being combined with water. The formulation is readily dispersible in water to produce a **suspension**. The typical active component content of this dry formulation is between 70% and 90%. Compared to WPs, dry flowables have a number of advantages because to their form, including the fact that they can be "poured" and measured with the same ease as liquid and the fact that very little dust is discharged into the air during the mixing and weighing processes. Such a solid particle is typically very finely powdered and suspended in a liquid, containing a high concentration or big amount of the active agent, and must be mixed with water before application.

Advantages	Disadvantages
• No dusting • No caking during storage • Simple to store, transport, and use • Less exposure of the applicator when mixing dry formulations • Concentrate spills are easiest to remove from porous surfaces	• Good agitation required • Residues may be visible • Abrasive to sprayers • May be slightly more expensive than other dry formulations • If you pour quickly from a big container, the product mass can settle to the bottom of the tank and make it hard to mix. • Expensive manufacturing cost in comparison to WP

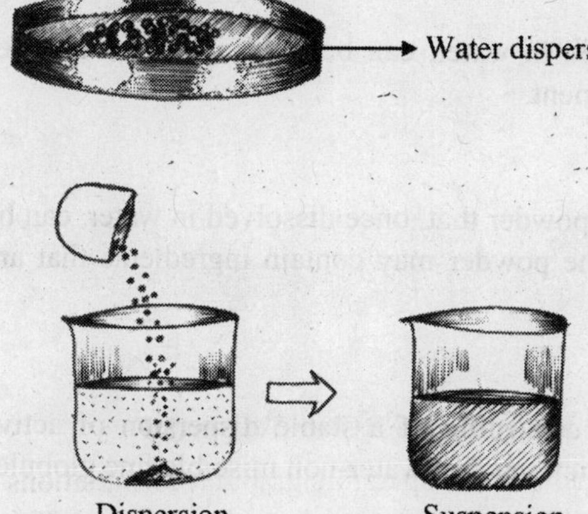

Eg: Thiamethoxam 25 % WG

Consitutents role	Name of the constituent	Contents (% w/w)
Active ingredient	Thiamethoxam Technical	25.00
Dispersing Agent	Sodium Lignosulfonate	5.00
Wetting Agent	Sodium Lauryl sulfate	3.75
Plasticiser	Butylated polyvinyl pyrrolidone	1.25
Carrier	Diatomaceous earth	5.00
Binder	Corn starch	60.00
Total		100.00

Eg: Fipronil 80 % WG

Consitutents role	Name of the constituent	Contents (% w/w)
Active ingredient	Fipronil Technical	80.00
Dispersing Agent	Sodium plycarboxylate	2.90
Dispersing Agent	Sodium dioctyl sulfosuccinate 85 %	1.90
Antifoam agent	Silicon oil emulsion	0.30
Wetting agent	Sodium lignosulfonate	Balance (Q.S)
Total		100.00

2. Water Dispersible Tablet (WT)

Tablets that, when broken up in water, will spread out the active ingredient. Each tablet is meant to be used by itself. Tablets are solids that are already made and have the same shape and size. They are usually round with flat or convex faces, and the distance between the sides is less than the diameter.

3. Water Dispersible Powder for Slurry Treatment (WS)

It consists of active ingredient, carrier, and necessary formulants, including colouring matter. It is free fine powder, free from visible extraneous matter and hard lumps.

4. Emulsifiable Granule (EG)

A granule formulation once dissolved in water, can be used as an oil-in-water emulsion to deliver the active component.

5. Emulsifiable Powder (EP)

An active substance in the form of a powder that, once dissolved in water, can be applied as an oil-water emulsion. The powder may contain ingredients that are insoluble in water.

6. Suspo-Emulsion (SE)

A fluid, heterogeneous formulation consisting of a stable dispersion of active ingredient(s) in the form of solid particles and of water-non miscible fine globules in a continuous water phase.

Eg: Pyriproxyfen 5 % + Diafenthiuron 25 % SE

Consitutents role	Name of the constituent	Contents (% w/w)
Active ingredient	Pyriproxifen Technical	5.00
Active ingredient	Diafenthiuron Technical	25.00
Nonionic surfactant	Ethoxylate octyl phenol	10.00
Nonionic surfactant	Tristyrylphenol ethoxylate	2.50
Anionic surfactant	Polyarylphenyl ether sulphate, ammonium salt	2.00
Nonionic surfactant	Copolymer butanol (Ethylene Oxide/ Propylene Oxide)	1.50
Stabilizing agent	Silica	0.80
Wetting agent	Glycol	4.00
Anti-foaming agent	Polydimethyl siloxane	0.50
Broad-Spectrum Antimicrobial Activity	1,2 benzisothiazol-3-one	0.20
Stabilizing agent	Poly saccharides (Xanthan gum)	0.20
Medium for Dispersion	Water	Q.S
Total		100.00

Solid Formulations For Dissolution

1. Wettable Powder (WP or W)

Spraying is done with a dry (powder) mixture that is mixed with water to make a solution. Unlike a liquid powder, it does not dissolve in water. Wettable powders are very finely ground solids, usually mineral clays that can absorb an active ingredient. When mixed with water a wettable powder forms a suspension and the mixture must be stirred in some way to avoid lumpy mixtures that can clog nozzles and lead to bad application. Most of the time, the active ingredient (a.i.) in a wettable powder is between 25% and 75%. The active ingredient is mostly a concentrated dust that has been ground into a powder and mixed with a fine clay-like substance to make it easier to use. Even though a WP is made from dry ingredients, it is mixed with water, which is the secondary agent. Most of the ingredients in a WP formulation don't mix with water, but this problem can be solved by adding a bipolar substance like a wetting agent and a dispersing agent. These are chemicals that help the powder get wet and spread the active ingredient all over the spray tank.

This WP suspension, however, is highly unstable and requires regular agitation to prevent settling and maintain the appropriate efficacy. One of the most common types of pesticide is a wettable powder. As long as agitation is possible, they can be employed for most pest problems and in most spray equipment.

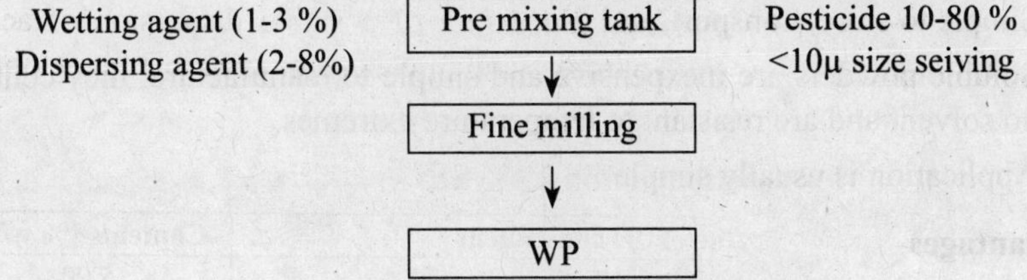

Advantages	Disadvantages
• Simple to handle, store, and transport • Relatively inexpensive • Less likely to harm treated plants, animals, and surfaces than ECs and other petroleum-based insecticides • Easily measured and mixed • Less absorbed by the skin and eyes than ECs and other liquid forms • Higher concentration of active ingredient can be incorporated • Solubility in organic solvent is not a criteria	• Possibility of inhalation while pouring the powder • Need good and steady agitation (usually mechanical) in the spray tank and quickly settle out if the agitation is turned off • Wettable powder is rough on sprayers, which makes them wear out quickly • Residues may be noticeable • When splashes of concentrate happen on porous surfaces, they can be hard to clean up • Hard to mix in hard water water • Dust explosion hazards during production • Handling problem • Requires expensive packaging, as volumes are high

Eg: Diafenthiuron 50 % WP

Consitutents role	Name of the constituent	Contents (% w/w)
Active ingredient	Diafenthiuron Technical	50.00
Wetting agent	Ethoxylated fatty alcohol	12.00
Dispersing Agent	Condensed alkyl naphthalene sulphonate sodium salt	15.00
Carrier	Precipitated silica	Q.S.
Total		100.00

2. Soluble Powder (SP)

This is a dry powder preparation with a high concentration of active component (15% to 95%) that dissolves in water (or another liquid) and creates a solution that may be applied. Soluble powders resemble wettable powders, but when mixed with water, they produce a real solution. A soluble powder entirely dissolves in water, but a wettable powder contains an emulsifier that creates a homogeneous suspension (colloid). These are finely ground, highly concentrated powders that are soluble in water and do not require a wetting agent.

Advantages

- Simple to mix
- Only minor agitation is required.
- Simple to store, transport, and use
- Soluble powders are inexpensive and simple to manufacture; they contain no solvent and are resistant to temperature extremes.
- Application is usually simple

Disadvantages

- Risk of inhalation while pouring powder spills can be difficult to clean up from porous surfaces
- In contrast to water-dispersible powders, the likelihood of sludge formation and blockages in line filters and spray nozzles is minimal.

Eg: Cartap Hydrochloride 50 % SP

Consitutents role	Name of the constituent	Contents (% w/w)
Active ingredient	Cartap Hydrochloride Tech. (Based on 98 % w/w)	51.20
Surfactant	Alkyl phenol ethylene oxide	0.70
Stabilizer	Isopropyl acid phosphate	0.30
Dye	Acid green dye	0.05
Diluent	4-0-B-D gatacto pyronosyl alpha-D glucophyranose/lactose	47.75
Total		100.00

Eg : Acetamiprid 20 % SP

Consitutents role	Name of the constituent	Contents (% w/w)
Active ingredient	Acetamiprid Technical	20.00
Primary Surfactant	Blend of sodium dodecyl benzene sulfonate	4.54
Inert filler	Sodium sulfate	
Co-Surfactant and Dispersing Agent	Sodium alkylate	
Buffering Agent and Chelating Agent	Sodium citrate dihydrate	0.10
Filler and Stabilizing Agent	Calcium hydrogen Phosphate dihydrate	
Colouring agent	Brilliant Blue	0.05
Diluent	Lactose monohydrate	Q.S.
Total		100.00

3. Water Soluble Powder for Seed Treatment (SS)

The powder should be dissolved in water prior to being applied to the seed. The composition consists of an active ingredient along with any required formulants, including coloring agents.

4. Water Soluble Tablets (ST)

The ST formulation is presented in tablet form, allowing for the administration of the active ingredient upon disintegration and subsequent dissolution in water.

5. Water-Soluble Packets (WSB or WSP)

Water-soluble packets effectively mitigate the risks associated with the mixing and handling of certain insecticides that possess high toxicity levels. The manufacturing process involves the precise packaging of specific quantities of wettable powder or soluble powder formulants within a specialized plastic bag.

Eg: In Indian market, Diafenthiuron 50 WP available in WSP (Water soluble packets). It can be easily soluble in water once a 20 g pouch is dropped in water.

6. Water Soluble Granules (SG)

A formulation consisting of granules to be applied as a true solution of the active ingredient after dissolution in water, but which may contain insoluble inert ingredients.

Eg: Dinotefuran 20 % SG

Consitutents role	Name of the constituent	Contents (% w/w)
Active ingredient	Dinotefuran Technical	20.00
Wetting agent	Dialkyl sulfosuccinate sodium salt	1.00
Surfactant	Alkylbenzene sulfonate sodium salt	3.00
Dispersing agent	Beta naphtysulfonate formaldehyde condensate sodium salt	3.00
Carrier or anti-caking agent	Amorphous silicone dioxide	2.26
Filler or diluent	Lactose monohydrate	Q.S.
Total		100.00

Eg : Emamectin benzoate 5 % SG

Consitutents role	Name of the constituent	Contents (% w/w)
Active ingredient	Emamectin benzoate Technical (Based on 95 % w/w min. purity of a.i)	5.27
Surfactant	Sodium N-methyl-N-oieoyl taurate	7.50
Dispersing agent	Sodium alkyl naphthalene sulfonate	1.00
Spreader	Polydimethylsiloxane	0.10
Filler or diluent	Lactose anhydrous	Q.S.
Total		100.00

7. Water soluble crystals

The crystals are packed in water-soluble plastic pre-packs. Once these are dropped in the spray solution they will be easily dissolved

Other Solid Formulations

Controlled release insecticide formulations

1. Polymer membrane- e.g. Microencapsulation
2. Reservoirs for insecticides
3. Rubber, polyvinyl chloride, gypsum-wax combination, polyester and acrylic resins, cellulose, starch, and gels like alginate and lignin are all examples of matrix systems that physically trap insecticides
4. Polymer systems containing covalently bound insecticides
5. Coated pesticide granules

1. Microencapsulated Materials (ME)

Microencapsulated formulations are made by covering pesticide particles (either liquid or dry) with plastic. Insecticides in microcapsules can be diluted with water and sprayed. The plastic coating degrades after spraying and gradually releases the drug.

Advantages	Disadvantages
• Longer-lasting effects can be achieved through gradual or delayed release of the active component • Lower mammalian toxicity • Reduced phytotoxicity • Environmentally friendly • User friendly • More secure handling • Greater residual management • Less plant harm	• Expensive than the conventional formulations • Polymer may have adverse impact on the environment • Due to variation in the release at different locations, the performance may not be uniform • The microencapsulation process may not be quantitative. Hence waste water will have considerable amount of active ingredient. • Require constant agitation • Bees take capsules back to hive • Slower decomposition may result in more residue at harvest

Microcapsules in agriculture

Lambda-cyhalothrin fast-acting capsules, sold under the brand name "Zeon Technology"

Slow / controlled release formulation

- Active ingredient is slowly released
- Increase size of particle thus reducing the specific surface/unit mass
- Advantage
- Extended bioactivity
- Reduced phytotoxicity
- Pesticide level in the environment is reduced
- Reduce mammalian toxicity

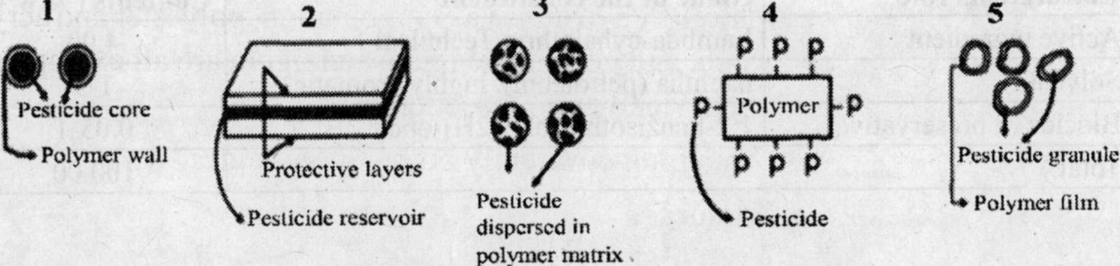

2. Impregnated Products

The active ingredient in some insecticides is embedded inside a plastic or other solid material. Pests are kept at bay as the pesticide evaporates or is gradually discharged, with the resulting fumes. Typical cases include:

- Livestock ear tags.
- Adhesive tapes and plastic pest strips.
- Pet collars.

Fertilizers may also be impregnated with insecticides. Over time, these chemicals dissipate and control pests in the area.

3. Pastes, Gels and other Injectable Baits

Ants and cockroaches are mostly controlled with pastes and gels. Cockroach control now relies on pastes and gels. Syringes and bait guns apply pastes and gels.

Eg : Fipronil 0.05 % GEL

Consitutents role	Name of the constituent	Contents (% w/w)
Active ingredient	Fipronil Technical	0.05
Phagostimulant	Fructose	5.00
Humectant	Glycerine	25.00
Thickener	Starch powder	32.00
Humectant, stabilizer, sweetner	Sorbitol	25.00
Thickener	Polysaccharide gum	0.90
Stabilizer	Butylated hydroxyl Toluene (BHT)	0.50
Colourant	Brown	0.10
-	Distilled water	Q.S.
Total		100.00

4. Capsule Suspension (CS): Normally diluted with water, a stable capsule suspension. A stable suspension of capsules in a fluid, normally intended for dilution with water before use.

Eg: Lamda cyhalothrin 4.9 CS

Consitutents role	Name of the constituent	Contents (% w/w)
Active ingredient	Lambda-cyhalothrin Technical	4.90
Solvent	naphtha (petroleum), highly aromatic	1-5
Biocide or preservative	1,2-benzisothiazol-3(2H)-one	0.05-1
Total		100.00

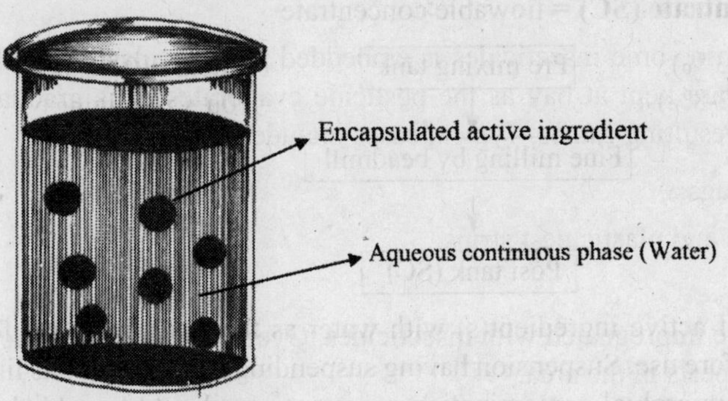

5. A mixed formulation of CS and SC (ZC)

A stable suspension of capsules and active ingredient(s) in fluid, normally intended for dilution with water before use.

Eg : Thiamethoxam 12.6 % + Lambda Cyhalothrin 9.5 % ZC

Consitutents role	Name of the constituent	Contents (% w/w)
Active ingredient	Lambda-cyhalothrin Technical	9.50
Active ingredient	Thiamethoxam Technical	12.60
Thickener	Octadecanoic acid,12-hydroxy homopolymer octadecanoate (polymer)	5.84
Surafactant and stabilizer	Styrylphenol polyethoxyester phosphate	7.76
-	Other ingredients	8.79
-	Water	Q.S.
Total		100.00

II) Liquid Formulations

However, liquid formulations varied greatly in properties that affect selection, pace and manner of application, and environmental impact.

1. Suspension Concentrate (SC)
2. Oil Dispersion (OD)
3. Emulsifiable Concentrate (EC)
4. Ultralow Volume Concentrates (ULV)
5. Flowable Suspensions or Aqueous Suspension (F, L or AS)
6. Flowable Concentrate for Seed Treatment (FS)
7. Dispersible Concentrate (DC)
8. Water Soluble Concetrate (Or) Water Soluble Liquids (or) Soluble Liquids (SL)

1. Suspension Concentrate (SC) = flowable concentrate

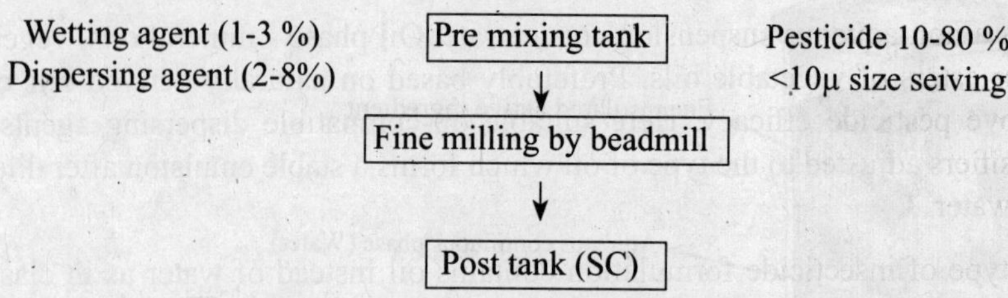

A stable suspension of active ingredient(s) with water as the fluid, intended for dilution with water before use. Suspension having suspending agent and looks like thick syrup. To avoid microbial contamination preservatives should be added to this formulation. Generally, while formulating in pre-mix and post tanks foaming will be formed to avoid an antifoaming agent will be added to SC formulation.

Eg: Fipronil 5 % SC

Consitutents role	**Name of the constituent**	**Contents (% w/w)**
Active ingredient	Fipronil Technical (Purity 90 %)	5.00
Carrier	Sunflower oil	5.00
Wetting agent	(Napthalene sulfonate)-3	3.00
Suspending agent	Sodium ligno sulfonate	8.00
Sticking/stabilizing agent	Carboxy methyl cellulose	4.00
Dispersing agent	Sodium salt of Dinaphthyl methanol sulphonic acid	3.00
Blend of emulsifiers	Proprietory blend of alkyl phenol ethoxylate, triglyceride ethoxylate, calcium alkyl benzyl sulfonate)	8.00
	Water	Q.S. to make 100
Total		100.00

Eg: Diafenthiuron 47.8 % w/w SC

Consitutents role	**Name of the constituent**	**Contents (% w/w)**
Active ingredient	Diafenthiuron Technical (Purity 90 %)	47.8
Wetting & Dispersing agent	Tristyrylphenol polyethylene glycol phosphoric acid, ester	1.90
Antifreeze	1,2 - propylene glycol	4.80
Antifoaming agent	Polydimethylsiloxane	0.48
Thickener	Polysaccharide derivative	0.09
Preservative	Formaldehyde (35 % aqueous solution)	0.09
Solvent	Water	Q.S.
Total		100.00

2. Oil Dispersion (OD)

OD is a non-aqueous suspension concentrate. Oil phase - mineral oils, vegetable oils or esters of vegetable oils. Preferably based on naturally derived oils could improve pesticide efficacy. Here suitable oil-compatible dispersing agents and emulsifiers adjusted to the type of oil which forms a stable emulsion after dilution with water.

This type of insecticide formulation contains oil instead of water as in classical suspension concentrate and typically has better retention and coverage.

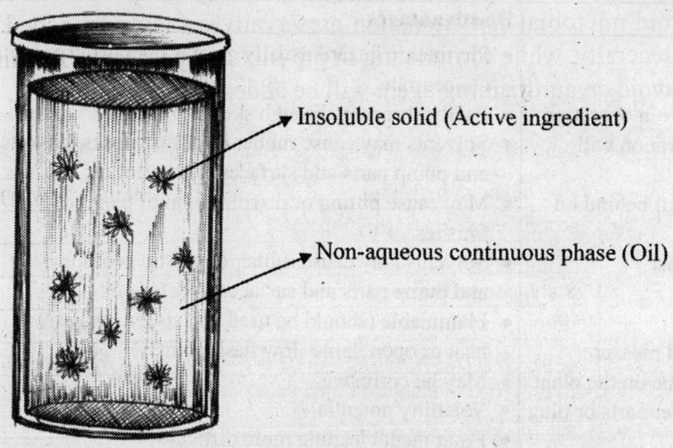

Eg: Cyantraniliprole 10.26 % w/w OD

Consitutents role	Name of the constituent	Contents (% w/w)
Active ingredient	Cyantraniliprole Technical	10.26
Wetting & Dispersing agent	Water	4.11
Emulsifier	Dodecyl benzene sulphonate in methylated and ethoxylated Seed Oil	25.00
Thickener	Fumed Silica	1.31
pH adjuster	Citric acid	0.02
Surfactant	Polyoxyethylene Sorbitol Fatty Acid	5.00
Carrier	Methylated Seed Oil	Q.S.
Total		100.00

3. Emulsifiable Concentrate (EC)

These liquid formulations contain toxicants and emulsifiers in organic solvent. These are leave no residue on fruits and veggies, unlike other wettable powders. Organic solvents (Petrolium ether) in the mixture may harm sensitive plants. Urban and industrial pest sprays use emulsifiable concentrates. The a.i. is dissolved in xylene and an emulsifier to combine with water. Inert components show the insecticide's emulsifying agent proportion. Wetting agents and emulsifiers have hydrophilic and hydrophobic ends. Long-chain emulsifying agents wrap around oil droplets and bind oil-water surfaces to prevent separation.

When mixed with water, the EC creates a milky-colored emulsion that is stable for several hours without agitation.

Emulsifier 5-10% Solvent → Mixing/Dissolving ← Pesticide 10-50%

↓

EC

Advantages	Disadvantages
• Low-agitation • Non-abrasive • While machinery is in operation, no settling or separation will occur • Minimal residue left behind on surfaces • Low production cost • Easy technology • Better bioefficacy • Easy to handle and measure • Little visible residue on the plant • Do not wear sprayer parts or plug screens or nozzles	• Phytotoxic hazard usually greater than water soluble concentrate • Easily absorbed through skin of humans or animals • Solvents may cause rubber or plastic hoses, gaskets, and pump parts and surfaces to deteriorate • May cause pitting or discoloration of painted finishes • Solvents may cause rubber or plastic hoses, gaskets, and pump parts and surfaces to deteriorate • Flammable (should be used and stored away from heat or open flame- low flash point) • May be corrosive • Volatility potential • Equipment cleaning more difficult • High volume of solvent • High dermal and eco toxicity • Problems in packaging disposal • May burn tender plant foliage

Answer this questions.

1. What is role of emulsifying agent in Emulsifiable concentrate (EC)?

A. Emulsifying agent convert insoluble toxicant to soluble form.

Eg: Dimethoate 30 % EC

Consitutents role	Name of the constituent	Contents (% w/w)
Active ingredient	Dimethoate Technical	35.50
Stabilizer	Epichlorohydrin	1.00
Emulsifying agent	Non-ionic poly oxy ethylene ether	8.00
Solvent	Aromax	10.50
Solvent	Xylene	30.00
Solvent	Cyclohexanone	15.00
Total		100.00

Eg: Lambda Cyhalothrin 5 % EC

Consitutents role	Name of the constituent	Contents (% w/w)
Active ingredient	Lambda Cyhalothrin Technical	5.00
Stabilizer	Epichlorohydrin	1.00
Emulsifying agent	Creslox AE 4 (Mixture of calcium salt of Alkyl Benzenesulphonate)	20.00
Emulsifying agent	Alkylphenol ethoxylate	57.00
Solvent	Methanol, Isobutanol and 23 % Aromex	7.00
-	Others	Q.S.
Total		100.00

4. Ultralow Volume Concentrates (ULV)

Ultra-low-volume concentrates have almost 100% active ingredient. These are both a formulation and an application technique, and are highly concentrated formulations designed to be used in specialized spray equipment that atomizes the concentrate droplets. The ULV spray equipment is used by most aerial applicators (airplane sprayers) that treat forested lands or large agricultural acreages. These kill the pest directly, usually by exposing it to lethal substances or unsuitable environmental conditions. These reduce the reproductive potential of a pest population often by modifying its environment (biotic or abiotic) or by restricting its movement. These modify the pest's behaviour to make it less troublesome (attract, repel, confuse, exclude and mislead it). These are usually sold as technical grade materials and not further diluted before application by special spray equipment. The extremely fine spray is applied at rates as low as one half pint to one-half gallon per acre.

Advantages

- Relatively easy to handle, transport, and store.
- Little or no agitation required.
- Not abrasive to equipment.
- Do not plug screens and nozzles.
- Leave little visible residue on treated surfaces.

Disadvantages

- High drift hazard due to small droplet size.
- Specialized equipment required.
- Easily absorbed through skin of humans or animals; high dermal and inhalation exposure risk (concentrated product applied as fine droplets).
- Products and/or solvents may cause rubber or plastic hoses, gaskets, and pump parts and other surfaces to deteriorate. Calibration and application must be performed with special care because ULV products are applied in concentrated form.

5. Flowable Suspensions or Aqueous Suspension (F, L or AS)

Some active ingredients are in-soluble solids: substances that will not dissolve in either water or oil. These may be formulated as flowables. (Most manufacturers use the letter "F" by the trade name to designate that the formulation is a flowable. However, some us the letter "L," meaning that an insoluble material is presented in "liquid" form). Most flowables are prepared by first impregnating them onto a dry carrier, such as clay. Then, the active ingredient plus carrier (or the active ingredients alone) are ground into a fine powder. Next, the fine powder is suspended in a very small amount of liquid (and perhaps other inert ingredients). The resulting product is a thick liquid suspension.

Earlier, it is stated that some insecticides are soluble in neither oil nor water, but are soluble in one of the expensive solvents, making the formulation quite expensive. To handle the problem, the technical material is blended with one of the dust diluents and a small quantity of water, leaving the insecticide-diluent mixture finely ground but in wet form. This 'wet blend' mixes well with water and can be sprayed with the same tank settling characteristic as wettable powder. Flowables combine many of the characteristics of liquid emulsifiable concentrates and dry wettable powders. They appear in the "Liquid Formulations" section because the end-use product is a thick liquid. Flowables are often used for the same types of pest control operations as ECs.

Advantages

- Easy to handle and apply; low exposure risk.
- Generally not phytotoxic.
- Splashes are less likely than with other liquid formulations.
- Can be mixed with water, and
- No inhalation hazard.

Disadvantages

- May settle; need shaking before measuring and mixing.
- Difficult to remove all of product from the container. Containers may be difficult to rinse.
- Require moderate agitation.
- May be abrasive; contribute to "wear and tear" of spray application equipment.
- Spills may be harder to clean up.
- May leave a visible residue on treated surfaces.

6. Flowable Concentrate for Seed Treatment (FS)

A stable suspension for application to the seed, either directly or after dilution.

Eg: Imidacloprid 48.0 % FS

Consitutents role	**Name of the constituent**	**Contents (% w/w)**
Active ingredient	Imidacloprid Technical	48.00
Azocolorant	Hydroxy naphthalene carboxamide	4.00
Emulsifying agent	Ethoxylated derivative of styrylated phenols	1.50
Emulsifying agent	Ethoxylated polymethacrylate in propylene	
Solvent	Glycol and water	4.50
Humectant	Trihydroxy propane	10.00
Thickening agent	Xanthan gum	2.50
Solvent	Phenyl methoxy methanol	0.10
Preservative	Blend of methyl isothiozolone and its chloro derivatives	0.10
Defoaming agent	Water solution of polydimethyl siloxane	0.10
Diluent	Water, demineralized	Q.S.
Total		100.00

Eg: Thiamethoxam 30 % FS

Consitutents role	**Name of the constituent**	**Contents (% w/w)**
Active ingredient	Thiamethoxam Technical	30.00
Film forming agent or adherent	Copolymer of Butanol	4.30
Humectant	Glycerine	5.10
Preservative	1,2-Benzisothiazole-3-one	0.20
Colouring agent	Monazo dye (Irgalith Red C2B)	3.40
Anti-freezing agent	Polyethylene glycol	2.60
	Water	Q.S.
Total		100.00

7. Dispersible Concentrate (DC)

A liquid homogeneous formulation to be applied as a solid dispersion after dilution in water. When added to water, insecticide completely mixes to produce a uniform dispersion containing microscopic crystals.

Eg: Afidopyropen 50 G/L DC

Consitutents role	Name of the constituent	Contents (% w/w)
Active ingredient	Afidopyropen Technical	50.00
Azocolorant	1,2 - Propylene Glycol	165.00
Emulsifying agent	Propylene Carbonate	1.50
Emulsifying agent	Propylene Carbonate	102.50
Solvent	Ethoxylated Tristyrylphenol	51.30
Solubilizer	N,N, N', N' – Tetrakis (2-hydroxypropyl) ethylene-diamine	2.05
Non-ionic surfactant	Alcohols, C 12-18, ethoxylated propoxylated	Q.S.
Total		1000.00

Eg: Isocycoseram 9.2 % w/w DC

Isocycoseram 10 % w/v DC

Isocycoseram 100 g/l DC

Consitutents role	Name of the constituent	Contents (% w/w)
Active ingredient	Isocycloseram Technical	9.20
Suspending agent	Solution of acrylic graft co-polymer in propylene glycol and water	4.60
Dispersing agent	Lignin derivatives	0.46
Emulsifying agent	Carboxylic acid, amine functionalised methyl ester	20.80
Solvent and penetration enhancer	Hydrocyalkyl dimethyl amide	Q.S.
Total		100.00

8. Water Soluble Concetrate (Or) Water Soluble Liquids (or) Soluble Liquids (SL)

These are liquids in their original state and are fully soluble in water and any other solvent. Solutions that are prepared in the right way will not leave unsightly residues or clog spray nozzles. It is a concentrated liquid insecticide formulation that may be used directly or require diluting. This is one of the formulation types that actually contain dissolved molecules, not suspended particles. They mix easily in water and require minimal agitation after dilution. These are forms clear liquid up on dilution in water.

Advantages	Disadvantages
• Mixes well with water • Equipment cleans readily • Non-volatile • Equipment-friendly • Not clog strainers • No agitation needed • Measureable and manageable • Rarely burns foliage • Easy to tank mix • Applicator dust-free	• Some salts cause eye discomfort • Some materials are corrosive in unlined steel tanks • There could be problems, if you mix concentrates together. • It's possible to settle down with a little shaking • Spray nozzles may be worn

Eg: Imidacloprid 17.8 % SL

Consitutents role	Name of the constituent	Contents (% w/w)
Active ingredient	Imidacloprid Technical	17.80
Non ionicEmulsifier	Ethoxylated alkyl aryl phenol derivatives	2.50
Dispersing agent	Polyvinyl pyrrolidone copolymer	1.00
Emulsifier	Dimethyl sulfoxide	38.40
Solvent and penetration enhancer	N-methyl pyrodlidone	Q.S.
Solvent	Aeromax	Q.S.
Total		100.00

III Gaseous formulation

1. Aerosols (A)

Aerosol formulations contain one or more active ingredients and a solvent. Most aerosols contain a low percentage of active ingredient. There are two types of aerosol formulations:

1. The ready-to-use type (often sold in pressurized sealed containers that serve as application devices).
2. Those made for use in electric or gasoline-powered aerosol generators that release the formulated product as a smoke or fog.

1.1. Ready-to-Use Aerosols (RTU)

Ready-to-use aerosol formulations are usually small, self-contained units that release pesticide when the nozzle valve is triggered. An inert pressurized gas pushes the pesticide through a fine opening when the gas is released, creating fine droplets. These products are effective in greenhouses, in small areas inside buildings, or in localized outdoor areas. Commercial models, which hold 5 to 10 pounds of pesticide, are usually refillable.

Advantages

- Easy to use; convenient.
- Portable.
- Easily stored.
- Convenient way to buy and apply a small amount of pesticide.
- Retain potency for some time.

Disadvantages

- Practical for only a few limited or specialized uses.
- Risk of inhalation exposure.

1. Eg: Imiprothrin 0.07 % + Cypermethrin 0.2 % Aerosol (HIT®)

Consitutents role	Name of the constituent	Contents (% w/w)
Active ingredient	Imiprothrin + Cypermethrin Technical	0.07 % + 0.2 %
Perfume	Mixture of nenzyl salicylate, Benzyl benzoate, Linalol & Rose crystal	0.20
Propellant	LPG	50.00
Solvents	Isopropyl Alcohol	0.50
Hydrocarbon solvent	Compounds with C_9 to C_{20} aliphatic branched and unbranched chain	Q.S.
Do not expose to temperature exceeding 50°C		
Total		100.00

2. Eg : Transfluthrin 0.08 % w/w Aerosol

Consitutents role	Name of the constituent	Contents (% w/w)
Active ingredient	Transfluthrin Technical (100 % purity)	0.08 %
Perfume	Citronello, Coumarin, Egenol, Giraniol, Linalool etc	0.30
Propellant	LPG	60.00
Hydrocarbon solvent	Compounds with C_9 to C_{20} aliphatic branched and unbranched chain	Q.S.
Total		100.00

4. Eg: Transfluthrin 0.88 % (w/w) Liquid vaporiser

Consitutents role	Name of the constituent	Contents (% w/w)
Active ingredient	Transfluthrin Technical	0.880
Antioxidant	Butylated Hydroxy Toluene (BHT)	1.000
Perfume	-	1.000
Carrier and Solvent	Deodorised kerosene	97.120
Total		100.00

The vaporizers release an insecticide called Allethrin. While it works on mosquitoes, it is also bothersome to people with allergies or asthma. In heavier concentrations, it can be harmful to children. So don't use a chemical mosquito vaporizer indoors. Research shows that prolonged action of the Pyrethroids commonly used in repellent vaporizers leads to hyperexcitation of the neurons system.

2. Formulations for Smoke or Fog Generators

Formulations for smoke or fog generators are not packaged and sold under pressure. They are used in machines that break the liquid formulation into a fine mist or fog (aerosol). Using a rapidly whirling disk or heated surface, the machines produce and distribute very fine droplets. These formulations are used mainly for insect control in structures such as greenhouses, barns, and warehouses and for outdoor mos quito and biting fly control.

Advantages

- Easy way to fill an entire space with pesticide.

Disadvantages

- Highly specialized use sites and equipment.
- Difficult to confine to target site or pest.
- Spills and splashes may be difficult to clean up and/or decontaminate.
- May require respiratory protection to prevent inhalation exposure.

A. Mosquito card (Good night – FAST CARD®) - 3 minutes instant action

Eg: Transfluthrin 1% FU

Consitutents role	Name of the constituent	Contents (% w/w)
Active ingredient	Transfluthrin	1.00
Oxidizing Agent	Potassium Nitrate	2.00
Bittering Agent	Denatonium Benzoate	0.005
Red dye	-	0.05
Perfume	-	2.00
Emulsifying Agent	Cetyl alcohol	0.01
Paper	-	74.00
Thickening agent/Humectant	Isopropyl myristate	4.50
Carrier and Solvent	Deodorised kerosene	Q.S.
Total		100.00

B. Mosquito Coil (Good night - Maha Jumbo®) – (Minimum 11 hours)

Consitutents role	Name of the constituent	Contents (% w/w)
Active ingredient	Prallethrin Technical	0.04
-	Other ingredients	99.96
Total		100.00

C. Eg: Prallethrin 1.2 % w/w Mosquito Mat, House hold insecticide

Consitutents role	Name of the constituent	Contents (% w/w)
Active ingredient	Prallethrin	1.20
Active ingredient	Prallethrin	12.00
Synergist	Piperonyl butoxide	24.00
Dye	-	0.60
Perfume	-	7.00
Carrier and Solvent	Deodorised kerosene	24.00
Thickening agent/Humectant	Isopropyl myristate	32.40
Total		100.00

Premix of prallethrin 100 mg will be injected on to the blank mat having approximately a weight at 900 mg resulting in a total weight of 1000 mg so that active ingredient wii be 1.2 % w/w. This mat contains 12 mg of prallethrin as active ingredient. Each prallwthrn mat is a ready to use mat which has to be heated using an electrical vaporiser.

Direction of use: For better results close the doors and windows for about 30 minutes initially after switching on the vaporiser.

Isopropyl Myristate is a synthetic oil used as an emollient, thickening agent, or lubricant in insecticide formulation. Its an ability to reduce the greasy feel caused by the high oil content of other ingredients in a product.

Deodorised Kerosene is used as an evaporating agent and as a solvent. Deodorised Kerosene is environment and human friendly unlike simple kerosene. Vapours from simple kerosene cause cancer due to the aromatic contents of oils (carcinogens) which are present in simple kerosene oil.

The liquid inside the refill is a a mixture of mostly 3 types of chemicals

Insecticide - this makes the mosquitoes fly away

Stabiliser / anti-oxidant - to prevent the insecticide getting oxidised due to heat.

A perfume - to prevent the humans from running away

All these chemicals are dissolved into deodorized kerosene so that they can be easily vaporized when subjected to temperature. The presence of deodorized kerosene (96.40%) as a solvent in vaporizers is dangerous when inhaled regularly. They cause nausea, increase in blood pressure, and bloodshot eyes. **BHT (Butylated**

hydroxytoluene) used commonly in all insect repellents can lead to long term enzyme changes that make the human body susceptible to cancer.

Prallethrin, the chemical ingredient if any insect repellent, acts on the central nervous system causing headaches, dizziness, and sometimes leads to asphyxiation. Prolonged exposure to these chemicals causes permanent brain and nervous system damage.

3. Aerosol dispensers (AE)

They are commonly used against the domestic pests (cockroaches, lice). The small droplets are more likely to be inhaled. The formulations often contain flammable propellant under pressure that poses potential hazards if the container is punctured.

Dictionary of pesticide formulation words

Pesticide Mixtures

Sometimes, product manufacturers combine insecticides with other insecticides or fertilizers for sale as premixes. There are two types of pesticide mixtures.

- Pesticide with fertilizer.
- Two or more insecticides or two or more fungicides or two or more insecticide and fungicides.

When insecticides are tank-mixed, all of the dosages must be at or below the label rate for each separate component of the mixture.

Adjuvants

An adjuvant is a chemical that can affect how a pesticide works. Adjuvants are improve the action of a pesticide. Change the characteristics of a insecticide formulation or a spray mixture (suspension or solution). Many adjuvants increase effectiveness and/or safety. Although they enhance the action of a pesticide or modify the properties of a spray solution, adjuvants alone have no pesticidal activity.

Types of Adjuvants

There are many types of adjuvants. Here are some that are commonly used:

1. **Antifoaming (defoaming) agents**-Reduce foaming of spray mixtures that may result from using some surfactants and/or from vigorous agitation.
2. **Buffers or pH modifiers**-Allow insecticides to be mixed with diluents or other insecticides of different acidity or alkalinity. Most pesticide solutions or suspensions are stable between pH 5.5 and 7.0 (slightly acidic to neutral). If you use a buffer, add it to the spray tank and mix well. The water must be pH neutral or slightly acidic before adding insecticides or other adjuvants.

3. **Compatibility agents**-Help combine insecticides (or insecticides and fertilizers) effectively; reduce or eliminate incompatibility.
4. **Drift control additives (deposition aids)**-Reduce drift; increase average droplet size and/or lower the number of "fines" (very small droplets) produced.
5. **Emulsifiers**-Allow petroleum-based insecticides (ECs) to mix with water.
6. **Extenders**-Keep insecticides active on a target for an extended period. Some adjuvant manufacturers use this name for stickers.
7. **Invert emulsifiers**-Allow water-based insecticides to mix with petroleum carrier.
8. **Plant penetrants**-Allow the pesticide to pass through (penpenetrate) the outer surface to the inside of treated foliage. Certain plant penetrants may increase penetration on some but not all plant species.
9. **Safeners**-Reduce the toxicity of a insecticide formulation to the pesticide handler or to the treated surface.
10. **Deflocculators/Spreader**-Allow pesticide to form a uniform coating layer over the treated surface Eg: Calcium casinate.
11. **Stickers**-Allow pesticide to stay on a treated surface. Some types of stickers increase adhesion of solid particles to a treated sur face. This reduces the amount of pesticide that washes off due to rain or irrigation. Others reduce evaporation and/ or slow photodegradation. (See "Extenders" above.)
12. **Thickeners**-Increase viscosity (thickness) of spray mixtures. Thickeners may reduce drift and/or slow evaporation. (Slowing evaporation is useful when applying systemic insecticides). It increases the time during which the active ingredient can be absorbed by or penetrate plant foliage.
13. **Wetting agents**-Allow wettable powders to mix with water.
14. **Pseudosynergist:** Stabilizes the droplet size and increases toxicity of the insecticides Eg: Glycerin.

15. Surfactants

Some of the most common adjuvants are **surfactants** (***sur***face ***act***ive ingredie***nts***), which alter the dispersing, spreading, and wetting properties of spray droplets. Examples of surfactants are wetting agents and spreaders. Surfactant improves the spray cover by reducing the surface rension of spray droplet.

Surfactants are classified by how they split apart into charged atoms or molecules, called ions.

1. **Anionic surfactants** have a negative charge. They are most often used with contact insecticides, which control the pest by direct contact instead of being absorbed systemically.

2. **Cationic surfactants** have a positive charge. Do not use them as "stand-alone" surfactants often, they are phytotoxic.
3. **Nonionic surfactants** have no electrical charge. They are often used with systemic products and help sprays penetrate plant cuticles. They are compatible with most pesticide products.

A pesticide can behave very differently in the presence of an anionic, cationic, or nonionic surfactant. For this reason, you must follow label directions when choosing one of these additives. Selecting the wrong surfactant can reduce efficacy and damage treated plants or surfaces.

Oftenly the words "adjuvant" and "surfactant" are used interchangeably indicating both does the same function. However, an adjuvant is any substance added to modify properties of a insecticide formulation or finished spray. A surfactant is a specific kind of adjuvant one that affects the interaction of a spray droplet and a treated surface. All surfactants are adjuvants, but not all adjuvants are surfactants.

Eg: Triton X-100, Teepol, Tween-85, and Soap.

Sucroglycerides/sucroglycerides derived from **vegetable oils** reported as novel biodegradable surfactants for plant protection formulations.

Insecticide formulations with different strength (low to high) available in Indian market

Type of formulation	Name of the products
Emulsifiable concentrate (EC)	azadirachtin 0.03% Min. Neem Oil Based E.C., azadirachtin 0.15% Neem Seed Kernel Based E.C.; azadirachtin 0.3% Neem Seed Kernel Based E.C. azadirachtin 1% (10000 ppm) Min. Neem Based E.C. ; abamectin 1.8% EC; bifenthrin 2.5% EC; bifenthrin 10% EC; chlorpyrifos 20% EC; carbosulfan 25% EC; Profenophos 50% EC; Propargite 57% EC
Soluble Powder (SP)	acetamiprid 20% SP; cartap Hydrochloride 50% SP ; acephate 75% SP
Water Soluble Granules (SG)	emamectin benzoate 5% SG ; dinotefuran 20% SG thiamethoxam 75% SG; acephate 95% SG
Wettable Powder (WP)	*Beauveria bassiana* 1.15% WP; Deltamethrin 2.5% WP; Diafenthiuron 50%WP
Suspension Concentrate (SC)	Spinosad 2.5% SC; Fipronil 5% SC; Bifenthrin 8%SC; Chlorfenapyr 10% SC; Chlorantraniliprole 18.5% SC; Buprofezin 25% SC; flubendiamide 39.35% SC; Spinosad 45% SC
Granules (GR/G)	fipronil 0.3% GR; Chlorantraniliprole 0.4% GR; Carbofuran 3% GR; Cartap Hydrochloride 4% GR; Chlorpyrifos 10% GR
Water Dispersible Granules (WDG)/(WG)	Flubendiamide 20% WG; thiamethoxam 25% WG Clothianidin 50% WDG; Flonicamid 50% WG; imidacloprid 70% WG
Dispersible powder (DP)	Chlorpyrifos 1.5% DP; Metaldehyde 2.5% DP

Codes For Formulations

AE	Aerosol dispenser	DT	Tablets for direct
CB	Bait concentrate	EC	Emulsifiable concentrate
MC	Mosquito coil	EG	Emulsifiable granule
CP	Contact powder	EO	Emulsion, water in oil
ME	Microemulsion	RB	Bait (ready fore use)
CS	Capsule suspension	EP	Emulsifiable powder
OD	Oil dispersion	SC	Suspension concentrate
DS	Powder for dry seed suspension	ES	Emulsion for seed treatment (= flowable concentrate)
EW	Emulsion, oil in water	SG	Water soluble granule
SP	Water soluble powder	GR	Granule
ST	Water soluble tablets	ULV	Ultra- low volume
TB	Tablet	RTU	Ready to use
WG/WDG	Water dispersible granule	SL	Soluble liquid
DP	Dustable powder	WP	Wettable Powder
FU	Smoke generator	ZC	Combination of SC+CS
ZW	Combination of CS+EW	ZE	Combination of CS+SC+EW
SE	Suspo emulsion	FU	Foam Unit
MC	Mosquito coil	LV	Liquid vaporiser

4

Insecticide Formulation Testing: Importance and Protocols

Introduction

Analysis of insecticide formulations are critical for the availability of quality insecticides in the market for ultimate benefit of farmers and ecosystem health. This chapter highlights the significance of comprehensive insecticide formulation analysis protocols, encompassing methods for determining active ingredient and assessing physical and chemical parameters. Employing advanced analytical techniques such as spectroscopy and chromatography for insecticide formulation analysis will eventually maintain the quality of insecticides for the sustainable management of pests.

In India, over 339 pesticides are registered that includes insecticides, fungicides, herbicides (DPPQS, 2024). The current market of non-genuine insecticide is INR 3,200 crores (USD 525 Million) which constitutes 25 per cent by value and 30 per cent by volume of the total domestic market of agrochemicals in India as per Industry reports, primary interviews, news articles and Tata Strategic analysis (FICCI, 2015). Monitoring registration and regulation of insecticides is governed by the Insecticide Management Bill 2020, which replaced the Insecticide Act 1968 (DPPQS, 2023). The Insecticide Act of 1968 established the Central Insecticide Laboratory, Faridabad, under section 16 with two regional insecticide testing laboratories at Chandigarh and Kanpur (DPPQS, 2023). Current Quality Control of Insecticides in States/UTs during the last five years shows that 68078 insecticide samples were analysed (DPPQS, 2023). In the beginning, these laboratories involved in analysing insecticide samples drawn by any officer or the body authorized by the Central or State Governments and submit certificates of analysis to the concerned authority. Regional Pesticide Testing Laboratories (RPTLs) had a huge target of analysis of 1550 samples per annum that necessitated the establishment of 71 State Pesticide Testing Laboratories (SPTL) spread across India of which 14 are NABL accredited (DPPQS, 2023). As per the Insecticide Management Bill 2020, the routine process of insecticide formulations monitoring includes drawing of insecticide samples randomly from the market and send it to SPTL for quality analysis. At SPTL, the insecticide analyst (Agriculture Officer/ Asst. Director of Agriculture) will perform the test and issue the certificate (Dileep

Kumar and Narasimha Reddy, 2021). If the sample fails to meet the required standard, the manufacturer faces the consequences in a court of law (Dileep Kumar and Narasimha Reddy, 2021).

Insecticide formulation analysis main moto is to find the substandard insecticide in the pesticide market. Substandard definition according to insecticide act, 1968 states that it does not conform to the active ingredient test approved for it by the Registration Committee and its active ingredient is within five percent of the nominal value when applied beyond the upper and lower limits prescribed for conforming to the test, provided that no tolerance limit shall apply in case of insecticides, which are registered on minimum purity basis (FICCI, 2015).

The nominal value for insecticide active ingredient decided by the bureau of Indian standards as if percent insecticide formulation up to 9 % the range should be between + 10 to -5 and above 9 and below 50 the range should be + 5 to -5 and 50 and above it should be + 5 and -3.

Eg: Acceptable active ingredient range for insecticides

S. No.	Insecticide	Nominal value of insecticide formulation	Per cent range	Acceptable range
1.	Emamectin benzoate 5 % SG	≤ 9 %	+10 to -5 %	4.75-5.50
2.	Imidacloprid 17.8 % SL	≤ 9 to ≥ 50 %	+ 5 to -5 %	16.91-18.69
3.	Acephate 75 % SP	≥ 50 %	+ 5 and -3 %	72.75-78.75

Insecticide formulations can be analysed in two ways viz., Instrumental insecticide formulation and volumetric insecticide formulation analysis (Table 1).

Table 1: Pesticide formulation testing methods

Instrumental methods	Volumetric methods
HPLC	Emulsion Stability
GC	Cold test
MS	Heat stability test
FTIR	Persistent Foam
NMR	Suspensibility test
UV-Vis. Spectroscopy	-
Particle Size Analyser	-
Thermal Analysis (Differential scanning calorimetry; DSC, Thermo gravimetric analysis; TGA)	-
X-ray Diffraction (XRD)	-
Automatic Potentiometric Titrator	-
Karl Fischer Titrator	-
Sieve shaker	-
pH meter	-
Flash point apparatus	-
Viscometer	-

The flow chart depicts the series of steps in insecticide formulation testing (Fig 1).

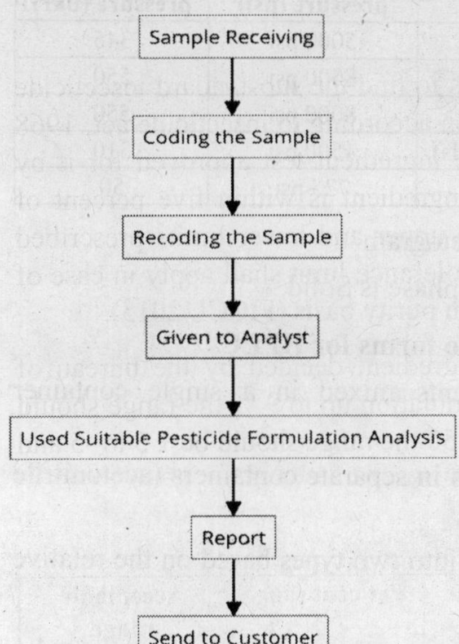

Fig. 1: Flow chart showing insecticide formulation analysis in the laboratory

1. Insecticide formulation Analysis

1.1 Instrumental Insecticide formulation Analysis

Instrumental insecticide formulation analysis involves the use of advanced analytical techniques and instruments to evaluate the composition, properties, and quality of insecticide formulations. These techniques provide highly accurate and sensitive measurements, allowing researchers, regulators, and manufacturers to gain insights into the chemical and physical characteristics of insecticide products. Some of the key instrumental methods commonly used in insecticide formulation analysis mentioned below

1.1.1 High-Performance Liquid Chromatography (HPLC): HPLC is a versatile technique used to separate, identify, and quantify individual components within a insecticide formulation. It is especially effective for analysing active ingredients, impurities, and degradation products. HPLC employs a liquid mobile phase to transport the sample through a stationary phase, and detectors viz., UV provide information about the separated compounds. The components of a chemical mixture can be separated using the HPLC process. These components are first dissolved in a liquid solvent, and then they are compelled to flow under intense pressure via a chromatographic column.

HPLC type	Maximum pressure (psi)	Maximum pressure (bar)
HPLC-High performance liquid chromatography	5000 psi	345
UFLC-Ultra flow liquid chromatography	8000 psi	550
UPLC-Ultra pressure liquid chromatography	8000 psi	550
Preparatory HPLC (used for purification of chemicals)	4500 psi	310
Flash HPLC (used for purification of chemicals)	725 psi	50

- Sensitivity of HPLC is nanogram to pictogram
- Mobile phase is Liquid and Stationary phase is Solid

The mobile phase has been prepared in two forms for HPLC

1. **Isocretic form:** two different solvents mixed in a single container (acetonitrile + water)
2. **Gradient form:** two different solvents in separate containers (acetonitrile and water separate containers)

In general liquid chromatography is divided into two types based on the relative polarities of stationary and mobile phases

1. **Normal phase (NP) chromatography:** stationary phase is more polar (e.g. silica gel), than the mobile phase (hexane, iso-octane). The least polar analytes elute first, then moderately polar and finally highly polar samples. Analyte retention.can be increased by decreasing mobile phase polarity.
2. **Reverse phase (RP) chromatography:** The mobile phase (often water, methanol, or acetonitrile) is more polar than the stationary phase (typically C_{18}). The more hydrophobic analytes tend to elute first. Boosting the polarity of the mobile phase improves analyte retention.

Detectors: UV-Visible detector, Fluorescence detector, Electrochemical detector, Photodiode array detector.

Note: When flowrate increases, Retention time (RT) will decrease

Eg: Acetamiprid 20 % SP (IS 16328: 2017), Imidacloprid 17.8 % SL (IS 15443: 2004).

List of HPLC column

SI.No.	Company	Colum type	Dimension	Name of the suitable equipment
1.	Shimadzu	Shim-pack GIST 5µm C8	4.6×150mm	HPLC Reverse phase
2.	Shimadzu	Shim-pack GIST 5µm C8	4.6×250mm	
3.	Shimadzu	Shim-pack GIST 5µm C18	4.6×150mm (HSS)	
4.	Shimadzu	Shim-pack GIST 5µm C18	4.6×250mm (HSS)	
5.	Shimadzu	Shim-pack UC-Sil 5µm	4.6×250mm	HPLC Normal phase
6.	Shimadzu	Shim-pack UC-Sil 5µm	4.6×250mm	

Column conditioning Procedure

For a typical C_{18} or C_8 column - Regular maintenance and proper cleaning will ensure the longevity and performance of HPLC C_{18} column.

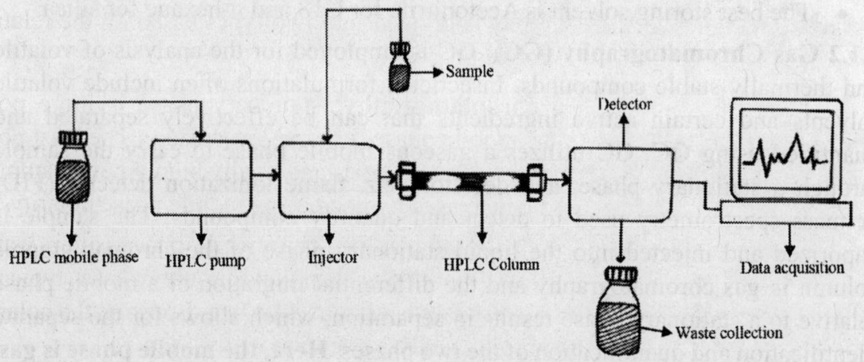

Column Conditioning

Changing the column from silica to C18 (Normal phase to Reverse phase)

- Change from mobile phase to n-Hexene and run for 1hr and remove the column and store it by placing end caps
- Place the union and run with DCM for 15 min
- This is followed by IPA for 15 min and methanol 15 min
- Then place the C_{18} column and run with methanol for another 15 min
- Finally run with the recommended MP for at least 30 min before injecting the sample.

Column Conditioning

Changing the column from C18 to silica (Reverse phase to Normal phase)

- Run the C_{18} column with Methanol before removing it for at least 1hr

- Then remove the column and store it in Methanol by keeping the end caps
- Place the union and run with IPA (15 min) – DCM (15 min) – n-hexane (15 min)
- Place silica column and run with n-Hexane for another 15 min followed by mobile phase for at least 30 min

Column Storage

- For short term storage, i.e., overnight, columns can be stored in the eluent used in last analysis.
- For Longer days store reverse phase columns in 100% Acetonitrile.
- Don't store column in buffered eluents.
- Make sure that all buffer are washed out of the column with 10 column volumes of HPLC grade water before flushing with Acetonitrile.
- Completely seal column to avoid sealing and drying out of the bed.
- The best storing solvent is Acetonitrile for C18 and n-hexane for silica

1.1.2 Gas Chromatography (GC): GC is employed for the analysis of volatile and thermally stable compounds. Insecticide formulations often include volatile solvents and certain active ingredients that can be effectively separated and quantified using GC. GC utilizes a gaseous mobile phase to carry the sample through a stationary phase, and detectors viz. flame ionization detector (FID) or mass spectrometry used to detect and quantify compounds. The sample is vaporized and injected into the liquid stationary phase of the chromatographic column in gas chromatography and the differential migration of a mobile phase relative to a stationary phase results in separation, which allows for the separate identification and quantification of the two phases. **Here, the mobile phase is gas, and the stationary phase is solid.**

Hydrogen, helium, nitrogen, or argon (chemically inert) makes up the mobile phase of the gas cylinder, while a non-volatile liquid used in the stationary phase. Due to its high thermal conductivity in comparison to most organic vapors, helium is favoured for thermal conductivity detectors despite its high cost. Nitrogen is the gas of choice when a high volume of carrier gas is required in a laboratory. In GLC, the partition coefficient difference is used to separate volatile chemicals. A substance needs to be sufficiently volatile and thermally stable to be analyzed by GC. In order for GC to analyze a component, that compound must exist in the gas or vapour phase between 400 and 450°C without decomposing.

Generally, packed columns are used in pesticide formulation analysis

Characteristics of packed columns

- Typically, a glass or stainless-steel coil.
- 1-5 m total length and 2-4 mm internal diameter.

- Filled with finely divided, inert, solid support material (commonly based on diatomaceous earth) coated with liquid stationary phase.

List of GC Packed column

SI. No.	Company	Colum type	Dimension	Equipment
1.	RESTEK	Packed Column	5% OV-101 DIATO-WHP 80/100 1m 2.1MMID 1/8INOD SS	GC-FID
2.	RESTEK	Packed Column	10%OV-101 DIA TO-WHP 80/100 2m 3.2MMID 3/16INOD SS	
3.	RESTEK	Packed Column	5%SE-30 DIA TO-WHP 80/100 1m 3.2MMID 3/16INOD SS	

Detectors used: Flame Ionization Detector (FID), Electron Capture Detector (ECD) & Nitrogen phosphorous Detector (NPD), Flame photometric detector (FPD), Mass Detector (MS).

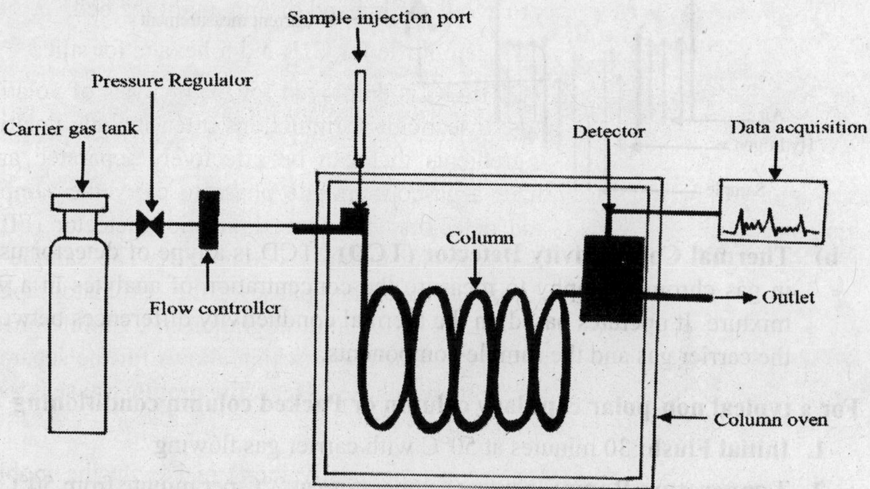

Eg: Chlorpyrifos 20 % EC (IS 8944: 2005), Fenpropathrin 10 EC (IS 15161: 2002), Fenvalerate 11 % EC (IS 11997: 1987), Organochlorine, Organophosphorus, Pyrethroids *etc*. can be analyzed by GC

Pesticide formulation analysis is performed using a flame ionization detector (FID) and a thermal conductivity detector (TCD).

Detectors

a) **Flame ionization detector (FID):** A most widely used detector. It is sensitive to organic compounds. The difference between PID (Photo ionization detectors) to FID results in an intense beam of ultraviolet radiation in the 105 to 150 nm range to absorb light of sufficient energy to cause an electron to leave and create a positive ion., rather than the flame, is used to ionize the molecules. Flame Ionization Detector (FID) responds to majority of pesticide compounds. The carrier gas sensitivity of GLC is in the range of 10 pg – 10 µg with usual analytical run time of 15-30 minutes. In the FID detector, the light source is Nickel.

Eg: Organochlorine, Organophosphorus, Pyrethroids etc. can be analysed by GC.

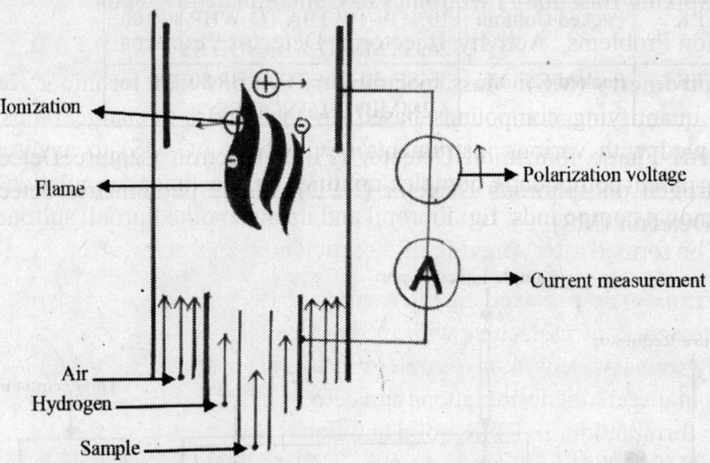

b) **Thermal Conductivity Detector (TCD)** : TCD is a type of detector used in gas chromatography to measure the concentration of analytes in a gas mixture. It operates based on the thermal conductivity differences between the carrier gas and the sample components.

For a typical non-polar capillary column or Packed column conditioning

1. **Initial Flush:** 30 minutes at 50°C with carrier gas flowing
2. **Temperature Ramp:** Increase temperature at 2°C per minute from 50°C to 300°C.
3. **Hold:** Hold at 300°C for 15 min.
4. **Cool Down:** Gradually cool to the operating temperature (e.g., 50°C).

Proper conditioning of GC column will ensure reliable and reproducible results, extending the column life and maintaining its performance.

Ways to overcome trouble shoot in HPLC/GC

- Peak Telling : Flow Path or Activity
- Bonus Peaks : In Sample or Back Flash (Carry Over)
- Split Peaks : Injector Problems, Mixed Solvent
- No Peaks : Wasn't Introduced, Wasn't Detected

- Response Changes : Activity, Injector Discrimination, Detector Problem
- Peak Forting : Overload or Solubility Mismatch, Injector Problems
- Shifting Retention : Leaks, Column Aging, Contamination or Damage
- Loss of Resolution : Separation Decreasing, Peak Broadening
- Baseline Disturbances : Column Bleed, Contamination, Electronics
- Noisy or Spiking Baseline : Electronics or Contaminated Detector
- Quantitation Problems : Activity, Injector or Detector Problems

1.1.3 Mass Spectrometry (MS): Mass spectrometry is a powerful technique for identifying and quantifying compounds based on their mass-to-charge ratios. MS can be coupled with various instruments (GC-MS or LC-MS) to analyse metabolites from parent compounds, complex mixtures in insecticide formulations and identify unknown compounds. Eg: Fipronil and its metabolite fipronil sulfone (metabolite will be formed after spraying of insecticide on leaf, soil, insect)

1.1.4 Fourier-Transform Infrared Spectroscopy (FTIR): FTIR spectroscopy measures the interaction of molecules with infrared light, providing information about functional groups present in the sample. It is useful for identifying different chemical groups in insecticide formulations and detect changes due to degradation or any change in formulation. In FTIR, solid and liquid solvents also can be used. Eg: Ethion (IS 10319: 1982)

1.1.5 Nuclear Magnetic Resonance (NMR) Spectroscopy: NMR provides detailed information about molecular structure and composition by analysing the interactions of nuclei with a magnetic field. It's particularly useful for elucidating the structure of complex molecules, confirming the identity of active ingredients, and studying the interactions between different components in the formulation.

1.1.6 UV-Visible Spectroscopy: UV-Vis spectroscopy measures the absorption of ultraviolet and visible light by molecules. It's used to quantify the concentration of active ingredients and assess the stability of formulations by monitoring changes in absorption spectra over time. Eg: Cartap hydrochloride (IS 14184: 1994), Monocrotophos, Glyphosate (IS 12502: 1988), Malathion EC (IS: 2567: 1978)

1.1.7 Particle Size Analyser (Dynamic Light Scattering, Laser Diffraction): Insecticide formulations manufactured in different types such as suspensions or emulsions, where particle size plays a critical role in stability and effectiveness. The techniques mainly dynamic light scattering and laser diffraction provide insights into particle size distribution, allowing manufacturers to optimize formulation properties. The optimum particle size recommended should be $<25\mu m$ but majority manufacturers maintain particle size $<10\ \mu m$ for better penetration of pesticide into the plant system. Eg: SC formulations

1.1.8 Thermal Analysis (DSC, TGA): Differential scanning calorimetry (DSC) and thermogravimetric analysis (TGA) provide insights into the thermal properties and stability of insecticide formulations. These techniques analyse melting point, decomposition temperatures, and potential thermal degradation pathways.

1.1.9 X-ray Diffraction (XRD): XRD is used to analyze the crystal structure of solid components in insecticide formulations. It's valuable for studying the arrangement of particles and assessing crystallinity changes that may occur during formulation or storage.

These instrumental methods offer a comprehensive understanding of insecticide formulations, including the identification, quantification, and characterization of active ingredients, inert components, and additives. The integration of these techniques ensures the quality, safety, and efficacy of insecticide products while aiding regulatory compliance and environmental protection efforts.

1.1.10 Automatic Potentiometric titrator: It is used to find the acidity/alkalinity of insecticide formulation. Eg: Chlorpyrifos – ≤ 0.05 percent by mass ((IS 8944: 2005)

1.1.11 Karl Fischer Titrator: This instrument is used to quantify the moisture present in pesticide formulation for example majority of technical and few formulation types (Imidacloprid Technical moisture content should be ≤1 % (IS 15443: 2004)

1.1.12 Sieve shaker with test sieves of 2 mm, 1.7 mm, 1.4 mm

Sieve test: Wet sieve test is a critical analytical technique in insecticide formulation as it ensures the particle size distribution is controlled and consistent. Eg: WP, SC and Granular formulations

Fipronil 5 % SC: Not less than 98 precent by mass of the material shall pass through a 45 micron IS sieve when determined by method described in 11.1 of IS 6940.

1.1.13 pH meter: An equipment used to analyse the pH in the formulation for example Acetamiprid pH should be between 7.0-9.0 (IS 16328: 2017) and Fipronil pH should be between 4.0-8.5 (IS 16145: 2013).

1.1.14 Flash point apparatus (Abel)

A flash point apparatus is a laboratory instrument used to determine the flash point of a flammable substance in EC and SL pesticide formulations. The flash point is the lowest temperature at which the vapour of a substance can ignite in the presence of an open flame or spark. It is an important safety parameter for handling and storing flammable materials (IS 1448: 1960)

The working of flash point apparatus:

Sample Preparation: A small quantity of the substance being tested (usually a few millilitres) is placed in a cup or container designed for the apparatus.

Test Procedure: The cup containing the sample is positioned in the apparatus, and it is typically exposed to an open flame or an electrical spark in a controlled environment.

Temperature Control: The temperature of the sample is gradually increased at a controlled rate (usually 1 or 2 degrees Celsius per minute) using a heating element.

Observation: An ignition source, such as a pilot flame, is directed towards the surface of the sample and the operator carefully observes the sample during the temperature ramp-up. The flash point is reached when a small flame or spark occurs at the surface of the sample, indicating that it has reached a temperature at which it can ignite.

Recording Data: The temperature at which the flash occurs is recorded as the flash point of the substance. The flash point of the material should be >24.5 °C.

1.1.15 Viscometer

Viscometers play a critical role duly measuring the viscosity of the formulation, a parameter important for ensuring the stability, efficacy, and ease of application of the pesticide. The viscosity range should be between 250-550 cp. **Centipoise (cp),** a unit of dynamic viscosity is commonly used to measure the liquids viscosity.

1 centipoise = 0.01, poise = **0.001** Pascal-second (Pa-s) in the SI unit system

Types of Viscometers

1. **Rotational Viscometers**: Commonly used in pesticide formulation. It operates by rotating a spindle in the fluid and measuring the resistance to rotation and one such example is Brookfield viscometers that normally spindle 63 number (60 rpm) rod (used in Brookfield viscometer).
2. **Capillary Viscometers**: This type measures the time it takes for a liquid to flow through a thin tube. They are less common in pesticide formulation but can be used for low-viscosity formulations.
3. **Falling Sphere Viscometers** measure the time it takes for a sphere to fall through the liquid.

1.2 Volumetric insecticide formulation analysis

Volumetric analysis is a branch of analytical chemistry that involves measuring the volume of a solution of known concentration (titrant) required to react completely with a solution of the analyte, thus allowing for the determination of the analyte's concentration. In the context of insecticide formulation analysis, volumetric methods are often used to quantify the concentration of active ingredients or specific chemical components within the formulation. Volumetric analysis methods are widely utilized due to their simplicity, accuracy, and relatively low cost. The details on few key volumetric techniques used in insecticide formulation analysis are,

1.2.1 Emulsion Stability: Emulsion stability is used for EC and EW pesticide formulations and is typically assessed using a 100 mL Crow receiver/Crow (Fig.1). The desired outcome is the observation of any phase separation, such as creaming at the top or sedimentation at the bottom, in a 100 mL emulsion created in standard hard water (342 ppm) with an EC volume **not exceeding 2.0 mL** (IS 6940: 1982). Emulsion stability.

1.2.2 Cold test- Pour 50mL of insecticide formulation (EC and SL formulations) in a 100mL transparent glass container and close it with a thermometer-equipped cork or stopper. Place the container in ice water to cool it to 10°C and allow the insecticide in the container to reach 10°C, then apply a little insecticide seeding crystal with minimal stopper opening in the shortest time. The formulation made from technical liquid insecticide does not need crystal seeding. Stir the material in the container gently at short intervals for one hour at 10°C. **After one hour, check for turbidity or solid/oily matter separation (IS 6940: 1982).**

1.2.3 Heat stability test: Initially, check the pesticide formulation (EC, SL) and its appearance in the transparent beaker. Place the sample in a bottle (one litre) until it is about three-fourths full. If the sample is volatile, seal it air-tight with sealing wax to avoid any loss of the volatile solvent. The filled bottle is kept in the oven at 54±1 □C for 24 hours. Remove the sample container from the oven and cool down to room temperature. Transfer to the beaker and observe the separation of any solid or oily matter; if homogeneous looking as in earlier observation, it is passed.

Eg: EC and SL formulations

1.2.4 Persistent Foam-The sample mass is the mass needed to make 200 mL of a suspension with the concentration suggested in the product's instructions. When multiple concentrations are advised, choose the highest. Pour 180 mL of standard hard water in the 250 mL measuring cylinder on a top pan balance and weigh 1g of suspension concentrate. Fill the volume to 200 mL with regular hard water. Stop the cylinder and invert 30 times. Start the stopwatch after placing the stoppered cylinder upright on the bench. Read foam production and remaining after 1 min.

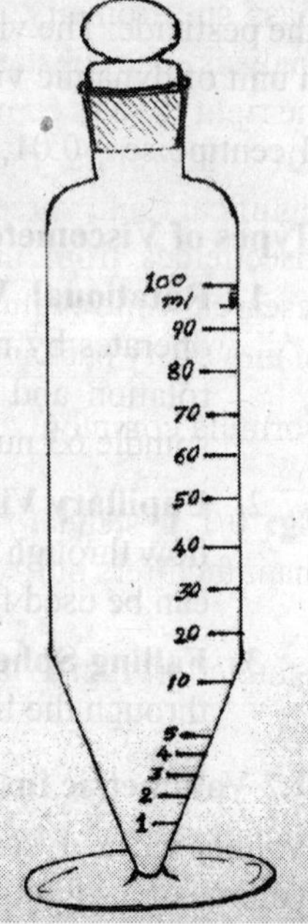

Fig. 1: Crow cylinder 100 mL

Eg: Fipronil 5 % SC (IS 16145: 2013) Annex D – The persistent foam shall not be more than 60 mL in 1 min.

1.2.5 Suspensibility test: The dry insecticide formulations (WP and SC) are slurred into a 50 ml test water in a 100 ml beaker. The slurry is quantitatively transferred to a 250 ml measuring cylinder and is stoppered and further inverted for 15 complete cycles. The measuring cylinder is allowed to stand for 30 minutes and later on the top 225 ml is drawn off and the remaining suspension is dried. The residue weight will determine the percent suspensibility.

Eg: Fipronil 5 % SC (IS 16145: 2013) Annex C-The suspensibility of the material shall not be less than 60 percent (m/m).

Conclusion

Pesticide formulation analysis (PFA) is a vital part of modern agriculture, assuring the efficacy, stability, and safety of the products. By employing advanced analytical techniques such as chromatography and spectroscopy, regulatory bodies can make decisions accordingly for detecting inferior quality insecticides. It is important to note that though volumetric methods are valuable for many analyses, they might not be suitable for all types of insecticide formulations or analyses. For complex formulations with multiple active ingredients and additives, more sophisticated instrumental techniques viz. chromatography, spectroscopy, and mass spectrometry might be necessary to achieve accurate and comprehensive analysis. The substandard pesticides can be detected using PFA that gives farmers the right to seek compensation from the producers as per consumer protection act, 2019. Therefore, the future implications of insecticide formulation analysis are significant and safeguarding the crop health. PFA ensures production of quality insecticides from industries, and thereby helps scientists to carry out quality research while evaluating insecticides against insect-pests before shortlisting them in university package of practices. It also creates trust among crop input dealers.

Formula commonly used in insecticide formulation analysis

Eg: 0.1 g standard weight of sample (chlorpyrifos 20% EC) is required and manufacturers purity is 95.37, then substitute this information in formula as …

$$\text{Standed weight to taken} = \frac{0.1 \times 100}{95.37}$$

$$= 0.1049$$

$$\text{Sample equivalent weight to standard be taken} = \frac{0.1049 \times 95.37}{\text{Purity of commercial formulation}}$$

$$\text{Sample equivalent weight to standard be taken} = \frac{0.1049 \times 95.37}{20}$$
$$= 0.5002$$

According to procedure the standard weight is 0.1049 and their equivalent weight of sample is 0.5002.

$$\text{Required ppm} = \frac{\text{Requiredppm} \times \text{Required volume (mL)}}{\text{Available ppm}}$$

Assuming a standard stock solution is 20000 ppm (200 mg/10 mL), and if linear concentrations viz., 14000 ppm, 12000 ppm, 10000 ppm, 8000 ppm, 6000 ppm, 4000 ppm, 2000 ppm are required, follow below method

Initially if required prepare 14000 ppm from 20000 ppm in 10 mL volumetric flask, follow below procedure and subsequently same method can be followed for concentrations required.

$$14000\text{ ppm} = \frac{14000 \times 10\text{ mL}}{20000} = 7\text{mL (Stock solution)} + 3\text{ mL (Distilled water)}$$

$$12000\text{ ppm} = \frac{12000 \times 10\text{mL}}{20000} = 6\text{mL (Stock solution)} + 4\text{ mL (Distilled water)}$$

$$10000\text{ ppm} = \frac{10000 \times 10\text{ mL}}{20000} = 5\text{ mL (Stock solution)} + 5\text{mL (Distilled water)}$$

$$8000\text{ ppm} = \frac{8000\text{ x}10\text{ mL}}{20000} = 4\text{mL (Stock solution)} + 6\text{mL (Distilled water)}$$

$$6000\text{ ppm} = \frac{6000 \times 10\text{ mL}}{20000} = 3\text{mL (Stock solution)} + 7\text{mL (Distilled water)}$$

$$4000\text{ ppm} = \frac{4000 \times 10\text{ mL}}{20000} = 2\text{mL (Stock solution)} + 8\text{mL (Distilled water)}$$

$$2000\text{ ppm} = \frac{2000 \times 10\text{ mL}}{20000} = 1\text{mL (Stock solution)} + 9\text{mL (Distilled water)}$$

Method validation parameters to be considered for pesticide formulation

Definition of Validation Parameters

1. **Specificity** -Specificity in method validation refers to the ability of an analytical method to distinguish and accurately measure the analyte of interest in the presence of other components, such as impurities, degradants, or matrix elements. It ensures that the method can specifically detect the target analyte without interference from other substances in the sample.
2. **Linearity:** Linearity is a measure of the method's ability (within a given range) to obtain test results that are directly proportional to the concentration (amount) of analyte in the sample.

3. **Accuracy:** The accuracy of an analytical procedure expresses the closeness of agreement between the value which is accepted either as a conventional true value or an accepted reference value and the value found.
4. **Precision:** The Precision of an analytical method is the degree of agreement among individual test results when applied repeatedly to multiple replicates of a homogenous sample.

Validation report

S/N	Validation Parameter	Results		Acceptance Criteria
1	**Specificity**			
	Chromatogram of Blank (Diluent)	Complies/Not complies		An injection of blank (diluent) does not exhibit any peak at the Retention time of Standard
	Chromatogram of pesticide sample preparation	Complies/Not complies		An injection of sample preparation should exhibit principal peak at the same retention time of Standard
2	**Linearity**			
	Correlation coefficient (r)	--		Not less than 0.99
	Slope Value	--		Report the value
	%y Intercept of response of nominal concentration	--		
3	**Accuracy**			
	Mean recovery at each level	Mean at 80%	--	97.00% to 103.00%
		Mean at 100%	--	
		Mean at 120%	--	
	Overall mean recovery (n=9)	--		97.00% to 103.00%
	Overall RSD	--		Not more than 2.0%
4	**Precision**			
4.1	**System Precision**			
	%RSD of peak area of six pesticide standard solution (n=6)	--		Not more than 2.0%
4.2	**Repeatability**			
	%RSD of repeatability (n=6) purity of pesticide sample	--		Not more than 5.0%
4.3	**Reproducibility**			
	%RSD of reproducibility (n=6) purity of pesticide sample	--		Not more than 5.0%
4.4	Cumulative %RSD of repeatability & reproducibility (n=12)	--		Not more than 5.0%

Pesticide	Formulation Type	Physical test parameters	Glassware and Equipment	Requirements	Reference Method
Chlorpyrifos	EC	Cold test	Thermometer	No turbidity or separation of-solid or oily matter or both shall occur when the material is subjected to the cold test at 10°C as prescribed in 13.1 of IS 6940	IS:6940-1982 (RA 2022)
		Flash point	Flash point Instrument	When determined by the method prescribed in IS 1448 [P: 20] the flash point of the material shall be minimum 24.5°C.	IS:1448 (P20) 1960
		Emulsion stability	Crow receiver (100 mL)	Any separation including creaming at the top and sedimentation at the bottom of 100 ml of emulsion prepared in standard hard water with 2.0 ml of EC shall not exceed 2.0 ml when tested by the method prescribed in 13.3 of IS 6940.	IS:6940-1982 (RA 2022)
Acetamiprid	SP	pH of 10 % Aqueous solution at 27 ±1°C	pH Meter	7.0 - 9.0	IS: 16328:2017
Imidacloprid	SL	Cold test	Thermometer	No turbidity or separation of-solid or oily matter or both shall occur when the material is subjected to the cold test at 10°C as prescribed in 13.1 of IS 6940	IS:6940-1982 (RA 2022)
		Flash point	Flash point Instrument	When determined by the method prescribed in IS 1448 [P: 20], the flash point of the material shall be above 24.5°C.	IS:1448 (P20) 1960
		Acidity	Automatic potentiometric titrator	When determined by the method prescribed in 13.5.4 of IS 6940, acidity (as H_2SO_4) shall not be more than 1.0 percent by mass.	IS:6940-1982 (RA 2022)
Hexaconazole	SC	Pourability	500 ml stoppered measuring cylinder	When determined by the method prescribed in Annex A, the rinsed residue shall not be more than 1 percent (m/m).	IS: 15692:2006

Pesticide	Formulation Type	Physical test parameters	Glassware and Equipment	Requirements	Reference Method
		Spontaneity of dispersion	Glass Suction tube (40 cm long, 5mm ID)	When determined by the method prescribed in Annex B, the spontaneity of dispersion of the material shall not be less than 70 percent (m/m)	IS: 15692:2006
			Density Bottle		
		Suspensibility	Glass Suction tube(40cm long,5mm ID)	When determined by the method prescribed in Annex C, the suspensibility of the material shall not be less than 80 percent (m/m)	IS:15692:2006
		Wet sieve test	45micron IS Sieve	Not less than 98 percent by mass of the material shall pass through a 45 micron IS sieve when determined by the method described in 11.1of IS 6940	IS:460 (P1)-1978
		Persistent foam	Graduate cylinder	When determined by the procedure prescribed in Annex D, the persistent foam shall not be more than 60 ml in 12 min	IS:15692:2006
			Stopwatch		
		Acidity/ Alkalinity	Automatic potentiometric titrator	When tested by the method prescribed in 13.5 of IS 6940 the acidity (as H_2SO4) or alkalinity (as NaOH) of the material shall not be more than 0.5 percent by mass, respectively	IS:6940-1982 (RA 2022)
Fipronil	SC	Pourability	A 500 ml stoppered measuring cylinder	When determined by the method prescribed in Annex A, the residue shall not be more than 5 percent (m/m) and rinsed residue shall not be more than 0.50 percent (m/m)	IS 16145 : 2013
		Spontaneity of Dispersion	Glass Suction tube (40 cm long, 5mm ID)	When determined by the method prescribed in Annex B, the spontaneity of dispersion of the material shall not be less than 60 percent (m/m)	
			Density Bottle		
		Suspensibility	Liquid Chromatography (HPLC) system equipped with UV-VIS detector	When determined by the method prescribed in Annex C, the suspensibility of the material shall not be less than 60 percent (m/m)	

Pesticide	Formulation Type	Physical test parameters	Glassware and Equipment	Requirements	Reference Method
		Wet Sieve Test	45micron IS Sieve	Not less than 98 percent by mass of the material shall pass through a 45 micron IS sieve when determined by method described in 11.1 of IS 6940	
		Persistent Foam	Graduate cylinder Stopwatch	When determined by the method prescribed in Annex D, the persistent foam shall not be more than 60 ml in 1 min	
		pH of Suspension	pH Meter	When determined by the method prescribed in Annex E, the pH of 1 percent solution in standard hard water (342 ppm) shall be between 4.0–8.5	
Fenvalerate	EC	Cold test	Thermometer	No turbidity or separation of-solid or oily matter or both shall occur when the material is subjected to the cold test at 10°C as prescribed in 13.1 of IS 6940	IS:6940-1982 (RA 2022)
		Flash point	Flash point Instrument	When determined by the method prescribed in IS: 1448 [P: 20]-1960 the flash point of the material shall be above 24.5°C	IS:1448(P20)1960
		Emulsion Stability	Crow receiver (100 mL)	Any separation including creaming at the top and sedimentation at the bottom of 100 ml of emulsion prepared in standard hard water with 2.0 ml of EC shall not exceed 2.0 ml when tested by the method prescribed in 13.3 of IS 6940	IS:6940-1982 (RA 2022)
		Acidity/ Alkalinity	Automatic potentiometric titrator	When tested by the method prescribed in 11.3 of IS: 6940-1982 acidity (as H_2SO4) shall be not more than 0.30 percent by mass. The product shall not be alkaline to litmus paper when tested according to 11.3.1.1 of IS : 6940-1982	IS:6940-1982 (RA 2022)
		5. Heat Stability	Glass Bottle Thermostat	After treating in accordance with the method prescribed in 13.4 of IS : 6940-1982 the material shall comply with the requirement Specified in 2.2.1 to 2.2.4 and 2.3.1	IS:6940-1982 (RA 2022)

Physical parameters needed as per BIS for Technical grade pesticides

Pesticide	Formulation Type	Physical test parameters	Instruments	Requirements	Reference Method
Chlorpyrifos	Technical	Acidity	Automatic Potentiometric titrator	0.1%	IS 6940: 1982 (RA 2022)
		Moisture	Karl Fischer titrator	0.1%	
Acetamiprid	Technical	Melting point	Thermometer (Long stem and short bulb)-0-100°C;100 to 200°C	98-100	IS 6940: 1982 (RA 2022)
			Bath and heating assembly-250 mL Kjeldahl flask		
		Moisture	Karl Fischer titrator	0.4	
Imidacloprid	Technical	Moisture	Karl Fischer titrator	1.0%	IS 6940: 1982 (RA 2022)
		Acidity/Alkalinity	Automatic Potentiometric titrator	0.5%	IS 6940: 1982 (RA 2022)
Hexaconazole	Technical	Moisture	Karl Fischer titrator	1.0%	IS 6940: 1982 (RA 2022)
		Acidity	Automatic potentiometric titrator	1.0%	IS 6940: 1982 (RA 2022)
Lambda-cyhalothrin	Technical	Moisture	Karl Fischer titrator	0.3%	IS 6940: 1982 (RA 2022)
		Acidity	Automatic potentiometric titrator	0.3%	IS 6940: 1982 (RA 2022)
Bifenthrin	Technical	Acidity	Automatic potentiometric titrator	0.20%	IS 6940: 1982 (RA 2022)
		Cis –isomer content	Reverse phase HPLC	97.0%	IS 6940: 1982 (RA 2022)
Fenpropathrin	Technical	Moisture	Karl Fischer Titrator	0.50%	IS 6940: 1982 (RA 2022)
		Acidity	Automatic potentiometric titrator	1.0%	IS 6940: 1982 (RA 2022)
Deltamethrin	Technical	Melting point	Thermometer (Long stem and short bulb)-0-100°C;100 to 200°C	98.0-101.0%	IS 6940: 1982 (RA 2022)
		Acid chloride	Erlenmeyer flask 200 mL	0.2%	IS 6940: 1982 (RA 2022)
		Acid + Anhydride	Erlenmeyer flask 200 mL	1.%	IS 6940: 1982 (RA 2022)
		Optical rotation	D-line	57±1.5%	IS 6940: 1982 (RA 2022)
		Acidity	Polari meter	1.0%	IS 6940: 1982 (RA 2022)
Fenvalerate	Technical	Acidity	Automatic potentiometric titrator	1.0%	IS 6940: 1982 (RA 2022)

5

Insecticide Classification

Insecticide classification majorly divided into three categories. A. based on chemical group, B. mode of entry in the insect, C. based on toxicity.

A. Classification of Insecticides based on chemical group

Historians have traced the use of pesticides to the time of Homer around 1000 B.C. earliest records of insecticide use pertain to the burning of "brimstone" (Sulfur) as a fumigant. Pliny the Elder (A.D. 23-79) recorded most of the earlier insecticide uses in his *Natural History* and included among these was the use of gall from a green lizard to protect apples from worms and rot. Later, a variety of materials used viz., extracts of pepper and tobacco, soapy water, whitewash, vinegar, turpentine, fish oil, brine, lye and many others. At the beginning of World War II (1940), insecticide selection was limited to several inorganic compounds, arsenicals, petroleum oils, nicotine, pyrethrum, rotenone, sulfur, hydrogen cyanide gas, and cryolite. It was during World War II that opened the *Chemical Era* with the introduction of a totally new concept of insect control.

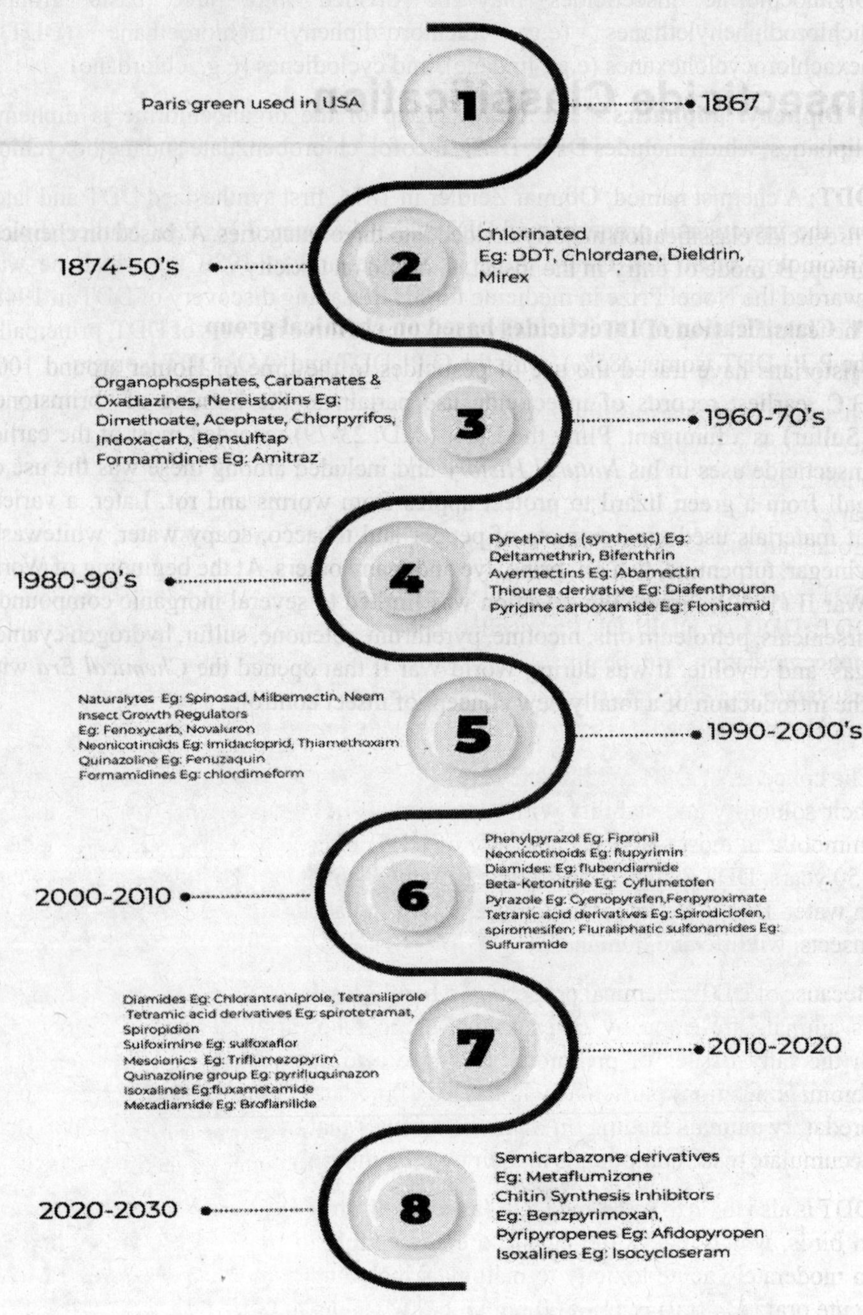
TIMELINE FOR EVOLUTION OF INSECTICIDES
1
Paris green used in USA
1867
2
1874-50's
Chlorinated
Eg: DDT, Chlordane, Dieldrin, Mirex
3
Organophosphates, Carbamates & Oxadiazines, Nereistoxins Eg: Dimethoate, Acephate, Chlorpyrifos, Indoxacarb, Bensulftap
Formamidines Eg: Amitraz
1960-70's
4
1980-90's
Pyrethroids (synthetic) Eg: Deltamethrin, Bifenthrin
Avermectins Eg: Abamectin
Thiourea derivative Eg: Diafenthouron
Pyridine carboxamide Eg: Flonicamid
5
Naturalytes Eg: Spinosad, Milbemectin, Neem
Insect Growth Regulators
Eg: Fenoxycarb, Novaluron
Neonicotinoids Eg: Imidacloprid, Thiamethoxam
Quinazoline Eg: Fenuzaquin
Formamidines Eg: chlordimeform
1990-2000's
6
2000-2010
Phenylpyrazol Eg: Fipronil
Neonicotinoids Eg: flupyrimin
Diamides: Eg: flubendiamide
Beta-Ketonitrile Eg: Cyflumetofen
Pyrazole Eg: Cyenopyrafen,Fenpyroximate
Tetranic acid derivatives Eg: Spirodiclofen, spiromesifen; Fluraliphatic sulfonamides Eg: Sulfuramide
7
Diamides Eg: Chlorantraniprole, Tetraniliprole
Tetramic acid derivatives Eg: spirotetramat, Spiropidion
Sulfoximine Eg: sulfoxaflor
Mesoionics Eg: Triflumezopyrim
Quinazoline group Eg: pyrifluquinazon
Isoxalines Eg:fluxametamide
Metadiamide Eg: Broflanilide
2010-2020
8
2020-2030
Semicarbazone derivatives
Eg: Metaflumizone
Chitin Synthesis Inhibitors
Eg: Benzpyrimoxan,
Pyripyropenes Eg: Afidopyropen
Isoxalines Eg: Isocycloseram

1. Organochlorines (Synthetic chlorinated hydrocarcons)

Organochlorine insecticides may be divided into three basic groups: dichlorodiphenylethanes (e.g., dichloro-diphenyl-trichloroethane (DDT)), hexachlorocyclohexanes (e.g., lindane), and cyclodienes (e.g., chlordane)

i) Diphenyl aliphatics:- The oldest group of the organochlorine is diphenyl aliphatics, which includes DDT, DDD, dicofol, chlorobenzilate and methoxychlor.

DDT: A chemist named, Othmar Zeidler in 1874, first synthesized DDT and later on, the insecticidal properties of DDT were discovered by Paul Muller (Swiss Entomologist) of J.R. Geigy, A.G. in Switzerland in 1939 for which he was awarded the Nobel Prize in medicine for his lifesaving discovery of DDT in 1948. The technical grade DDT is actually a mixture of three isomers of DDT, principally the P, P'-DDT isomer (85%), with the O,P'-DDT and O,O'-DDT isomers

This insecticide was useful in the control of malaria, yellow fever, typhus, body lice, bubonic plague and many other insect-vector diseases i.e. mainly used for public health purposes. In addition to public health uses, growers used on a variety of food crops *viz*., cotton, beans, soybeans, cabbage, tomatoes, cauliflower, groundnut and even used in pest control in buildings.

DDT and its metabolites accumulate in body fat and other tissues, as either DDT, DDD or DDE. DDT cancelled because of concern over carcinogenicity, bioaccumulation and health effects on wildlife. In addition to these concerns, resistance to DDT occurs in some insects (housefly) that develop the ability to quickly metabolize DDT into the lower toxicity breakdown product DDE.

The concerns of DDT are that they are highly persistent in the environment due to their solubility and stability with reported half-life of between 2-15 years and is immobile in most soils. The half-life of DDT in an aquatic environment is about 150 years. DDT is highly fat soluble (dissolves in fat easily), but is poorly soluble in water. Due to its fat loving nature, it tends to accumulate in the fatty tissues of insects, wildlife and humans.

Because of DDTs chemical properties, it has the tendency to accumulate in animals, as animals are eaten by other animals higher up, DDT becomes concentrated in the fatty tissues of predators. They are also accumulated in the food chain (biomagnification) particularly in tissues of predatory and parasitic insects. Thus, predatory animals feeding on many individuals that have DDT in their fat tissues accumulate these compounds in their own fat multiplying the accumulation effect.

DDT is also toxic to birds and DDE (a metabolite of DDT) causes eggshell thinning in birds, which makes the eggs more susceptible to fracturing. DDT is slightly to moderately acute toxicity to mammals including humans, when ingested with acute oral LD_{50} (rat) is 113-800 mg/kg body weight and acute dermal LD_{50} (rat) is 2500-3000 mg/kg.

Dicofol is an organochlorine pesticide, specifically classified as a miticide. This group includes compounds that are structurally similar to DDT, sharing a common chemical structure characterized by diphenyl and aliphatic components.

Note: **The compound, DDT has banned for production and use in India**.

ii. Benzene hexa chloride (BHC): Hexchlorocyclohexane (HCH) also known as BHC was first prepared in 1825 by Michael Faraday, who did not recognize its insecticidal properties. The insecticidal properties of HCH were discovered in 1940 by French and British entomologists. In its technical grade, there are five isomers, *alpha, beta, gamma, delta and epsilon*. α isomer constitutes 65-70%, β isomer constitute 5-6%, γ isomer constitutes 13%, δ isomer constitutes 6%, and ε isomer constitutes 3-4%. Surprisingly, only the *gamma* isomer has insecticidal properties and consequently, the gamma isomer was manufactured and sold as odorless insecticide, ***lindane***. The technical HCH has a characteristic musty odor.

iii. Cyclodienes: The cyclodienes group of insecticides are chlordane, heptachlor, aldrin, and diéldrin which were used for the control of termites, soil-borne insects etc

a) **Lindane:** In 1912, Van der Linden discovered four isomers and the insecticidal properties lie in γ isomer. So pure form of γ HCH is named as Lindane in honor of Van der Linden, a colorless crystal compound and was marketed under the brand name Gammexane®

b) **Polychloroterpenes:** Only two polychloroterpenes were developed i.e. toxaphene and strobane. These act on the neurons causing an imbalance in sodium and potassium ions, similar to that of the cyclodiene insecticides.

c) **Endosulfan:** Endosulfan is a chlorinated hydrocarbon insecticide and acaricide of the cyclodiene subgroup. It is practically insoluble in water, moderately soluble in most organic solvents. In India, it was available as Endosulfan 35EC (Thiodan®) and was the widely used broad spectrum insecticide/acaricide for control of insects in various crops but now the insecticide has been banned as per the CIB regulations. All the compounds belonging to cyclodiene subgroup has banned for production and use in India including Endosulfan.

2. Organophosphates (OP)

Organophosphates (OPs) are chemical compounds formed by the interaction of alcohols with phosphoric acid. They were originally synthesized in 1854, but their extreme toxicity was not discovered until the 1930s. Organophosphates were utilized as pesticides in the 1930s, but the German military developed them as neurotoxins during WW-II. **Tetraethyl pyrophosphate (TEPP)**, the first synthetic OP pesticide, was developed in Germany prior to World War II to replace nicotine, an extremely poisonous and lipid-soluble plant insecticide. Sarin and tabun, two highly lethal nerve gas agents produced about the same time, were not employed throughout the conflict. The mode of action of OP compounds is the same as nerve gases" (sarin, soman, and tabun), due to the similarities of chemical structures. During World War II, Gerhard Schrader of Germany discovered the insecticidal characteristics of the first OP pesticide in 1937. The insecticide was called after its discoverer, **Schradan.**

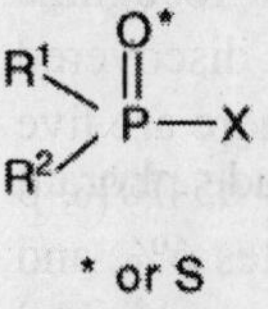

O*
R1 ‖
P—X
R2

* or S

An organophosphorous insecticide's Schradan formula, contains the thioate prodrugs have a sulphur atom, the ultimate reactive compounds, the "oxons," have an oxygen atom at the double bond to phosphorus. R1 and R2 are identical in most widely used pesticides and either O-methyl or O-ethyl groups.

Classification of OPs

Organophosphates typically contain two alkyl groups and a third group known as the leaving group, which is frequently an aryl or heterocyclic group. The leaving group is more sensitive to hydrolysis than the alkyl groups. Organophosphate insecticides with P=S configuration have little or no inhibitory activity on acetylcholine esterase enzyme unless activated by enzymatic or non-enzymatic oxidative desulfuration to the equivalent oxon, i.e., P=O configuration (Table 1). For example, Malathion to Malaoxon, Parathion to Paraoxon, and similar chemicals are known as "indirect inhibitors of AChE." Following the initial exposure of the effector junction and organophosphate, the enzyme-phosphoryl bond is strengthened by the removal of one alkyl group from the phosphoryl adduct. This is referred to as aging.

OP compounds may also be classified based on mode of entry and practical uses:

Subgroup (SG)-I: These compounds are soluble in water and have limited chemical stability; however, they are rapidly hydrolyzed and are primarily employed as contact insecticides. Eg: TEPP, tetrachlorvinphos, and mevinphos

SG-II: Soluble in oils but poorly soluble in water; moderate to high chemical stability. Persistent contact pesticides like these can soak into the leaf's waxy coating but remain unable to penetrate the plant itself. The toxicant is rendered active prior to its arrival at the activation site.

Eg. Malathion, methyl parathion, diazinon, fenitrothion and trichlorophon (Dipterex).

SG-III: Chemically stable molecules with a moderate to high partition coefficient between oil and water are able to move around inside plants. These pesticides work systemically, activating them before they reach the target area. These chemicals have a systemic character due to their water solubility.

Eg: Dimethoate, phosphatidyliodone, disulfoton, formothion, phorate, and demeton methyl are other examples.

SG-IV: Compounds with a high vapour pressure (0.012 mm Hg at 20°C) and low chemical stability can be utilized as fumigants

Eg: dichlorvos

SG-V: OP insecticides are used as soil insecticide. Examples include phorate, chlorfenvinphos, bromophos, and diazinon.

Table 1: Martin classification of organophosphates

Type of OP	Structure	Example
Phosphates	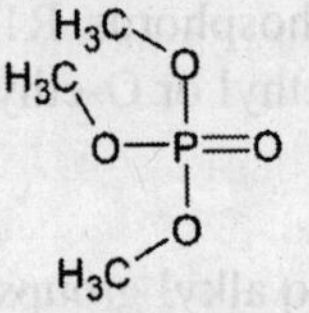 Phosphate	Monocrotophos Chlorfenviphos Chlorpyrifos-methyl Dichlorvos Tri-o-cresyl phosphate
Phosphonates	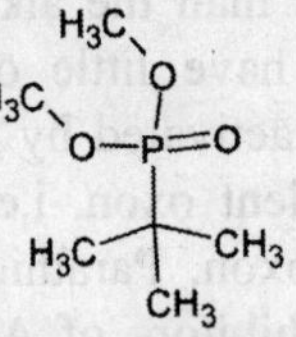 Phosphonate	Trichlorfon
Phosphorothionates (Thiophosphates)	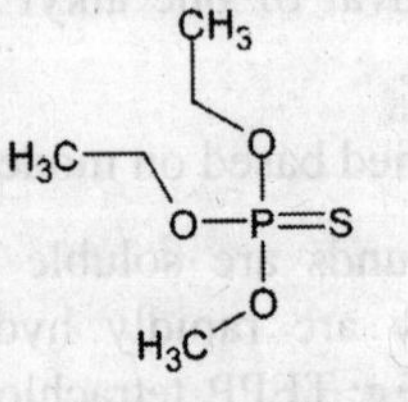 Phosphorothionate	Pirimiphos-methyl Bromophos Diazinon Traizophos Fenitrothion Fenthion Methyl parathion Chlorpyrifos
Phosphorothiolates (S-substituted)	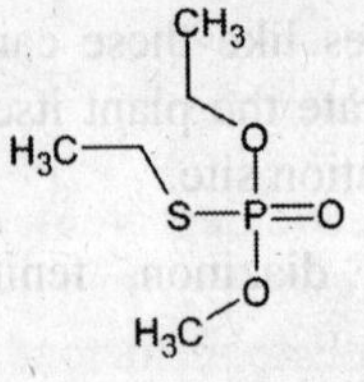 Phosphorothiolate	Demeton – S- methyl Ecothiopate

Type of OP	Structure	Example
Phosphorothiolothionate (S-substituted) (Dithiophosphates)	Phosphorothiolothionate	Dimethoate Malathion Phorate Terbufos Azinphos-methyl
Phosphoroamidate	Phosphoroamidate	Acephate Fenamiphos Methamidophos

Phosphorus esters have unique combinations of oxygen, carbon, sulfur, and nitrogen linked to the phosphorus atom. The names of the phosphor molecules listed above are part of considerably longer chemical names for the compounds that comprise these building blocks (moieties). There are three chemical classes of organophosphorus insecticides:

- Derivatives of amino acids with a carbon chain structure
- Phenyl derivatives having a benzene ring in which the phosphorus moiety replaces one hydrogen atom
- Heterocyclic derivatives having a phosphorous group and ring structure as well. On the other hand, one or more carbon atoms are replaced by oxygen, nitrogen, or sulfur in a heterocyclic carbon ring, which can have three, five, or six rings.

Aliphatic organophosphates

- Malathion is the most well-known and traditional insecticide in this class. This insecticide has been widely used for about 40 years in agriculture, public health, and the management of storage and household pests. It has low mammalian toxicity and is effective against a wide variety of insect pests.
- A number of aliphatic OPs, which are effective against sucking insect pests, have a systemic activity. These include dimethoate, demeton, and monocrotophos. Certain insecticides are favored for use on vegetable crops because, despite their extreme toxicity, they have a brief residual activity. The insecticides acephate, demeton, demeton-methyl, dichlorvos, dimethoate, malathion, methamidophos, mevinphos, monocrotophos, phorate, phosphomidon, and trichlorfon are included in this category.

Phenyl Organophosphates

- These substances are more persistent and generally more stable than aliphatic ones. Methyl parathion is used extensively in horticulture, public health initiatives, and agriculture. It is somewhat less harmful than only other older members of this class, like fenitrothion and tetrachlorvinphos. Bromophos, fenthion, fensulfothion, fenitrothion, parathion methyl, phosalone, and triazophos are among the compounds in this group.

Heterocyclic Organophosphates

A
‖
O—P—X
|
O

(Scheme 2)

The general formula for all organophosphate pesticides is as follows (Scheme 2). Where RO and R'O are alkoxy groups, which are typically either O, O diethyl or O, O dimethyl, and A is oxygen or sulfur. These are the same in the majority of commercial pesticides. The acidic group X is particularly vulnerable to chemical attack because the phosphorus atom is starting to take on a positive charge. It is frequently referred to as "the leaving group" because of this. Phosphorus's electrophilic behavior is dependent on how it exits group X.

Organophosphorus insecticides work because of the nature of group X, but if this group is electrophilic, it tends to pull electrons away from P and form a positive site close to the phosphorus atom. The compound's hydrolysis is facilitated by the establishment of positive nature at this location. The other radicals cannot be disregarded, even if group X's nature is mostly to blame for the organophosphorus pesticides' effects. For instance, even though an analog of parathion has the same acidic or leaving group as parathion, it is biologically far less active than parathion.

This means that the biological activity of organophosphorus insecticides is not just based on its chemical qualities, but also on its steric factors. This is why the molecules' biochemical interactions are stronger. The molecule stops the middle of the esterase from working by fitting into it in a way that can't be undone.

Table 2: Organophosphorus insecticides and their groups

Group	Insecticide
Diphosphoric acid	TEPP
Phenylphosphonothionic acid	Leptophos
Phosphonothioic acid	EPN(*O*-ethyl *O*-*p*-nitrophenyl phenylphosphonothionate)
Phosphonic acid	dichlorvos, trichlorfon,
Phosphoric acid	chlorfenvinphos, dicrotophos, Naled, monocrotophos, phosphamidon, tetrachlorvinphos
Phosphorothioic acid	chlorpyrifos, coumaphos, demeton, diazinon, famphur, fenthion, fenitrothion, fensulfothion, methyl parathion, oxydemeton methyl, parathion, ronnel, temephos,
Phosphoramidic acid	Cyolane
Phosphorodithioic acid	azimphos-methyl, dimethoate, dioxathion, disulfoton, ethion, ethoprop, imidan, methidathion, phorate
Phosphoramidothioic acid	Methamidophos

Chemistry

Most organophosphates are moderately soluble in water, have a high oil-water partition coefficient, and have low volatility (with the exception of dichlorvos). Additionally, they have a low vapor pressure. The vapour pressure and solubility of a pesticide determine whether it will be removed from the atmosphere by surface water or rain, or if it will be volatilized from a surface. The absorption of organic chemicals by plants is typically attributed to the properties of the chemical, particularly those that determine the logarithm of the n-octanol–water partition coefficient ($\log K_{OW}$). KH and K_{OW} are the properties that determine the persistence of a chemical in the environment. A high KH value indicates that the pesticides will be more likely to volatilize. Koc is an essential input parameter for the estimation of the environmental distribution and exposure level of a chemical substance.

Table 3: Physicochemical properties of organophosphates

Insecticide	Solubility (20–25 °C) (mg/L)	V_p (Pa) (20–25 °C)	Log K_{ow}	Koc (cm³/g)	Half-Life $T_{1/2}$ (Days)
Acephate	650	2.26×10^{-4} 1.7×10^{-6} (23–25 °C)	−1.87	0.88	13
Chlorfenvinphos	145	1.0×10^{-3}	3.8	-	-
Chlorpyriphos	1.4	2.7×10^{-3}	4.96	-	94
Diazinon	60	1.2×10^{-2}	3.3	4981	23
Dichlorvos	18,000	**2.1**	1.9	272	-
Dimethoate	23	1.1×10^{-3}	0.7	20	7
Fenamiphos	700	0.12×10^{-3}	3.3	267	16
Fenitrothion	30	18×10^{-3}	-	-	-
Fenthion	4.2	7.4×10^{-4}	4.84	15,000	34

Insecticide	Solubility (20–25 °C) (mg/L)	V_p (Pa) (20–25 °C)	Log K_{ow}	Koc (cm^3/g)	Half-Life $T_{1/2}$ (Days)
Malathion	145	5.3×10^{-3}	2.75	1800	1
Methamidophos	90,000	2.3×10^{-3}	0.8	1.7	>2.6
Mevinphos	Miscible	1.7×10^{-3}	0.13	44	3
Monocrotophos	Miscible	2.9×10^{-4}	−0.22	1	30

V_p = vapour pressure; **Log Kow**, also known as the octanol-water partition coefficient, is a measure of a compound's hydrophobicity and ideal range of 2 to 4 for pesticides; **Koc**, or the soil organic carbon-water partition coefficient, is a measure of a chemical's tendency to adsorb to soil organic matter; The **half-life** of a substance is the time required for its concentration to decrease by half.

Phosphamidon: Phosphamidon is a systemic insecticide and effective against pests infesting rice, cotton, vegetables.

Demeton (Systox): Demeton is a systemic insecticide and has a long residual life.

Oxydemeton methyl (Metasystox-R)-It is a systemic insecticide and nothing but a methyl analogue of the thiol isomer of demeton. It is mainly used as foliar spray to control sucking insects.

Phorate: It is marketed as Thimet. Phorate is a systemic insecticide and miticide. It is used for cotton pests, rice pests, thrips, leaf hoppers etc.

Malathion: It is one of the safest organophosphate insecticide used today, It has both contact and stomach action, It is wid

Dichlorvos: Dichlorvos is a very volatile insecticide and has a vapor pressure of 1.2 x 10^{-2} mm Hg at 20°C. It gives rapid knockdown and kill of insects, the brand names available are DDVP and Vapona.

DDVP (dimethyl 2,2dichlorovinyl phosphate) is a contact and stomach insecticide with fumigant and penetrant properties, commonly known as dichlorvos. Despite its low vapor pressure (Table 3) and high boiling point (120°C/14 mm), dichlorvos is unable to penetrate materials. **Consequently, it is of no value as a commodity fumigant**. Rather, it is dispersed as a true gas to control insects in the open spaces of structures. It has been successfully employed against moths and the cigarette insect in tobacco warehouses, and it is recommended for use in free spaces. Dichlorvos exhibits fumigant properties in the field through the direct evaporation of liquid concentrate through the application of heat.

The residues of dichlorvos decrease to extremely low levels during storage and shipment due to its rapid degradation following application. The rate of disintegration increased as the temperature and moisture content of the material or its environment increased. It is possible to infer that the rapid hydrolysis of dichlorvos vapour in enclosed spaces containing foodstuffs results in the rapid disappearance of substantial residues of this chemical within a brief period of time. Dichlorvos is rapidly removed from plant leaf surfaces by volatilization and hydrolysis, with a half-life of within a few hours. A limited quantity enters the waxy layers of plant tissues, where it may remain for longer (FAO/WHO, 1971). Dichlorvos is highly soluble in water and is unlikely to bioconcentrate in fish or other aquatic animals.

Parathion and malathion are poor cholinesterase inhibitors in vitro, but quite effective in vivo. The explanation is that parathion and malathion are transformed into paraxon and malaxon. This process is commonly referred to as oxidation, but because there is no change in valency, it should be called desulfuration or bioactivation. In the body, sulfur atoms form the molecules parathion and malathion. This molecule is replaced by oxygen, resulting in the production of P=O instead of P = S. Because =O is more electrophilic than =S (i.e. attracts more electrons), P in P=O molecule gains enough positive charge to engage quickly with cholinesterase enzyme.

One example of isomerization from the thiono type to the thiolo type is the change from malathion to malaxon. This is the reason why phosphorothionates are indirect cholinesterase enzyme inhibitors that don't work until they change to the P=O form. At room temperature, chemical isomerization takes a long time, but it may be sped up by enzymes inside living things. Phosphorrothionate is likely to contain thiolo isomers because OP pesticides are made at high temperatures. These isomers may directly block the cholinesterase enzyme.

Malathion is better than parathion because it is safer. The World Health Organization (WHO) says that parathion is three to six mg/kg of acute oral toxicity and malathion is fourteen hundred mg/kg of acute oral toxicity. Malathion is used in different types of products and public health use i.e. in the form of sprays, protein hydrolysates and yeast baits for fruit flies (Matthews, 2018).

Reaction between Acetylcholine and Cholinesterase

In 1920, Otto Loewi provided evidence that acetylcholine (ACh) functions as a chemical mediator, facilitating the transmission of nerve signals between neurons at synapses. Acetylcholine (ACh) is a neurotransmitter that is synthesized from acetyl-coenzyme A (acetyl-CoA). Acetylcholine (ACh) is a neurotransmitter that is synthesized from acetyl-CoA, which is derived from glucose and choline through the enzymatic activity of choline acetyltransferase. Acetylcholine (ACh) is stored in presynaptic membranes in vesicles and is released upon activation. Acetylcholinesterase (AChE) employs a hydrolytic mechanism to break down the neurotransmitter acetylcholine (ACh) into choline and acetate, so effectively ending its influence on the muscarinic and nicotinic receptors. Organophosphates have the ability to form a permanent bond with acetylcholinesterase (AChE) and hinder the process of acetylcholine (ACh) breakdown. The release of ACh leads to excessive activation of both muscarinic and nicotinic receptors, which are extensively dispersed throughout the body.

The hydrolysis of acetylcholine by the enzyme cholinesterase occurs in three distinct steps, resulting in the recovery of both acetylcholine and the enzyme.

Stage 1: The reaction between acetylcholine and cholinesterase results in the formation of a complex called the "Michaelis complex."

Stage II: The enzyme complex produces choline and enzyme that has been acetylated. The process being referred to is commonly known as an acetylation reaction.

Stage III: The final stage involves a deacetylation reaction in which the acetylated enzyme is hydrolyzed to produce free enzyme and acetic acid.

$CH_3CO . E \longrightarrow CH_3COOH + EH$

(Acetylated enzyme) (acetic acid) (enzyme)

In order to prevent the accumulation of acetylcholine across the synapse of the neuromuscular junction, these reactions occur rapidly in living organisms. The choline portion is eliminated, and the acetic acid recombines to produce new acetylcholine, thereby closing the cycle.

Reaction Between op and Che Enzyme

The reaction between OP compounds and AchE is analogous to that of the standard substrate, acetylcholine. Consequently, the initial stages of the reaction between OP compounds and ChE enzyme are essentially analogous to the reaction between Ach and ChE enzyme. However, in the final stages of the deacetylation, "Ach hydrolysis" occurs at a rapid pace, and the enzyme is then recovered. The OP compounds are potent ChE inhibitors due to the fact that dephosphorylation occurs at an incredibly sluggish rate in the ChE inhibition reaction.

The ChE enzyme contains two active centers: the anionic site and the esteratic site. The hydrolysis of linkage is catalyzed by the esteratic site, while the anionic site is negatively charged and bonds the trimethylammonium group. The esteratic site is composed of three groups: basic (histidine imidazole), hydroxyl (serine), and acidic (tyrosine hydroxyl) (Fig. 1).

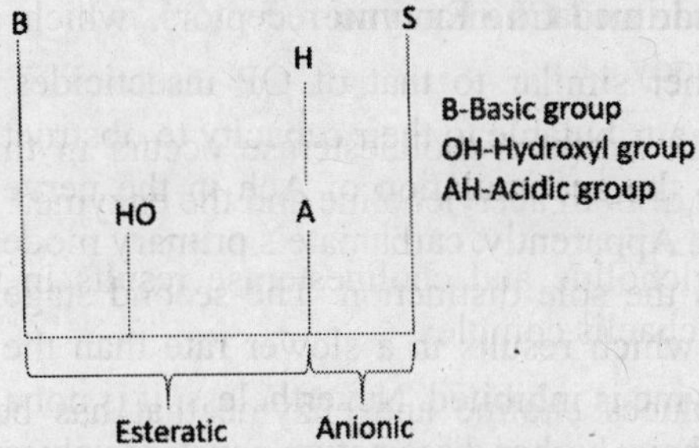

Fig. 1: Basic structure of esteric site

A reversible complex is generated by the enzyme's action on paraxon. The phosphorus atom of paraxon is attacked by the hydroxyl group of the enzyme during this process. The p-nitrophenol is produced by the transfer of the hydrogen atom from the acidic group to the paraxon. The remaining product is an enzyme that has been phosphorylated.

Phosphorylation is the term used to describe the process. Two factors determine the reversibility of the reaction: the affinity of the inhibiting compound and the rate of phosphorylation. The reversible complex undergoes immediate conversion to a phosphorylated enzyme, a process known as dephosphorylation, due to the high phosphorylation constant. During the typical deacetylation reaction of an acetylated enzyme, the rate of deacetylation is significantly higher (295,000 molecules/active center/min), whereas the rate of dephosphorylation is extremely sluggish (3 molecules/active center/min). Consequently, the amount of enzyme recovery is extremely negligible.

The reaction is represented by the following: EH is the enzyme, AB is the insecticide, and A is the phosphorylating group, while B is the leaving group. (The non-phosphoryl or non-carbamyl portion of organophosphorus or carbamate insecticides is referred to as the "leaving group").

K+1 K+2 K+3

EH + AB → EH.AB → EA → EH+AOH

K-1 BH HO

The Ach will not be promptly removed from the receptor surface of the muscle if the ChE enzyme is eliminated or a complex is formed with a slower release of the enzyme, as in the aforementioned reaction. This would result in the muscle remaining depolarized for an extended period of time than is typical, which will trigger multiple action potential trials to pass through the muscle. This will cause the muscle to twitch, which in turn leads to tetanus and ultimately paralysis of the muscle tissue.

Reaction Between Carbamate Insecticide and Che Enzyme

Carbamates react with AchE in a manner similar to that of OP insecticides. Therefore, the carbamates' toxic nature is attributable to their capacity to obstruct the nerve enzyme AchE, which leads to the accumulation of Ach in the nerve synapse and a reduction in nerve function. Apparently, carbamate's primary mode of ChE inhibition is reversible, which is the sole distinction. The second stage reaction in carbamates is carbamylated, which results in a slower rate than the OP. Consequently, a small amount of enzyme is inhibited. Nevertheless, it is not a general rule, as there are carbamates that have higher dissociation constant values than OP compounds and are therefore effective inhibitors of ChE.

All organophosphate compounds have two characteristics: they are often far more poisonous to vertebrates than other types of insecticides, and the majority are chemically unstable or non-persistent. This distinguishing trait led to their application as agricultural alternatives for persistent organochlorines. One characteristic distinguishing these OP insecticides is that they inhibit the cholinesterase enzyme.

Why OPs are toxic to insects?

Certain nerves in insects work by releasing acetylcholine (ACh) into the intracellular space (i.e., the synapse) where the nerve cell contacts a muscle cell (or another nerve cell). Acetylcholine causes muscle cells to contract, which is prevented by acetylcholinesterase, an enzyme that destroys the produced ACh signal molecules. Organophosphates are acutely poisonous because they chemically attach to the acetylcholinesterase enzyme, preventing it from destroying ACh, and the insect dies with its muscles contracted and nervous system stimulated. It is important to note that ACh is a neurotransmitter found in most animal species, not just insects. As a result, organophosphate insecticides can be hazardous to a wide range of non-target species, including people.

Toxicology

Organophosphates are readily absorbed by inhalation and eating. The extent to which substances can penetrate the skin and be absorbed into the bloodstream varies depending on the specific agents. There is significant variety in the extent to which these different pathways absorb substances. For example, the lethal dose (LD_{50}) of parathion when administered orally to rats ranges from 3-8 mg/kg. This dosage is very toxic and essentially the same as the lethal dose when absorbed via the skin, which is also 8 mg/kg. However, when it comes to toxicity, phosalone is less harmful when absorbed via the skin compared to when ingested orally. The LD_{50} values for rats are 1,500 mg/kg and 120 mg/kg for dermal and oral exposure, respectively. Typically, extremely toxic substances are more likely to have a greater level of toxicity when they come into contact with the skin compared to moderately poisonous substances.

Many organothiophosphates convert from thions (P=S) to oxons (P=O) in the environment under oxygen and light and in the body mostly by liver microsomal enzymes. Oxons break down faster than thions but are more poisonous. Thions and oxons hydrolyze at the ester bond to produce low-toxicity alkyl phosphates and leaving groups. They are excreted or altered before excretion. Loss of one alkyl group from the phosphoryl adduct strengthens the enzyme-phosphoryl link after effector junction and organophosphate exposure. This is aging. The bond becomes permanent. Aging might take minutes to days depending on the agent. An oxime can de-phosphorylate (reactivate) some phosphorylated acetylcholinesterase enzymes depending on their age. The only FDA-approved oxime is pralidoxime. Europe and Asia employ obidoxime and HI-6 oximes. Since the 1950s, oximes have been employed to poison OP.

Insecticide	**Oral LD_{50} for rats {mg/kg}**
Terbufos	2-5
Mevinphos	3-12
Ethyl parathion	3-13
Azinphos ethyl	12-17
Chlorfenvinphos	10-40
Methyl parathion	14-24
Triazophos	57-68
Oxydemeton methyl	65-80
Dichlorvos	56-108
Quinalphos	62-137
Chlorpyrifos	135-165
Fenthion	190-315
Diazinon	300-400
Dimethoate	500-600
Fenitrothion	800
Malathion	2800

*[#]Eyer, P. The Role of Oximes in the Management of Organophosphorus Pesticide Poisoning. *Toxicol Rev* **22**, 165–190 (2003).

Some interesting facts

Why only monocrotophos is recommended/widely used for root feeding?

The efficiency of three systemic insecticides (methamidophos, monocrotophos, and dicrotophos) in coconut was investigated using root absorption technique. Of the three systemic insecticides tested, dicrotophos was the least effective. Despite having the same efficacy as monocrotophos, methamidophos is phytotoxic and a contact insecticide, making it dangerous to workers and palms. Monocrotophos, which digests without phytotoxicity, is better for root absorption treatments. Root absorption was used to analyze residues from old and young nuts at 7, 26, 5, 7, and 17 days after treatment. Monocrotophos remains in old and young nut water and albumen 117 days after treatment. (WHO/FAO insecticide residue tolerance is 200 ppb). So, wait at least two months before consuming nuts.

Insecticide	**Tradename/Brand name**
Acephate	Acephate 75SP (Orthene®)
Chlorpyrifos	Chlorpyrifos 20EC(Dursban®), Chlorpyrifos 50EC (Predator®)
Chlorpyrifos-methyl	-
Dichlorvos	Dichlorvos (DDVP) 76EC (Nuvan®)
Dimethoate	Dimethoate 30EC (Rogor®)
Ethion	Ethion 50EC (Fosmite®)
Fenitrothion	Sumithion (Sumitomo)
Malathion	Malathion 50EC (Cythion)
Methyl Parathion	Methyl Parathion 50EC (Metacid®)
Monocrotophos	Monocrotophos 36SL (Nuvacron®)
Oxydemeton-methyl	Oxydemeton-methyl 25EC (Metasystox®)
Parathion	-
Phenthoate	Phenthoate 50EC (Phendal®)
Phosphamidon	Phosphamidon 40SL (Kinadon Plus®)
Phorate	Phorate 10G (Thimet®)
Profenophos	Profenophos 50EC (Curacron®)
Quinalphos	Quinalphos 25EC (Ekalux®)
Triazophos	Triazophos 40EC (Hostathion®)

3. Carbamates

Carbamates are synthetic derivatives of physostigmine, a key alkaloid of Calabar bean (*Physostigma benenosum*). Carbamates resemble organophosphate pesticides physically and mechanistically. Carbamates are amino formic acid-derived N-methyl carbamates. Ester derivatives from simple carbamic acids are unstable, especially in basic circumstances. Ester derivatives of carbamates are crystals with low Pv and variable water solubility. Water-soluble carbamates have different vapor pressures. KH represents Henry's law's balancing of these two processes. They are poorly soluble in chloroform and toluene but substantially soluble in acetone.

CH_3 N CH_3 N O H_3C O H_3C NH

Physostigmine

Carbamates have contact and stomach action, while few insecticides have systemic action. The main worry with this group is that insects can recover if the dose is minimal because enzyme inhibition is easier to reverse than with OP insecticide categories. Due to their lower environmental half-lives, carbamates have replaced OP and OC insecticides.

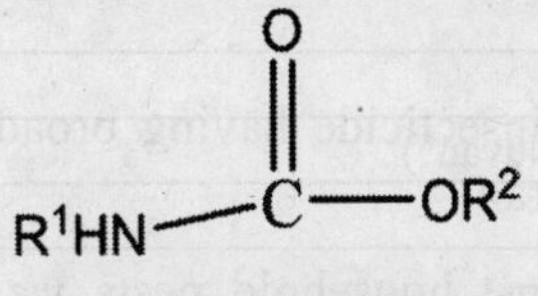

$$R^1HN-\overset{\overset{\displaystyle O}{\|}}{C}-OR^2$$

Fig. 2: General Structure of carbamate pesticide

R^2 is an aromatic or aliphatic moiety. Where R^1 is a methyl group (Fig.2)

Important carbamate compounds commonly used are

- The first successful carbamate insecticide, carbaryl was introduced in 1956
- The carbamate compound used as insecticide and nematicide is Carbofuran 3G
- One public health insecticide registered under this group is Propoxur commonly used against cockroaches and other household insect pests.
- Other carbamate compounds are aldoxycarb, bendiocarb, oxamyl, methiocarb, promecarb

Heterocyclic methylcarbamates: The insecticides in this group are Mobam and Carbofuran.

Carbofuran: It is the widely used carbamate insecticide next to carbaryl. It is systemic insecticide and mainly effective in cotton, borers in rice and sugarcane. It is widely sold under brand name Furadan.

Heterocylic dimethyl carbamates

The compounds belonging to this group are Pyrolan, Isolan, Dimetilan. These are used to control houseflies and sucking insects infesting various crops

Oximes

1. Aldicarb: It is highly toxic carbamate insecticide with acute oral LD_{50} in rats of 1-30mg/kg. It is a systemic insecticide and also acts as nematicide to control the soil nematode
2. Methomyl: It is commercially available as Lannate. It is broad spectrum insecticide and used on various crops viz. vegetables (Chilli, tomato, Cauliflower, cabbage) and cotton. It is effective as ovicide.

Napthyl carbamates

1. Carbaryl: Carbaryl compounds with a hydroxyl group attached directly to a phenyl or napthyl ringand was the most widely used carbamate insecticide today under the trade name Sevin. It is a wide spectrum insecticide used to control pests infesting cotton, vegetables etc.

Phenyl carbamates

Mexacarbate: Mexacarbate is a contact and systemic insecticide having broad spectrum of action and was commonly available as Zectran

Propoxur or Arprocarb (Baygon): It is effective against household pests *viz.* cockroaches, flies, mosquitoes, etc. It is a contact insecticide with rapid knockdown effect and also shows systemic action when applied to soil

There are 3 major sub groups of carbamates:

1. N-methyl carbamate esters of phenols- that is the e.g. Carbaryl
2. N-methyl and N-dimethyl esters of heterocyclic phenols e.g. Carbofuran.
3. Oximes- e.g. Aldicarb

Why OPs and Carbamates are preferred over OC insecticides?

As compared to OC, OPs and Carbamates are

- Less environmentally persistent
- More biodegradable
- Less subject to biomagnifications
- Usually unstable in the presence of sunlight
- More acutely toxic to nontarget species

Difference between Organophosphates and Carbamates

Organophosphates	**Carbamates**
• Derivatives of phosphoric acid	• Derivatives of carbamic acid
• Generally, much more toxic to vertebrates than carbamates	• very low mammalian and oral toxicity
• Cholinesterase (ChE) inhibition is irreversible	• Cholinesterase (ChE) inhibition is reversible
• When cholinesterase (ChE) enzyme is inhibited by an OP compound, it is said to be phosphorylated	• when cholinesterase (ChE) enzyme is inhibited by an carbamate compound, it is said to be carbamylated
Eg: chlorpyrifos	Eg: carbofuran

Carbamate insecticides used for pest control

Insecticide	Formulations
Aldicarb	Aldicarb (Temik®)
Carbosulfan	Carbosulfan 25EC (Marshall®)
Carbaryl	Carbaryl 75%WP (Sevin®)
Carbofuran	Carbofuran 3G (Furadan®)
Methomyl	Methomyl 40%SP (Lannate®)
Thiodicarb	Thiodicarb 70WP (Larvin®)
Propoxur	Propoxur (Baygon®)

4. Synthetic Pyrethroids

Pyrethrum is extracted from the flowers of Chrysanthemum, *Chrysanthemum cineriaefolium* flower contains on an average 1.3 percent pyrethrins. The dried flowers were probably used as insecticides in ancient China as long ago as the 1st century AD and by the Middle Ages were to be found in Persia. They became known in Europe via Armenian traders around 200 years ago as 'Persian dust', which was produced from dried flowers of *Chrysanthemum roseum* widely grown in Kenya and Ecuador. It is one of the oldest and safest insecticides available and was first used as powder in around 1851 as the original louse powder to control body lice in the Napoleonic Wars.

The full-scale commercial production of pyrethrins from Chrysanthemum flowers began in the mid-19th century, the chief ingredients in the extract being pyrethrin I and II which are still in use today in household sprays. However, their general use in agriculture was limited by their low stability in air and light, and the cost of production.

Pyrethroids are a major class of neurotoxic insecticides. They are synthetic analogues of the naturally occurring insecticidal esters of chrysanthemic acid (pyrethrins I) and pyrethric acid (pyrethrins II), originally found in the flowers of *Chrysanthemum cinerafolis*. The alcohol moiety of the pyrethrins has three natural variations giving rise to the pyrethrin series pyrethrin I and II, jasmolin I and II, and cinerin I and II. Subsequent structural modification of pyrethrins during the period from 1924 to 1970 produced the first synthetic pyrethroids, the most successful of which was allethrin which closely duplicated the structure of cinerin I.

Fig. 3: Basic structure of natural pyrethrins

Compound comprise the natural pyrethrins (pyrethrin I and II, jasmolin I and II, and cinerin I and II)

$R^1 = H$, $R^2 = CH = CH_2$ - Pyrethrin I; $R^1 = H$, $R^2 = CH_3$ – Cinerin I; $R^1 = H$, $R^2 = CH_2CH_3$ – Jasmolin I; $R^1 = CO_2$, $R^2 = CH = CH_2$ - Pyrethrin II; $R^1 = CO_2$, $R_2 = CH_3$ - Cinerin II; $R^1 = CO_2$, $R^2 = CH_2CH_3$ - Jasmolin II (Fig.3)

Why pyrethroids (pyrethrins analogus products) were synthesized though plant-based extracts, Pyrethrins were safer to environment.

Pyrethrins do not accumulate in the body, their excretion is rapid, and they do not persist in environmental media, as they degrade rapidly. However, their cost, high biodegradability and lack of photostability limit the use of pyrethrins, which require piperonyl butoxide (with low oral toxicity) as a synergistic agent to enhance their insecticidal effectiveness. For this reason, analogous products were synthesized with structure and properties similar to pyrethrins, the so-called pyrethroids.

In simple terms, synthetic pyrethroids have been developed because the natural pyrethrins tend to break down quickly when exposed to air, light and heat. There are six compounds that comprise the natural pyrethrins: pyrethrin I and II, jasmolin I and II and cinerin I and II. The synthetic pyrethroids can be classified as first and second generation.

Photostable compounds, with high insecticidal activity, low mammalian toxicity and limited soil persistence, were achieved between 1968 and 1974 by Michael Elliott and colleagues in the laboratories at Rothamsted. The first of these compounds was permethrin and this was soon followed by cypermethrin and deltamethrin. At the time of its discovery, deltamethrin was the most active insecticide ever known. The first non-cyclopropane pyrethroid, fenvalerate developed by the Sumitomo Chemical Company were also synthesized during this period. From 1975 to 1983, compounds with increased efficacy against ticks and mites were developed, including fluvalinate and flumethrin.

Chemistry and Discovery of pyrethroids

First generation pyrethroids are esters of chrysanthemic acid and an alcohol, containing a furan ring and terminal side chain moieties Eg: allethrin. Second generation pyrethroids have 3-phenoxybenzyl alcohols derivatives in the alcohol moiety and have had some of the terminal side chain moieties replaced with a dichlorovinyl or dibromovinyl substitute and aromatic rings. The addition of the alpha-cyano group to the 3-phenoxybenzyl alcohol group in the second generation pyrethroids has increased the insecticidal potency and stability of these compounds. Initially esters were produced using the same cyclopropane carboxylic acids, with variations in the alcohol portion of the compounds. The first commercial synthetic pyrethroid, allethrin (Fig. 4), was produced in 1949, followed in the 1960s by dimethrin, tetramethrin, resmethrin, prothrin, and proparthrin. 3-Phenoxybenzyl esters were also found to be active as pesticides (phenothrin, permethrin) (Fig. 5).

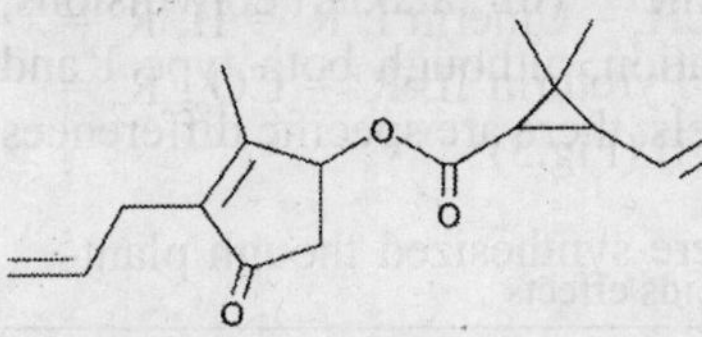

Fig. 4: Allethrin

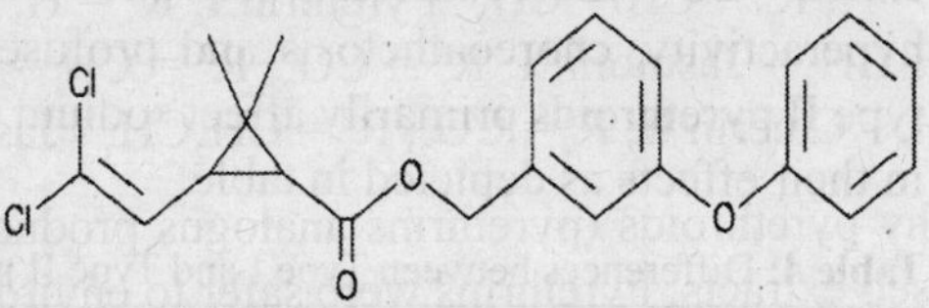

Fig. 5: Permethrin

Similar insecticidal activity was found in a group of phenylacetic 3-phenoxybenzyl esters, despite the lack of the cyclopropane ring. This led to the development of fenvalerate, an alpha-cyano-3-phenoxy- benzyl ester, and other related compounds. All pyrethroids are lipophillic in nature and are practically insoluble in water. Examples are allethrin and permethrin.

There were interesting discoveries of pyrethroids as depicted below.

Interesting discovery of pyrethroid insecticides based on generations

Generations	**Examples**
First generation	allethrin
Second generation	tetramethrin, resmethrin, bioresmethrin, bioallethrin, and phonothrin.
Third generation	fenvalerate, permethrin
Fourth generation	bifenthrin, lambda-cyhalothrin, cypermethrin, cyfluthrin, deltamethrin, fenpropathrin, flucythrinate, fluvalinate, prallethrin, tau-fluvalinate, tefluthrin, tralomethrin, zeta-cypermethrin, Transfluthrin

Synthetic pyrethroids with this basic cyclopropane carboxylic ester structure (no cyano group substitution) are known as type I pyrethroids (Fig.6). The insecticidal activity of synthetic pyrethroids was enhanced further by the addition of a cyano group at the benzylic carbon atom to give alpha-cyano (type II) pyrethroids and examples of type II pyrethroids include cyphenothrin and cypermethrin (Fig.6). The synthetic pyrethroids contains both Cl and F is cyfluthrin.

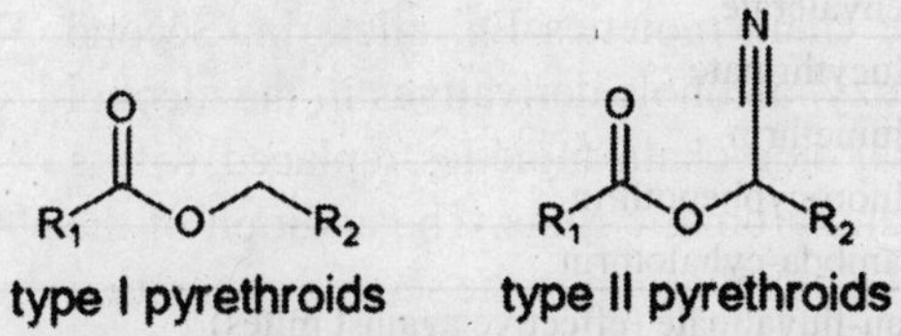

type I pyrethroids **type II pyrethroids**

Fig. 6: Basic structure of type I and type II pyrethroids

Animal studies suggest that the two structural types of pyrethroids give rise generally to distinct patterns of systemic toxic effects. Type I pyrethroids produce the so-called "T (tremor) syndrome", characterized by tremor, prostration and altered "startle" reflexes. Type II (alpha-cyano) pyrethroids produce the so-called "CS (choreoathetosis/salivation) syndrome" with ataxia, convulsions, hyperactivity, choreoathetosis and profuse salivation, although both type I and type II pyrethroids primarily affect sodium channels, there are specific differences in their effects as depicted in table.

Table 4: Differences between Type I and Type II pyrethroids effects

Type I pyrethroids	Type II pyrethroids
• Absence of alpha-cyano group (Fig.6)	• Presence of alpha-cyano-3-phenoxybenzyl esters (Fig.6)
• Keep sodium channels open	• Cause depolarization of myelinated nerve membranes without repetitive discharges
• Produce repetitive firing of sensory nerve endings	• Associated with a decrease in action potential amplitude
• Prolong sodium channel opening only long enough to cause repetitive firing of action potentials (repetitive discharge) following a single stimulus	• Prolong sodium channel opening for such long periods that the membrane potential becomes depolarized to the point that the generation of an action potential is impossible (depolarization-dependent block).
• Greater insecticidal activity when the ambient temperature decreases	• Type II pyrethroids become more effective with increase in temperature.
• Produce effects on cultured neurons that are easily reversed by washing with a pyrethroid-free solution	• Produce effects on cultured neurons that are largely reversible after washing cells with a pyrethroid-free solution

Table 5: Type I and type II pyrethroids

Type I	**Type II**
Allethrin	Alpha-cypermethrin
Bioallethrin	Cyfluthrin
Bromophenothrin	Cyhalothrin
Bifenthrin (effective against termites)	Cypermethrin
Bioresmethrin	Deltamethrin (storage pests)
Cismethrin	Esfenvalerate
Kadethrin	Fenpropathrin
Permethrin	Fenvalerate
d-Phenothrin	Flucythrinate
Prallethrin	Flumethrin
Resmethrin	Fluorocyphenothrin
Sumethrin	Lambda-cyhalothrin
Tefluthrin (effective against soil pests)	Tau-fluvalinate (effective against mites)
Tetramethrin	
Transfluthrin	

Table 6: List of Pyrethroid Isomers

Pyrethroid Isomers
Allethrin d-allethrin, bioallethrin, s-bioallethrin
Cyhalothrin lambda-cyhalothrin
Cypermethrin alpha-cypermethrin
Fenvalerate esfenvalerate
Phenothrin d-phenothrin
Resmethrin bioresmethrin, cisresmethrin

Structure activity studies of Pyrethroids

Synthetic pyrethroids are categorized as hydrophobic compounds having log octanol-water partition coefficient near to 6.0 along with less environmental persistence ranging from 12 to 197 days (Feo et al. 2010)

Pyrethroids possess two to three chiral carbon atoms (Fig.4) and toxicity is highly dependent on **stereo chemistry**. Both cis and trans chrysanthemic pyrethroids have substantial insecticidal activity and depends on the alcohol moiety used and for all of the chrysanthemic acid analogues, the IR configuration of the cyclopropane Cl is an absolute requirement for insecticidal activity while most pyrethroids are sold as racemic mixtures, whereas deltamethrin is marketed as single IR cis, S alpha cyano 3-phenoxybenzyl stereospecific isomer. The acid moieties of some pyrethroids also bear a close resemblance to DDT and even in 1970s, a DDT-pyrethroids hybrid structure namely "cycloprothrin" which was insecticidally active were characterized.

Fig. 7: *mark indicates two chiral carbon atoms in pyrethroids

All pyrethroids have at least four stereoisomers, with different orientation of the substituents on the cyclopropane ring (or the equivalent part of the phenylacetate). The isomers have different biological activities. In general, only the cyclopropane carboxylic acid esters with the R absolute configuration at the cyclopropane C_1 atom, and alpha-cyano-3-phenoxy benzyl esters with the S absolute configuration at C-alpha are toxic to man or insects (R and S refer to variations in the three dimensional structure of the molecule) (Fig.7).

Characteristics of pyrethroids

- Low toxicity to mammals and birds
- High toxicity to fish if applied directly to water
- Require very low doses to kill insects (high arthropod toxicity)
- Fast-acting
- Especially effective against chewing insects
- Bind tightly to soil and organic matter (therefore not as effective in penetrating soil to kill underground pests)
- Dissolve very poorly in water.

Mode of Action

The pyrethrin and pyrethroid insecticides affect peripheral and central nervous systems of insects. They initially stimulate nerve cells to produce repetitive discharges and eventually cause paralysis, an effect similar to, but more pronounced than that of DDT. As with DDT only a small fraction of the sodium channel population needs to be modified by pyrethroids for the generation of repetitive discharges. After modification by pyrethroids, the channels remain open as the insecticide impedes channel closing either by inactivation or deactivation, and the sodium channels retain the ability to conduct Na^+. However, the membrane potential is shifted so that the nerve cells function in a new, and relatively stable, state of abnormal hyperexcitability. In insects this produces an incapacitating, but sublethal effect, known as 'knockdown'.

The amplitude of the sodium current continues undiminished until the level of hyperexcitability overwhelms the capacity of the cell to maintain the activity of the sodium pump. Higher lipophilicity gives better knockdown rates as the pyrethroid penetrates to the target more quickly. However, such compounds may not give good 'kill' due to a tendency to dissociate from the target. Type I pyrethroids (e.g., permethrin) are generally good knockdown agents due to their ability to induce repetitive firing in axons, resulting in restlessness, un-coordination and hyperactivity followed by prostration and paralysis.

Type II compounds, as typified by deltamethrin, have a cyano group at α-benzylic position (α -carbon of the 3-phenoxybenzyl alcohol) and cause a pronounced convulsive phase that results in better kill because depolarization of the nerve axons and terminals is irreversible. The differing physiological effects are explained by the fact that the duration of modified sodium currents by Type I compounds lasts only tens or hundreds of milliseconds, whilst those of Type II compounds last for several seconds or longer.

Why pyrethroids show low toxicity to mammals (vertebrates) than insects invertebrates)

- Pyrethroids are 2250 times more toxic to insects than mammals
- Differences in their potency as neuronal toxins to insects than mammals
- Rates of detoxification differ between invertebrates and vertebrates
- Sensitivity of invertebrate neuronal sodium channels to pyrethroids is ten times greater than in mammals
- Invertebrates typically have body temperatures some 10°C lower than mammals
- Pyrethroid hepatic metabolism (detoxification) is faster in mammals.
- Small insect size increases the likelihood of end-organ (neuronal) toxicity prior to detoxification.
- low toxicity of pyrethroid insecticides in mammals is due to poor dermal absorption (the main route of exposure) and metabolism to non-toxic metabolites

If natural pyrethrins were effective why synthetic pyrethroids are developed?

Pyrethrins are natural insecticides derived from chrysanthemum flowers, while synthetic pyrethroids are man-made versions designed to mimic the properties of pyrethrins. Both are used for insect control, but synthetic pyrethroids tend to be more stable and potent, with potentially higher toxicity and longer environmental persistence compared to pyrethrins.

Effect of Temperature on pyrethroids toxicity

Pyrethroid insecticides are more effective in respect of both toxicity and knockdown rates when they are applied at lower ambient temperatures (i.e., they display a negative temperature activity coefficient), e.g., permethrin activity against housefly is considerably greater at 18°C than at 32°C (LD_{50} 28.3 vs 101.3, respectively). This negative correlation with temperature may be one factor that contributes towards their relatively low mammalian toxicity.

The acute mammalian toxicity of pyrethrins is generally low. The most common adverse effect is their great sensitizing power, related to the sesquiterpene lactones contained in the plant extract (pyrethrum). These substances cause allergic rhinitis and contact dermatitis (O'Malley, 1997). Natural pyrethrins are more toxic by skin contact than by ingestion, while synthetic pyrethroids are more potent when ingested. However, the latter are not skin sensitizers or irritants, although may cause inflammation and paresthesia after coming into contact with the skin.

Metabolism of pyrethroids in mammals

There is some stereospecificity in metabolism, with *trans*-isomers being hydrolyzed more rapidly than the *cis*-isomers, for which oxidation is the more important metabolic pathway. Although the alpha-cyano group reduces the susceptibility of the molecule to hydrolytic and oxidative metabolism, the cyano group is converted to the corresponding aldehyde (with release of the cyanide ion), followed by oxidation to the carboxylic acid, sufficiently rapidly for efficient excretion by mammals.

During metabolism of the pyrethroids, the chrysanthemic acid ester is usually cleaved via esterase or mixed function oxidase activity and any resulting alcohol moieties are converted to their corresponding acids. These metabolites are partly conjugated to glucoronide and both the conjugates and free acids are excreted in the urine. 3-phenoxy benzoic acid (3-PBA) is a metabolite that is common to as many as 20 synthetic pyrethroids.

Target Site Resistance to Pyrethroids/ DDT

Busvine first recognized the most common form of resistance against pyrethroids, so called Knockdown resistance (kdr) in houseflies in 1951. The Kdr factor is now known to be a recessive allele conferring cross-resistance to the entire class of pyrethroids including its botanical derivative, pyrethrins. Besides this type of resistance, a second recessive resistance character designated as Super-Kdr is observed which confers much greater resistance to pyrethroids.

The kdr resistance factor studies show that, the modification is not at the primary site where the insecticide binds, but perhaps in a region responsible for relaying the perturbation whereby the ion conductance properties of the sodium channel are modified. Super-kdr is clearly sensitive to structure and is more likely to be present at the actual binding site for the pyrethroid insecticides, and it shows greater resistance to compounds possessing cyclic side chains in the alcoholic component in the pyrethroid structure but changes in acid component have little effect. The resistance studies also show that a number of aminoacid substitutions takes place in the pyrethroid resistant insect, thereby reduce their sensitivity to this group of insecticides.

Advantages and Concerns Associated with Pyrethroids

Advantages	Concerns
• Safe and effective • Contact activity • Act as a repellent also • Low toxicity for mammals • Dosages can be kept low	• Biodegradable. • Short persistence • Toxic for fish and non-target invertebrates

Their principle disadvantage is their very broad insecticidal activity, which tends to eliminate many beneficial organisms including insects.

What is the most common route of pyrethroid exposure- Dermal Exposure? yes

Pyrethroids are commonly regarded as relatively safe insecticides. However, the acute toxicity of pyrethroids to mammals varies with structure and oral toxicity.

Table 7: Pyrethroids used for pest control

Pest type	Pyrethroids
Publich health	Bifenthrin 10%WP, Cyfluthrin 10 % WP Alphacypermethrin 05 % WP, Deltamethrin 25% WDG
Impregnation of Bed nets for mosquitoes	Lambda-cyhalothrin 2.43 % CS, Alphacypermethrin, Deltamethrin
House hold insecticides	Allethrin, S-Bioallethrin, Bifenthrin, Cypermethrin, Cyfluthrin, Lambda-cyhalothrin, Lambda-cyhalothrin, Permethrin, Prallethrin
Termites control	Bifenthrin 2.50 % EC
Mites control	Bifenthrin 8 % SC
Storage pests and godowns	Deltamethrin 2.50 % WP

Use of Deltamethrin 2.50 % WP against Stored pests

Commodity/structure	Target stored pest	Dilution in water (liter)
wheat & rice (grain & seed in stacks)	rice weevil, lesser grain borer, khapra beetle, red flour beetle, saw toothed grain beetle, rice moth, almond moth	1 litre/30 m^2
Walls, ceilings floors of Godowns	rice weevil, lesser grain borer, khapra beetle, red flour beetle, saw toothed grain beetle,rice moth, almond moth	1.5-2.5 litre/50 m^2

Table 8: Acute toxicity of pyrethroids to rats

Compound	Oral LD$_{50}$ (mg/kg)	
	Males	Females
S-bioallethrin	370	320
Bifenthrin	70	54
Cyfluthrin	155	160
cypermethrin	297	372
deltamethrin	95	87
esfenvalerate	87	87
fenpropathrin	71	67
lambda cyhalothrin	79	56
permethrin	1200	1200
pyrethrins	710	320
resmethrin	1695	1640
Tefluthrin	22	35

Table 9: Commercial insecticides in market containing pyrethroids

Active ingredients	Product trade names
Allethrin	Pyanmin
Alphamethrin	Alphaguard
Bifenthrin	Talstar
Beta-Cyfluthrin	Bulldock
Cypemethrin	Cymbush
Deltamethrin	Decis
Fenpropathrin	Meothrin
Fenvalerate	Sumicidin
Lambda cyhalothrin	Karate

5. Oxadiazines

Indoxacarb

The choice of candidate DPX-JW062 was made by considering its demonstrated effectiveness in killing insects, its lack of harm to non-target creatures such as predatory insects, fish, birds, and mammals, and its ability to quickly break down in the environment.

Physico-Chemical Properties of Indoxacarb

Melting point	88.1°C
Solubility	Water-0.20ppm; Acetone: >250g l^{-1}
Partition coefficient (log kow)	4.65
Vapour pressure	9.8 x 10^{-9} Pa (20°C); 2.5 x 10^{-8} Pa(25°C)

Insecticidal activity and Properties of Indoxacarb

Indoxacarb is an extremely effective agricultural pesticide that decomposes rapidly in the environment and causes minimal harm to non-target organisms. It was classified as a reduced-risk pesticide by the U.S. Environmental Protection Agency (EPA), which indicates that it provides both health and environmental benefits in comparison to other pesticides that are currently available.

Indoxacarb is a broad-spectrum lepidopteran insecticide that is also effective against other pests from a diversity of insect orders. Indoxacarb is responsible for the management of the majority of the globally significant lepidopteran pests, such as *Helicoverpa* (bollworms), *Spodoptera* (armyworms), *Trichoplusia* (loopers), *Plutella* (diamond back moth), *Ostrinia* (borers), *Lobesia* (berry moths), *Cnaphalocrocis* (leaf folders), *Pandemis* (leaf rollers), *Tuta* (pinworms), and *Agrotis* (cutworms). Additionally, indoxacarb modulates a spectrum of parasitic insect pests, including plant bugs, flea hoppers, leaf hoppers, beetles, sawflies, and apple maggot flies.

Indoxacarb is also highly effective fire ant product for use on public land, golf courses, and homes. It is used in lure matrix to control a range of cockroach and ant species, and it has been shown to have commercial activity against termites, fleas, mosquitoes, flies, and silverfish. Indoxacarb is mostly ingested by the target insects, but it can also be absorbed via the cuticle. Indoxacarb is a highly effective adulticide and ovilarvicide because it kills developing larvae in the egg and prevents hatching, despite the fact that the larval life stage is the major target of control. Even at sublethal rates, indoxacarb produces a strong eating inhibition. Insects exposed to a sublethal dose of indoxacarb tend to consume less food, develop at a slower rate, pupate, and emerge later than untreated larvae.

The quick suppression of insect feeding results in immediate crop protection with live insects partially paralyzed, smaller, dehydrated, and shrunken, with no defense against environmental threats. The other affected behaviors are reduced egg-laying, mating disruption, inability to metamorphosis, trouble emerging from the pupal case, inability to excavate soil for pupation, repellency, and uncoordinated F1 progeny. In contrast to synthetic pyrethroids, indoxacarb control is positively correlated with high temperatures, hastening insect population declines.

Indoxacarb protects crops for 5 to 14 days, depending on application rate and crop. It is very lipophilic and penetrates into the leaf waxy cuticle which inturn aids residual control, translaminar activity, and remarkable rainfastness.

Oil-based treatments can penetrate the leaf and repel a wide range of sucking insects. The dry formulation is also translaminar; however, tank mixing with an oil-based surfactant can increase activity. The temperature and pH of a pesticide in a spray tank are the most important factors determining its chemical stability. Indoxacarb compositions have high tank stability across a wide pH range (5-9). Furthermore, spray tank temperatures have no effect on indoxacarb stability between 12 and 32 degrees Celsius. When used in the correct order, indoxacarb formulations have been demonstrated to be compatible with tank mix partners. All indoxacarb formulations are rainfast and have excellent ultraviolet stability.

Indoxacarb is the most recent commercial pesticide to target voltage-gated sodium channels (VGSCs). These channels are critical for the intercellular transmission of electrical impulses in both vertebrate and invertebrate nervous systems. Insect VGSCs, like vertebrate channels, have three basic states: resting (closed), in which the channel is nonconductive. ii) an activated open state, in which an inward flow of Na^+ via the channel depolarizes the cell and eventually creates an action potential; and iii) an inactivated state (closed), in which the channel is nonconductive and cannot be activated. The restoration of inactivated channels to their resting state is a voltage-dependent process in which cells remain refractory until the cell membrane is repolarized.

Several species, including spiders, scorpions, and carnivorous marine molluscs, have developed highly selective neurotoxins that immobilize their prey by targeting VGSC. Nine distinct VGSC binding sites were discovered using neurotoxins, synthetic insecticides, and local anesthetics. Pyrethroids and DDT bind to site 7, causing channel activation to be voltage dependent.

Proinsecticides Action of Indoxacarb

Indoxacarb is a proinsecticide that requires bioactivation to have a strong insecticidal effect. In metabolic studies with pest insects treated with 14C-DPX-JW062 (a 50:50 mixture of the active and inactive enantiomers of indoxacarb), rapid conversion into the N-decarbomethylated metabolite DCJW was observed (Fig.8).

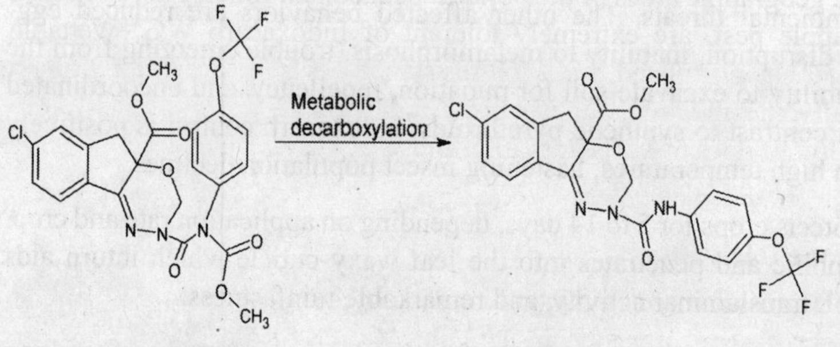

Metabolic decarboxylation

Indoxacarb DCJW

Fig. 8: Indoxacarb converted to DCJW

This bioactivation was associated with hydrolytic esterase and amidase metabolism in the midgut and fat bodies, and cytochrome 450 inhibitors had no influence on it. When a preparation from the central nervous system of the Lepidoptera, *Manduca sexta*, was employed, DCJW inhibited nerve conduction > 25 times more effectively than DPX-JW062 did and further insecticidal action was due to the (S)-enantiomer, which was twice as effective as the racemic mixture.

Effect of indoxacarb on mammal voltage gated sodium channels

Indoxacarb and DCJW both inhibited sodium currents after binding to the inactivated state, albeit their efficacy against rat sodium channels was substantially lower than that of insects. Despite its potency against mammalian VGSCs, indoxacarb has a high safety profile.

Indoxacarb's safety is heavily impacted by its differential sodium channel affinity. In insects, DCJW is extraordinarily potent, with an IC_{50} value of less than 30nM, as opposed to the low micromolar IC_{50} value of rat VGSCs. Furthermore, indoxacarb is the most frequent oxadiazine in mammals, while being ten times less powerful than DCJW against rat VGSCs. In humans, indoxacarb is turned into DCJW at a slow rate, however in insects, it is swiftly converted into DCJW.

Table 10: Acute toxicity of Indoxacarb

Oral LC_{50}	1730 mg kg^{-1} (rat male)
Oral LC_{50}	268 mg kg^{-1} (rat female)
Dermal LD_{50}	>5000 mg kg^{-1} (rat)
Inhalation LC_{50} (4h)	>5.5mg l^{-1}
Dermal Irritation	Non-irritant
Eye irritation: moderate eye irritant	Moderate eye irritant
Ames test	negative

Resistance

After many years of consistent use on the Hawaiian island of Oahu, a population of diamond back moth (*Plutella xylostella*) has developed resistance to indoxacarb. Similarly, some geographic areas of the oblique-banded leaf roller, *Choristoneura rosaeana*, an apple pest, are extremely tolerant of Indoxacarb WG (Wettable Granule). Ahmad has postulated that such resistance could be achieved through enhanced oxidative degradation, a mechanism initially selected by field exposure to azinphosmethyl. It does not appear that failing to activate the chemical has any effect on the resistance mechanism.

6. Semicarbazone Derivatives

Metaflumizone was discovered by Nihon Nohyaku in the early 1990s and belongs to the new class of semicarbazone insecticides. It is now being globally co-developed as the animal health product, ProMeris®, in cooperation with Fort Dodge Animal Health and as an agricultural and cooperation with BASF. Metaflumizone was developed in a synthesis program initiated from a pyrazoline insecticide lead.

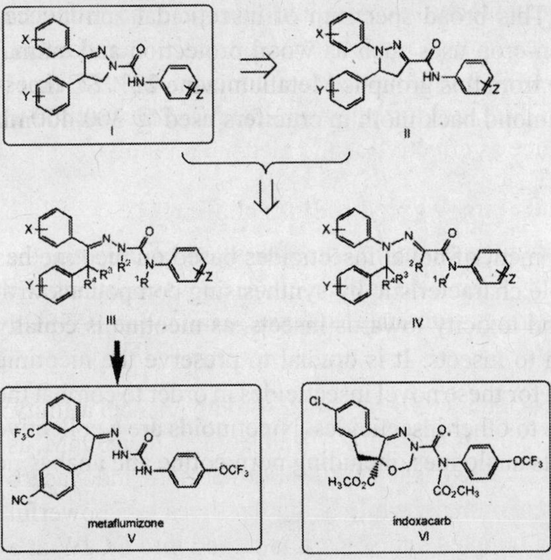

Fig. 9: The evolution of chemical design from pyrazoline insecticides to semicarbazone insecticides and oxadiazine insecticides.

These compounds showed insecticidal activity similar to the parent pyrazolines (Fig.9. I), but rapidly lost activity when exposed to light and were too labile for effective commercial use. Subsequently, semicarbazone insecticides (Fig.9. III) and their isomers (Fig.9. IV) Were designed by opening the central ring structure of the pyrazoline insecticides (Takagi et al., 1996). The semicarbazone compounds (Fig.9. III) Provided excellent insecticidal activity and possessed the appropriate stability. The isomers (Fig.9. IV) Were more stable than their parent compounds (Fig.9. II) but showed only moderate insecticidal activity. Optimization of the semicarbazone insecticide leads (Fig.9. III) led to the discovery of metaflumizone (Fig.9. V). Metaflumizone is an ideal candidate insecticide having activity against a variety of insects and a degradation profile that is consistent with environmental friendliness. At around the same time, oxadiazine insecticides, in which an oxygen atom was introduced into a pyrazoline ring, were designed and indoxacarb (Fig.9.VI) was discovered (Annis et al., 1992; McCann et al., 2001). One can see that these structures (Fig. I, II, V, VI) share a common structural backbone with two phenyl rings connected with five atoms, including three nitrogen atoms and one carbonyl group located at similar position

Metaflumizone derivatives showed excellent activity against insect pests in the order Lepidoptera, Coleoptera, Diptera and Hymenoptera, and demonstrated a lack of cross-resistance with other commercial insecticides such as organophosphates, carbamates, pyrethroids and benzoylphenylureas, suggesting a new mode of action (Salgado and Hayashi,). Initially, metaflumizone had been developed for agricultural crop protection. In addition to the broad insecticidal spectrum on crop pests, metaflumizone also showed excellent activity against termites, e.g. *Coptotermes formosanus* (Takagi et al., 2005), and fleas, e.g. *Ctenocephalides felis* (Yamaguchi et al., 2005). This broad spectrum of insecticidal activity has allowed for the extension to non-crop uses such as wood protection and animal health. One compound available from this group is Metaflumizone 22% SC that is found to be effective against diamond back moth in crucifers used @ 300-400ml/ acre (Waiting period -3 days)

7. Neonicotinoids

The primary goal of the development of novel insecticides based on nicotine has been to address this unfavourable characteristic by synthesising compounds with a higher degree of selectivity and toxicity towards insects, as nicotine is equally or more toxic to mammals than to insects. It is crucial to preserve the nicotinic acetylcholine receptor as a target for these novel insecticides in order to combat the development of insect resistance to other insecticides. Nicotinoids are a collective term that refers to nicotine and its analogues, including nornicotine and anabasine (Fig. 10).

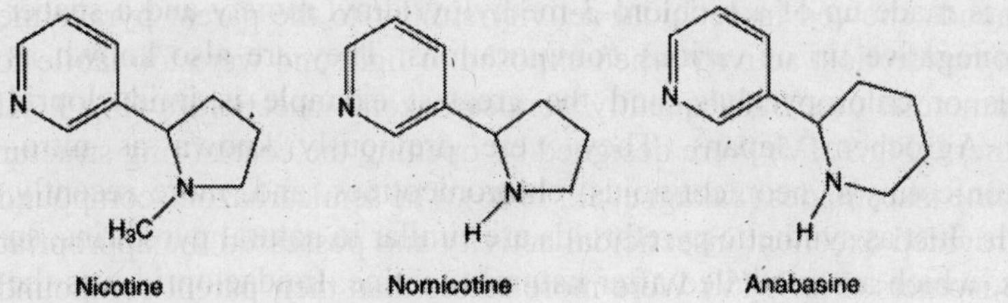

Nicotine Nornicotine Anabasine

Fig. 10: Nicotine and its analogue

In contrast to nicotinoids, neonicotinoids are synthetic, newly developed insecticides that target the nicotinic acetylcholine receptors. However, they exhibit a high degree of selectivity directed at insects. Shell (Modesto, USA) developed the first neonicotinoid in the 1970s: nithazine (Fig. 11). Nithiazine is a 2-nitromethylene tetrahydro-1,3-thiazine that is selected from a series of nitroalkyl heterocyclic compounds. The molecular models of nithiazine are distinct from those of nicotine, but they function on the nicotinic acetylcholine receptors the same way as nicotine. Nithiazine poses a selective toxicity to insects; however, its field application is restricted due to its low photostability.

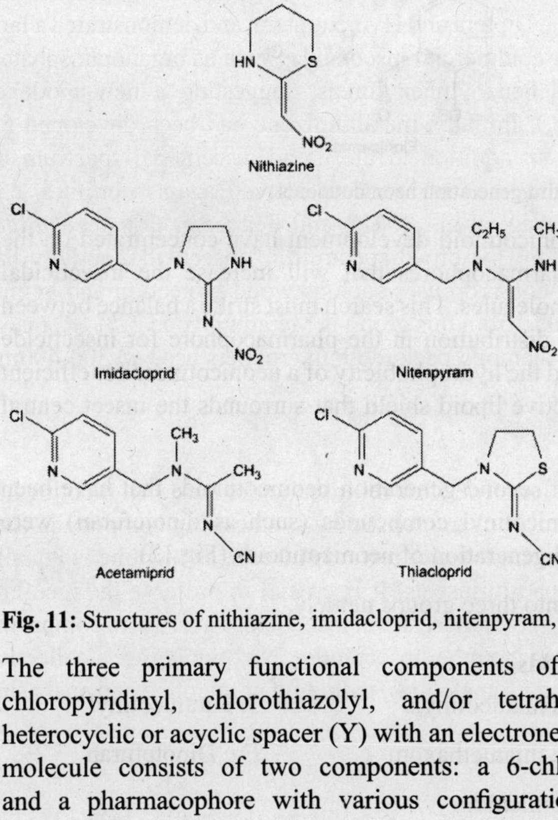

Nithiazine

Imidacloprid Nitenpyram

Acetamiprid Thiacloprid

Fig. 11: Structures of nithiazine, imidacloprid, nitenpyram, acetamiprid, thiacloprid

The three primary functional components of neonicotinoid (X-YZ) are chloropyridinyl, chlorothiazolyl, and/or tetrahydrofuryl moieties (X), a heterocyclic or acyclic spacer (Y) with an electronegative tip (Z). A neonicotinoid molecule consists of two components: a 6-chloro-3-methylpyridine moiety and a pharmacophore with various configurations. The first generation of neonicotinoids is made up of a 6-chloro-3-methylpyridinyl moiety and a spacer with an electronegative tip of various configurations. They are also known as chloronicotinyls or chloropyridyls, and the greatest example is imidacloprid (Nihon Bayer Agrochem, Japan). They were previously known as nitro-guanidines, neonicotinyls, neonicotinoids, chloronicotines, and, more recently, chloronicotinyls. Just as synthetic pyrethroids are similar to natural pyrethrins, so are nicotinoids, which are modelled after natural nicotine. Imidacloprid was the first commercially available neonicotinoid chemical in 1991 (foliage treatment/ seed dresser/soil treatment), and it has contact and systemic activity.

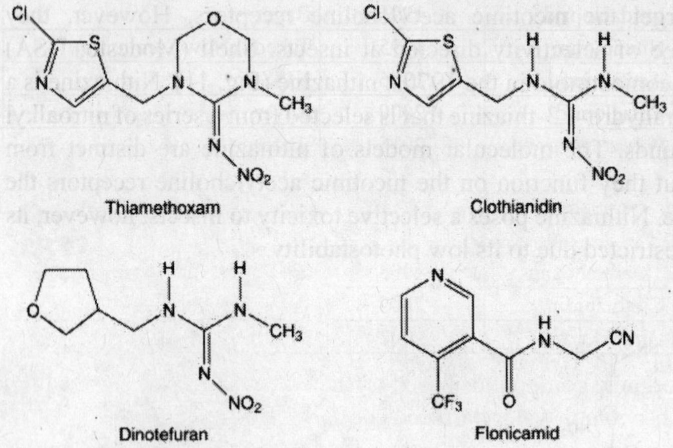

Thiamethoxam Clothianidin

Dinotefuran Flonicamid

Fig. 12: Structures of second and third generation neonicotinoids

The most recent efforts in neonicotinoid development have concentrated on the hunt for heterocycles and pharmacophores that will increase the insecticidal characteristics of the existing molecules. This search must strike a balance between the need for optimal electron distribution in the pharmacophore for insecticide binding to receptor subsites and the hydrophobicity of a neonicotinoid for efficient penetration through the protective lipoid shield that surrounds the insect central nervous system.

Figure 12 depicts examples of second-generation neonicotinoids that have been released to the market. Furanicotinyl compounds (such as dinotefuran) were recently developed as the third generation of neonicotinoids (Fig.12).

Neonicotinoids are classified into three groups namely,

Classification of Neonicotinoids

1. Chloronicotinyls	2. Thianicotinyls	3. Furanicotinyl
Eg: Imidacloprid	Eg: thiamethaxom	Eg: Dinotefuran
Acetamiprid	Clothianidin	

Neonicotinoids, the fourth generation of insecticides in the pesticide business, are further categorised into five-membered rings, six-membered rings, and fourth-generation products such as chlorinated, thiolated, and furanyl derivatives based on cyclic structures and substituents. This classification not only indicates the technological advancement of neonicotinoid insecticides, but also demonstrates the pesticide industry's ability to innovate.

Table 11: Neonicotinoids Development

Generation	Product	Year of development	Developing Company
First Generation	Imidacloprid	1991	Bayer
	Nitenpyram	1995	Takeda (Now Sumitomo)
	Acetamiprid	1996	NISSO
	Thiacloprid	2000	Bayer
	Flupyradifurone	2015	Bayer
	Flupyrimin	2013	Meiji Seika
	Cycloxaprid	2018	Shanghai Shengnong/ ECUST(China)
Second Generation	Thiamethoxam	1998	Novartis (Now Syngenta)
	Clothianidin	2002	Bayer/ Takeda
Third Generation	Dinotefuran	2002	Mitsui Chemicals Agro

Plants often absorb organic compounds based on their characteristics, particularly the logarithm of the n-octanol-water partition coefficient (log KOW).

Neonicotinoids, like imidacloprid, act by binding to nicotinic acetylcholine receptors, resulting in a proven novel insecticidal target site. Thiamethoxam's mode of action has been studied in aphids as target insects and in the functional assay of electrophysiological studies on locusts as a laboratory model insect. All tested neonicotinoids behaved as antagonists of acetylcholine (and nicotine) at concentrations ranging from the threshold concentration to the low micromolar range. Overall, the findings indicate that thiamethoxam and imidacloprid have different routes of action, as do neonicotinoids in general, albeit they all appear to operate on nicotinic acetycholine receptors.

Imidacloprid

The five-membered neonicotinoid imidacloprid was discovered in 1984 as a result of modifying the structure of the previously announced six-membered nithiazine in quest of enhanced action. Because of the photolabile 2-nitromethylene group, nithiazine was never commercialised for agriculture. During the 1980s, scientists at Nihon Tokushu Noyaku, a subsidiary of Bayer in Japan, researched synthesis based on the lead structure nithiazine. Insecticidal action was significantly increased by substituting an N-containing heteroaryl-methyl group (N-(6-chloro-pyrid-3-ylmethyl)-N-ethyl amine; CPM) into the 5-membered ring system of 2-nitromethylene-imidazolidine (X-Y= CH-NO2). Imidacloprid was chosen for this study because of its insecticidal potential, photostability, and long-lasting action, as well as its superior systemic characteristics in both protected and field environments.

Imidacloprid, the first member of the chloronicotinyl insecticide (CNI) family, was an efficient pesticide used in many countries to control agricultural pests, termites, and veterinary pests. Imidacloprid's water solubility and low partition coefficient in octanol-water are unaffected by pH values ranging from 4 to 9 at 20°C. Imidacloprid's low partition coefficient suggests that it has no potential for accumulation in biological tissues or, as a result, in the food chain. Imidacloprid has good fast absorption and translaminar transport properties. While imidacloprid exhibits acropetal mobility in the xylem, its penetration and translocation in cotton leaves are less significant, as demonstrated by phosphor-imager autoradiography. Imidacloprid's mobility in the xylem makes it particularly suitable for seed treatments and soil applications, however it is also effective as a foliar spray. Because imidacloprid contains no acidic hydrogen, its pKa is greater than 14, making phloem transport unlikely.

Effect on insect pests

The imidacloprid has a high intrinsic potency and great systemic qualities. The active ingredient must be taken up by the roots before it can be applied to the soil through drip or spray irrigation, furrow application, or seed treatment. This imidacloprid can be used to cover seeds (Gaucho), spray on leaves (Admire), treat the soil (by watering) or apply it to plant parts (stem, trunk, etc.) as grains. These ways must be cheaper and better for the earth to use.

Imidacloprid has a wide range of activity, a good long-term impact, and is plant-friendly. Imidacloprid is primarily used to control sucking insects (aphids, whiteflies, plant and leafhoppers, thrips, scales, mealybugs, plant bugs, and psyllids), as well as pests that are resistant to conventional pesticides. Apart from its direct insecticidal efficacy, imidacloprid has various sublethal side effects. Such effects, which can vary in dose, include repellency, a reduction or stoppage of feeding, a reduction or cessation of reproductive activities, an overall reduction in movement or activity, and enhanced susceptibility to biological control. Imidacloprid's anti-feeding actions result in less transmission time and a shorter period of xylem contact, resulting in a significant reduction in the number of infections per time unit.

Furthermore, field studies have shown that multiple foliar spray administrations of imidacloprid resulted in improved health and increased plant development even in the absence of insect infestation, and it also aids plants in mitigating the impacts of abiotic and biotic stress. Imidacloprid can be applied to the soil in a variety of ways, allowing it to be carried to the pest within the plant while avoiding harm to beneficial species.

Though imidacloprid is totally destroyed, its metabolism is heavily impacted by the technique of administration, as well as the period and plant species involved. Based on soil metabolism tests, imidacloprid was found to be totally metabolised to carbon dioxide and does not stay in the soil. Imidacloprid has a half-life of 156 days when degraded aerobically (DT_{50}) under typical laboratory conditions.

Nitenpyram

Takeda Chemical Industries Ltd., now Sumitomo Chemical Takeda Agro Company Ltd., discovered this compound in 1995 by optimising the substituents of an open chain nitroethene. It debuted in the Japanese market in 1995 under the brand name Bestguard. Novartis Animal Health launched nitenpyram as a systemic, fast-acting adult flea control medication for cats and dogs in 1999 under the brand name Capstar.

Physico-chemical properties

The nitromethylene is distinguished by its extraordinarily high-water solubility (840 gl^{-1}) and low partition coefficient (-0.64). The compounds decompose rapidly in water under light and in soil under aerobic circumstances, owing to their weak photostability.

The remarkable feature of nitenpyram is its insecticidal activity via translaminar action and translocation. Because of its high water solubility, it has good systemic action and no phytotoxic effects. It can be used to manage pests in soil utilising various soil treatment methods, such as seedling dip and application paired with fertiliser via irrigation systems (insectigation). The polar nitenpyram has a brief persistence in soil, with a half-life (DT_{50}: 1-15 days), which likely offsets the relatively weak sorption that could lead to mobility across the soil.

Efficacy against target pests and application rates

Nitenpyram suppresses homopterous insect pests such as leaf hoppers and plant hoppers in rice, whiteflies and aphids on vegetables, and thrips attacking fruit trees, vegetables, and tea, as well as pests attacking crops grown under protected agriculture. It is more active by ingesting than contact, and even at low concentrations, nitenpyram prevents pest feeding. When used as a foliar spray, the substance exhibits both translaminar and ovicidal activity. In Japan, granular formulations applied to soil showed promising results for controlling *Liriomyza trifolii* in tomatoes and chrysanthemums cultivated in greenhouse conditions.

Acetamiprid

The CPM moiety is present in N-cyano-acetamidine acetamiprid (1995, Nippon Soda Co., Ltd), which is similar to nitenpyram and the five-membered ring systems imidacloprid and thiacloprid. Acetamiprid was found by Nippon Soda as part of its nitromethylene research program while optimising unique 2, N-cyanoimino compounds containing an imidazoline five-membered ring obtained from Nihon-Bayer Agrochem. Methyl and ethyl were discovered as favourable substituents R1 on the secondary amino group, and both compounds demonstrated strong efficacy against cotton aphids.

Physico-chemical properties

Acetamiprid has both systemic and translaminar insecticidal action due to its high water solubility (4.25 gl^{-1} at 25°C). It is stable in water at various pH levels (4.0, 5.0, and 7.0) and temperatures (pH 9 at 22°C), but it hydrolyses at higher temperatures (35°C and 45°C) and high pH (9) to produce two primary hydrolytic transformation products. This pesticide likewise undergoes photo transformation at pH 7.0 and has a half-life of 34 days. Acetamiprid is moderately mobile (based on Koc values) and fast biodegradable in soils (DT_{50}-1-2 days in clay loam/light clay soils), as opposed to field collected soil (DT_{50}-12 days).

There are many types of sucking insects that can be killed with acetamiprid. These include aphids, termites, beetles, and thysanopteran adults. It works on contact and throughout the body for a long time, and it's only mildly effective against peach fruit moths, oriental fruit moths, and DBM. Even though insecticides don't get very far through the skin, some bad bugs viz., *Bemisia tabaci*, *Planococcus citri,* and *Thrips palmi,* are also very vulnerable. This pesticide is commonly used on many types of crops, including cereals, cotton, tobacco, tea, and horticulture plants. It can be applied to the soil, leaves, or the seeds. It also worked well against pests that are resistant to organophosphate and pyrethroids. Due to toxic and repellent nature, it also gets good rid of termites attack. This insecticides has to be avoided in fields where honeybees are present as it is poisonous to them.

Thiacloprid

Bayer Crop Science found and developed thiacloprid, a second member of the CNI family that is made up of five neonicotinoid molecules. This neonicotinoid has the CPM residue linked to the cyclic2-(N-cyanoamino)-thiazolidine (CIT) moiety, similar to imidacloprid.

Physico-chemical properties

Thiacloprid, once applied to the leaves, is resistant to hydrolysis even under conditions of heavy rain and sunlight, allowing for adequate plant uptake of the chemical by continuous penetration of the active ingredient into the leaf. Thiacloprid photolysis in water at pH 7 has a half-life of more than 100 days. It is stable under sunshine on oil surfaces, and its photostability is superior than that of other neonicotinoids.

The way it penetrated and moved through the plant was similar to imidacloprid, and it was clear that it could move through the whole system. In fact, the pattern of C_{14}-labeled thiacloprid's movement showed that it could move up through the xylem just one day after it was applied to cabbage.

Formulations

Thiacloprid has been specifically designed for foliar applications in the form of Suspension concentrate (480 SC Calypso or Alanto) and 30% and 70% water dispersible granules under the trade name Bariard. A robust spray solution has been achieved through the development of a novel oil dispersion (OD) formulation, Biscaya O-TEQ 240.

Efficacy and Safety: This pesticide exhibits great stomach and contact properties in addition to its outstanding systemic activities. Aphids, whiteflies, a few species of thrips, beetles (Colorado potato beetle, cotton boll weevil), lepidopteran moths (*Cydia pomonella*-ovicidal effect) that attack temperate fruits (pome and stone fruits), and dipteran species (*Rhagoletis, Ceratitis, Dacus*) that attack peaches and olives are among the pests that it controls. Another significant benefit is that it can be applied prior to, during, and following the flowering of fruit crops because it doesn't affect pollinating insects i.e. parasitic wasps, honey bees, or bumble bees. Furthermore, thiacloprid does not upset the balance between predator and prey.

Clothianidin

In 2002, Takeda Chemical Industries Ltd. released Clothianidin. The N-nitroguanidine pharmacophore in this open chain structure is comparable to the imidacloprid, a five-membered neonicotinoid, except that the CTM moiety has taken the place of the CPM group.

Physico-chemical properties

Clothianidin has a comparatively low water solubility ($0.327 g^{-1}$ at 20°C), vapour pressure (1.3×10^{-10} Pa at 25°C), and volatility when compared to other neonicotinoids with an N-nitroguanidine pharmacophore. Clothianidin has no acidic or alkaline qualities at the necessary pH, hence the pH of the aqueous solution has no effect on the compound's physical and chemical properties. Although clothianidin is stable in the pH range of 4-9, photolysis contributes significantly to its degradation in the environment, and breakdown in water/sediment systems was shown to be significantly faster in anaerobic than aerobic settings. This is also shown by the fact that clothianidin's octanol-water partition coefficient is 0.7 at 25°C, which means it absorbs well into soil.

Clothianidin is taken up by plants through their cotyledons and roots when they are new seedlings and their roots when they are already established seedlings. The tests of translocation show that clothianidin can move sidewards and up and down. The active ingredient that the root takes in is usually quickly moved to the leaves. Translaminar activity makes this possible by moving the compound across the leaf tissues, from one side to the other. Because of these physical and chemical features, clothianidin doesn't build up in living things and due to non-volatile nature, not a lot of it is likely to be in the air.

Efficacy on target pests

Clothianidin can be used to treat the soil, leaves, and seeds of plants to get rid of sucking and chewing insects like planthoppers, stink bugs, aphids, and whiteflies that attack citrus, fruit, veggies, rice, and rapeseed (Tomlin, 2009). It works on both the translaminar and root systems. Because of these qualities, clothianidin can be used on leaves, seeds, and grounds. Because it works so well on the roots, clothianidin is very good at getting rid of a wide range of insects that eat roots, stems, and leaves, as well as pests that live in the soil and live in the halo around the seed. Bayer Crop Science sells clothianidin under the brand names Dantop for application to leaves, Dantotsu for application to soil as a WSG, and 600 FS Poncho for flowable concentrate for seed treatment (FS: flowable concentrate for seed treatment).

It is possible for thiamethoxam to change into clothianidin, an open chain neonicotinoid, by cutting the ring either by hydrolysis or metabolism in living things. The pharmacokinetic studies with thiamethoxam in the Colorado potato bug showed that it quickly changed into clothianidin when it was put on the skin or eaten. Not long ago, it was discovered that thiamethoxam changes into clothianidin inside tomato fruits. Looking at neonicotinoid tolerance also supports the idea that clothianidin is the active ingredient. According to the IRAC MoA categories, all neonicotinoids are in group 4A.

Dinotefuran

The N-nitroguanidine dinotefuran was found by adding an N-nitroimino group to the structure of acetylcholine as a lead chemical in 2002 by Mitsui Chemicals. Dinotefuran is different from other neonicotinoids in the market because it has a racemic and alicyclic TFM (6-tetrahydro-fur-ylmethyl) moiety inplace of halogenated CPM and CTM moieties.

Physico-chemical properties

There is a lot of dinotefuran in water (54.3g l^{-1}) but not much partitioning (-0.644). It stays stable in water at four different pH levels (4, 7, and 9 at 20°C).

Dinotefuran also works very well in many plants, both systemically and translaminarly. Changing the structure of dinotefuran by changing the nitrogen substituents, the pattern of substitutions in the TFM group, or the pharmacophore leads to big changes in how well it kills insects. Dinotefuran works on nicotinic acetycholine receptors (nAChRs) in the same way that other open chain neonicotinoids act.

Efficacy against target pests and application rates

Dinotefuran, a systemic pesticide with contact and oral action, is easily assimilated into plants and carried acropetally. It is used in agriculture to suppress a variety of sucking insects, including Coleoptera, Diptera, and certain Lepidoptera (Tomlin, 2009). Dinotefuran is effective against bugs, hoppers, fruit moths (Lepidoptera), flea beetles (Coleoptera), leaf miners (Diptera), and thrips in cotton, vegetables, fruits, and ornamentals by consumption and contact.

Melting Point (°C)	94.5-101.5
Partition coefficient (log Pow at 25°C)	-0.644
Vapor pressure (mPa at 20°C)	Not described
Solubility in water (g l^{-1} at 20°C., pH 7.0)	54.3± 1.3 (purified water)
Solubility in organic solvents (g l^{-1} at 20°C)	Not described
Dissociation constant pKa (at 20°C)	No dissociation in range pH value 1.4-12.3

It is capable of being administered through foliar application and as a soil application (drenching), which includes root systemic activity. The environmental and toxicological profile is favourable, with minimal mammalian, avian, and aquatic toxicities. The substance is available in four different forms: 2% granule (OSHIN) for use in paddy nurseries, 0.5% dust for foliar rice application (Starkle, Phantom), 1% granule for soil incorporation in vegetables (Aluibarin), and 20% granule for foliar application in fruit and vegetables (Safari).

Thiamethoxam

The discovery of thiamethoxam, a second-generation neonicotinoid belonging to the thionicotinyl class, was the result of a research initiative on neonicotinoids that Ciba Giegy (Since 1996, Novartis; Now Syngenta) initiated in 1985. The biological properties of thiamethoxam were observed to be influenced by the unique combination of an oxadiazine ring and an N-methyl group as a pharmacophore substituent.

Physico chemical properties

Thiamethoxam's physico-chemical characteristics facilitated its rapid and efficient absorption in plants, as well as its xylem transport. In fact, the efficient production of all plant parts located acropetally from the thiamethoxam application site is a direct result of this systemic activity of thiamethoxam.

Similar to other neonicotinoids, thiamethoxam binds very strongly to nAChRs. However, these neonicotinoid molecules may have different ways of binding depending on the species of insects they are found in, and thiamethoxam and dinotefuran may also have different ways of binding to receptors.

Biological mode of action

Thiamethoxam works very quickly on the insects it is meant to kill. In aphids and the Colorado potato beetle, the first signs show up 15 to 30 minutes after they take it in, while in whiteflies, they show up 1 hour later. When the sucking insects stop eating, they pull out their stylets, stretch their legs, and move their heads forward. It doesn't matter if the insects die 24 hours later; the effects are the same as knockdown chemicals because the insects can resume feeding and won't try to get in again.

Thiamethoxam is a crystalline substance that has no smell and a melting point of 131.9°C. At 25°C, the molecule has a low molecular mass and dissolves in water at a rate of 4.1 g liter[1]. It also has a low partition co-efficient (10.13 at pH 6.8). There is no separation seen between pH 2 and 12. These features help plants take them in quickly and efficiently, and they also help xylem movement. All parts of the plant that are above the application sites can be protected by this systemic action. This pest killer can be applied on leaves or in the ground to get rid of flies, aphids, whiteflies, thrips, rice hoppers, rice bugs, and some lepidopterous species (Tomlin, 2009).

However, thiamethoxam has a relatively low affinity for the [3H] imidacloprid binding site compared to the other neonicotinoid pesticides. These other insecticides have nanomolar affinity for housefly and other insect nAChRs. Because it is a proneonicotinoid, it didn't bind very well because it was shown to bind to clothianidin in both animals and plants (Nauen et al., 2003).

Thiamethoxam is marketed since 1998 under the trade names Actara (for foliar and soil treatment) and Cruiser (for seed treatment) to control aphids, whiteflies, thrips, rice hoppers, beetles (Colorado potato beetle, flea beetle), leaf miners, and a few lepidopterans. Thiamethoxam demonstrated good to exceptional effectiveness against homopteran, coleopteran, and some lepidopteran pests following foliar, drench, and seed treatment applications. The field testing was conducted in many green and fruity vegetable crops around the world, and this insecticide displayed high activity against sucking pest's complex. It has also shown remarkable efficacy in controlling thrips following soil treatment. Following the initial seed treatment, it is recommended to use subsequent foliar sprays to keep pest pressure below the threshold. Thiamethoxam is an extremely effective seed treatment against a wide variety of soil-dwelling insects.

The half-life of thiamethoxam is more than one year at room temperature and it is also stable at pH 7.0 (half-life: 200–300 days). It is more easily broken down at pH 9.0 (half-life: a few days). Thiamethoxam breaks down quickly when exposed to light. Even though no breakdown was seen in the active ingredient or mixtures after being stored at 54°C for two months, a breakdown using heat was seen when the temperature rose above 150°C. Thiamethoxam breaks down at mild to slow rates in lab soils. Under normal circumstances, the half-life will be between 34 and 75 days. In the field, the breakdown usually happens faster, mostly because grounds in the field have more microbes living in them and another important way that insecticide break down is by being exposed to light.

Soil metabolism

Thiamethoxam is hydrolytically extremely stable at pH 5.0 (half life, >1 year at room temperature) and at pH 7 (estimated half life, 200-300 days at room temperature). The compound is more labile at pH 9 (half-life of a few days). The half-life of thiamethoxam in laboratory soils typically ranges from 34 to 75 days, whereas it degrades more rapidly in field conditions due to the elevated microbial activity and exposure to light, which are both significant degradation pathways.

Safety profile

Thiamethoxam and other neonicotinoid insecticides exhibit favourable safety profiles as a result of their low application rates, preferential affinity for nicotinic receptor (nAChR) subtypes in the central nervous system (CNS) of insects, and weak penetration of the mammalian blood–brain barrier (Sheets et al., 2016).

Mammalian toxicology, Ecotoxicology, and natural enemies: Thiamethoxam is rapidly absorbed and removed in the urine, mostly unaltered. Thiamethoxam is in WHO hazard class III because it is low in mammalian toxicity whether given to rats orally, safe to skin and eye non-mutagenic, non-genotoxic, and non-teratogenic.

Ecotoxicology and effect on natural enemies: Thiamethoxam shows low-toxicity profile for water vertebrates, invertebrates, avians, soil invertebrates, and beneficial arthropods, except for honey bees and bumble bees. Thiamethoxam is mildly to moderately toxic to most beneficial insects but safe to predatory mites in the field.

Acute toxicity	**Result**
LD_{50} (rat acute oral)	1563 mg kg^{-1}
LD_{50} (rat acute dermal)	>2000 mg kg^{-1}
LC_{50}(rat inhalation, 4h)	>3720 mg kg^{-1}
Skin irritation(rabbit)	Non -irritant
Eye irritation (rabbit)	Non -irritant
Skin sensitization (guinea pig)	Non sensitizing
Ecological toxicity	
Avian oral (Bobwhite quail)	1552 mg kg^{-1}
Freshwater fish (Rainbow trout)	>125 mg kg^{-1}
Freshwater invertebrate (*Daphnia magna*) EC_{50} (48h)	>100 mg l^{-1}
Earthworm EC_{50} (14 day)	>1000 mg kg^{-1}
Bee contact LD_{50} (Honeybee)	0.024 μg/bee (highly toxic)

Thiamethoxam phytotonic effect- Thiamethoxam seed treatment resulted in increased plant vigour, and this "vigour effect" (faster germination, early seedling emergence, greener canopy, healthier and more roots, etc.) is especially noticeable under abiotic stresses (drought, heat, cold, soil salinity).

Octanol-water partition coefficient of neonicotinoids

Insecticide	near to 6.0 along with less environmental persistence ranging from 12 to 197 days (Feo et al. 2010)		
	A	B	C
Dinotefuran	−0.549	−0.549	−0.549
Imidacloprid	0.57	0.57	0.57
Clothianidin	0.70	0.70	0.905
Thiacloprid	0.74	1.26	1.26

Source: Values obtained from The Pesticide Manual (16th ed.). b) Values obtained from the 2011 Pesticide Handbook. c) Values obtained from the Footprint Pesticide Database of IUPAC.)

The biochemical mode of action of neonicotinoid pesticides has been widely explored and characterised throughout the last decade. They act preferentially on insect nAChRs, a family of ligand-gated ion channels found in insects' central nervous systems that are responsible for fast neural transmission. The nAChR is a pentameric transmembrane complex made up of four transmembrane domains and an extracellular domain with the ligand binding site (Nauen et al., 2001; Tomizawa and Casida, 2003). Neonicotinoid insecticides target the acetylcholine binding site on the hydrophilic extracellular domain of α-subunits. The ability to remove tritiated imidacloprid from its binding site corresponds with insecticidal efficacy (Liu and Casida, 1993a; Liu et al., 1993b).

Table 12: Physical-Chemical Properties of Neonicotinoids

Insecticide	Melting Point (°C)	Partition coefficient (log Pow at 21°C)	Vapor pressure (mPa)	Solubility (g l^{-1} at 20°C)		Dissociation constant pKa
				Water	Solvents *	
Imidacloprid	144	0.57	hpa-4x10^{-12}	0.61 (no influence of pH value)	Dichloromethane:67	n.d.
Acetamiprid	98.9	0.80 (25°C)	<1.0 x 10^{-6} (25°C)	4.25(distilled water) 2.95(pH 7 buffersystem)	Ethanol >200 Dichloromethane: >200	0.7 weak base (25°C)
Clothianidin	176.8	0.70 (25°C)	1.3 x 10^{-10}	0.327	Methanol-6.26	11.09
Dinotefuran	94.5-101.5	-0.644 (25°C)	n.d.	54.3 ±1.3 (Purified water)	n.d.	No dissociation in range pH value of 1.4-12.3
Thiacloprid	136	1.26 (20°C)	3 x 10^{-12}	0.185(not influenced by pH in the range of pH 4-9)	Dichloromethane-160 Dimethyl sulfoxide-150	No acidic or basic properties in aqueous solutions
Thiamethoxam	139.1	-0.13 (25oC)	6.6 x 10^{-9} Pa at 25°C.	4.1mgl^{-1}	-	No dissociation constant within the pH range of 2-12.
Nitenpyram	83-84	-0.64	1.1 x 10^{-6} (20°C)	840 (pH 7.0)	Methanol-670	3.1 & 11.5

n.d- not determined, *solubility in solvents were given only for higher values

Use of neonicotinoids in the Indian market

S. No	Insecticide	Formulations used in pest management
1	Acetamiprid	Acetamiprid 20%SP (Pride®)
2	Clothianidin	Clothianidin 50%WDG (Dantop®)
3	Dinotefuran	Dinotefuran 20% SG (Osheen®)
4	Imidacloprid	Imidacloprid 17.8SL (Confidor®)
5	Nitenpyram	-
6	Thiacloprid	Thiacloprid 21.7% SC (Calypso®)
7	Thiamethoxam	Thiamethoxam 25%WG (Actara®)

Note: Two compounds are exclusively used for seed treatment i.e. imidacloprid 48%WS and imidacloprid 70%WS (Gaucho®) and thiamethoxam 70%WS (Cruiser®) and thiamethoxam 30%FS (Tamma®).

Compound	Manufacturer (Year introduced)	Group	Remarks
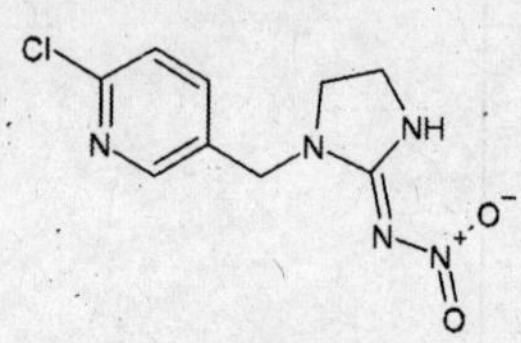 Imidacloprid	Bayer Crop science (1991)	Neonicotinoid	First commercial CNI with highest turnover
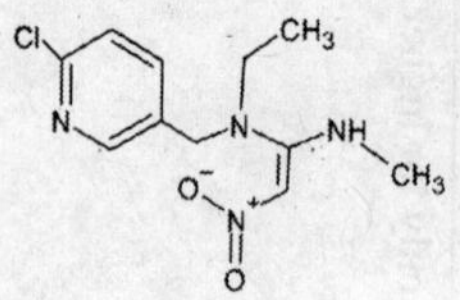 Nitenpyran	SumiTake (1995)	Neonicotinoid	Open chain nitro methylene
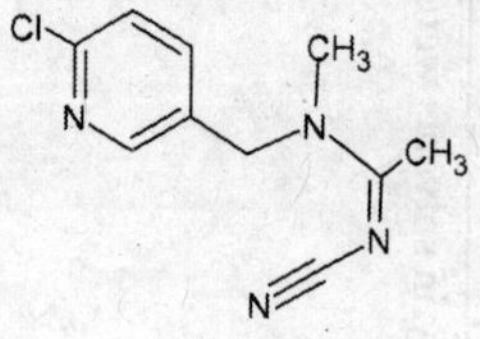 Acetamiprid	Nippon Soda (1995)	Neonicotinoid	Open chain N-cyano acetamidine
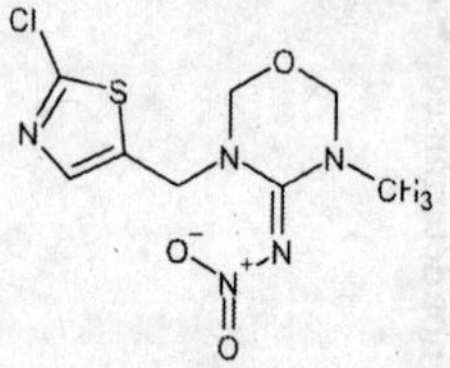 Thiamethoxam	Syngenta (1998)	Neonicotinoid	Six membered heterocyclic nitroguanidine

Compound	Manufacturer (Year introduced)	Group	Remarks
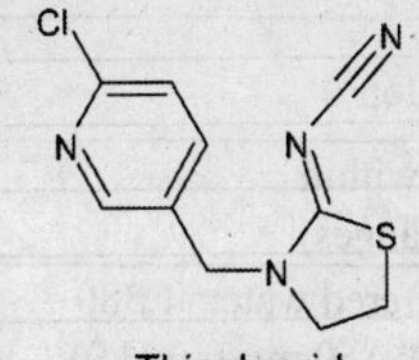 Thiacloprid	Bayer Crop science (2000)	Neonicotinoid	Five membered N-Cyano amidine
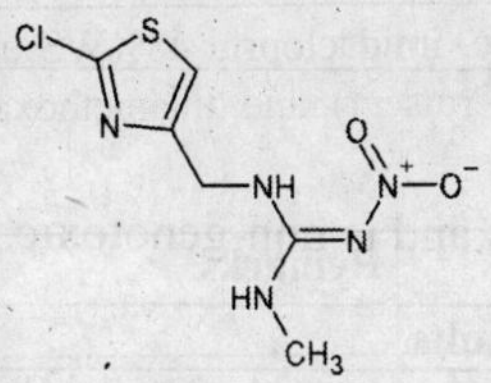 Clothianidin	SumiTake/Bayer Crop Science (2000)	Neonicotinoid	Open chain N-nitroguamidine
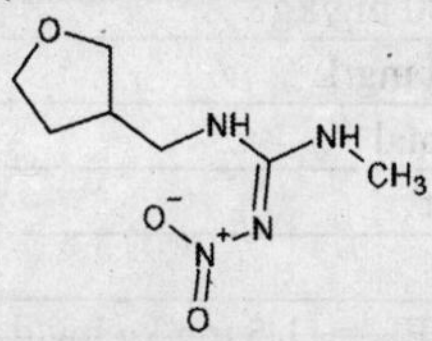 Dinotefuran	Mitsui Toatsu (2002)	Neonicotinoid	Racemic open chain neonicotinoid

8. Sulfoxamines (Sulfoxaflor)

Sulfoxaflor, a compoound discovered by Dow Agro sciences, though not classified as a neonicotinoid, but closely related and is a high-efficacy nAChR agonist with low affinity for the imidacloprid binding site.

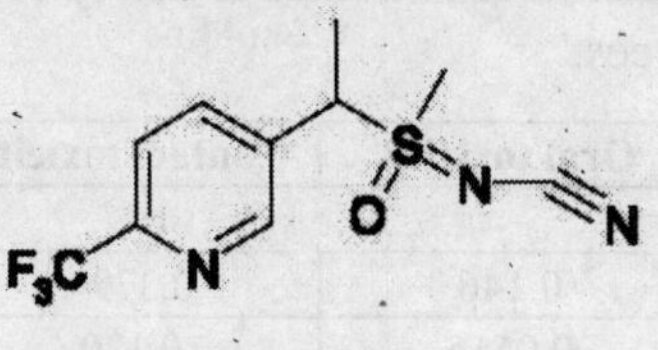

Fig. 13: General structure of Sulfoxaflor

Table 13: Physical and Chemical Properties of Sulfoxaflor Active

Melting point	112.9°C
Vapor pressure	≤1.4 x 10^{-6} Pa at 20°C
Octanol/water partition coefficient (log KOW) at 19°C	pH 7: Log Kow = 0.802
Dissociation constant (pKa)	>10 (does not fully dissociate within environmentally relevant pH ranges)
Solubility in water (mg/L @ 20°C)	Purified water=670 mg/L; Buffered water=1,380 mg/L (pH 5); 570 mg/L (pH 7); 550 mg/L (pH 9)
Organic solvent solubility (g/L @ 20°C)	Acetone- 217 g/L

Mammalian Toxicology

Sulfoxaflor active exhibits low acute mammalian toxicity, and is non-genotoxic.

Study	Animal or test system	Results
Acute oral LD_{50}	Rat	1,000 mg/kg
Acute dermal LD_{50}	Rat	>5,000 mg/kg
Acute inhalation LC_{50}	Rat	>2.09 mg/L
Dermal irritation	Rabbit	Minimal
Eye irritation	Rabbit	Slight
Skin sensitization	Mouse	None
Developmental toxicity	Rat	NOAEL = 11.5 mg/kg bw/d
Genotoxicity	Ames test Chromosomal aberration Mouse micronucleus (in vivo)	Negative Negative Negative
Acute neurotoxicity	Rat	NOAEL = 25 mg/kg bw/d

Acute Toxicity to bees (laboratory studies)- It exhibits acute toxicity to bees when the bees were exposed by oral or contact routes of administration. Sulfoxaflor technical and formulated products had similar toxicities to honey bees. The primary metabolite was not toxic to honey bees.

Test material	Oral toxicity	Contact toxicity
(*Apis mellifera*) (48-hr LD_{50}) (µg a.i./bee)		
Technical (95.6% a.i.)	0.146	0.379*
SC formulation	0.0515	0.130
WG formulation	0.08	0.244
Bumble bee (*Bombus terrestris*) (72-hr LD_{50}) (µg a.i./bee)		
SC formulation	0.027	7.554

*72-hr LD_{50}

Sulfoxaflor effect on other Non-Target Organisms

Acute toxicity to birds	Oral LD_{50} = 676 mg/kg body weight (bobwhite quail)
Dietary toxicity to birds	5-day dietary LC_{50}>5,620 mg/kg diet (bobwhite quail, mallard duck)
Reproductive toxicity to birds	NOAEL = 81.2 mg/kg bw/d (bobwhite quail) NOAEL = 25.9 mg/kg bw/d (mallard duck) No reproductive effects were observed at any dosage
Acute toxicity to fish	96-hour LC_{50} >387 mg/L (rainbow trout)
Acute toxicity to invertebrates	*Daphnia magna*-48-hour EC_{50} >399 mg/L Earthworm-14-day LC_{50} = 0.885 mg/kg soil
Chronic toxicity to invertebrates	*Daphnia magna*-21-day NOEC = 50 mg/L Earthworm-56-day NOEC = 0.1 mg/kg soil

Sulfoxaflor active exhibits very low acute toxicity to fish, freshwater crustaceans (*Daphnia magna*), slightly to moderately toxic to birds in acute oral toxicity studies and did not exhibit any effects on reproduction in birds.

Environmental Fate and Fate in Animals

Microbial degradation is the predominant mechanism of degradation of Sulfoxaflor in the environment. Sulfoxaflor bio-degrades very rapidly in soil. Sulfoxaflor degrades slowly by photolysis in water. The metabolism of Sulfoxaflor and one of its metabolites was studied in rats, ruminants, and poultry. In these animals, Sulfoxaflor is rapidly absorbed and rapidly eliminated with negligible metabolism and not accumulate in the animals' fatty tissues.

Sulfoxaflor 21.8 %SC is recommended against sucking pest complex in rice and cotton

Crop	Target pest	Dosage/acre	
		a.i (gm)	Formulation (ml)
Rice	BPH, WBPH	36	150
Cotton	whitefly, cotton mealy bug	36	150
	jassids, aphid	30	125

9. Nereis Toxin Analogues

Over 80 years ago, a neurotoxin was isolated from a large (~40 cm) annelid worm (***Lumbriconereis heteropoda***) that occurs along the coasts of Japan. The discovery of this toxin resulted from use of the worm as a fish bait. It was noticed by fisherman that flies eating the dead worms would often become paralyzed and die. Fishermen who had handled this annelid worm would occasionally develop a headache, nausea, and respiratory difficulties. Nereistoxin (NTX, Figure 14), the active substance, was isolated by Nitta and 28 years later its structure was reported by Okaichi and Hashimoto. It was the first example of an animal toxin that was successfully used as a basis for analog synthesis.

It was found to be a nicotinic acetylcholine receptor (nAChR) antagonist when applied to frog skeletal muscle. Nereistoxin insecticides are effective for lepidopteran, hemipteran and coleopteran pest control, but relatively toxic to insects, especially lepidopteran larvae such as that of the rice stem boring insect. Consequently, numerous analogs of NTX including cartap (Figure.14) have been used as insecticides. Konishi synthesized and studied the insecticidal activity of many nereistoxin derivatives and established the structure-activity relationship (SAR) of nereistoxin derivatives. However, there are only five insecticide products (Figure 15) currently on the market that use nereistoxin as their active ingredient.

The myriad nAChRs of insects are primarily expressed within their central nervous systems. NTX suppresses electrical signaling within the insect central nervous system and acts as an antagonist at insect nAChRs. A more recent investigation also suggested that NTX may largely exert its insecticidal action by blocking insect nAChR ion channel.

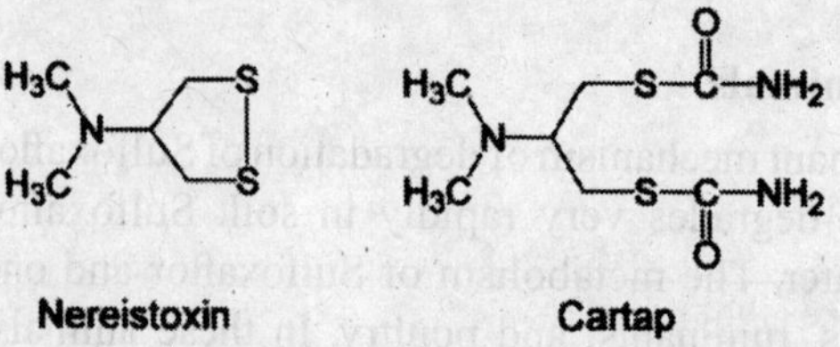

Fig. 14: Structures of nereistoxin (NTX) and cartap, one of the nereistoxin-based insecticides. Cartap primarily acts as a pro-insecticide, being converted into the more active NTX

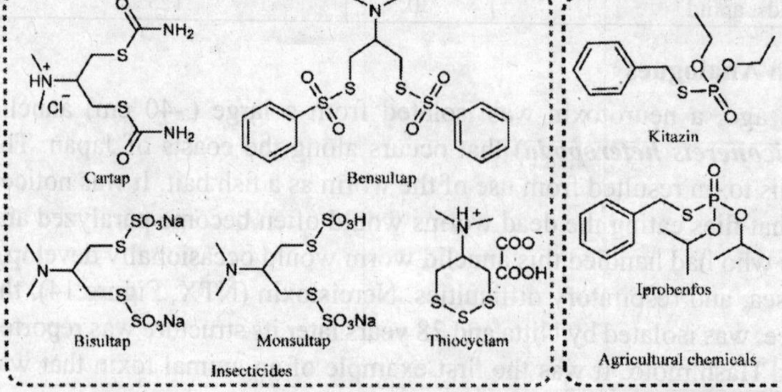

Fig. 15: Insecticides developed based on nereistoxin as their active ingredient

Table 14: Evolution of nAChR agonists

Compound	Manufacturer (Year introduced)	Group	Remarks
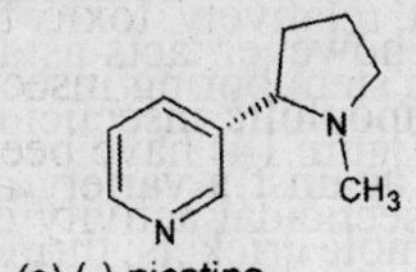 (s).(-)-nicotine	1814	Nicotinoid	Natural product, extract of tobacco
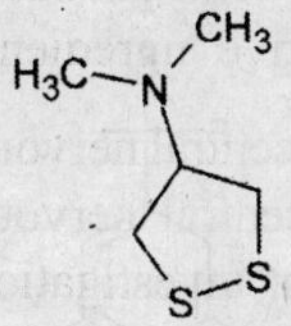 Nereistoxin	-	Nereistoxin	Marine annelid
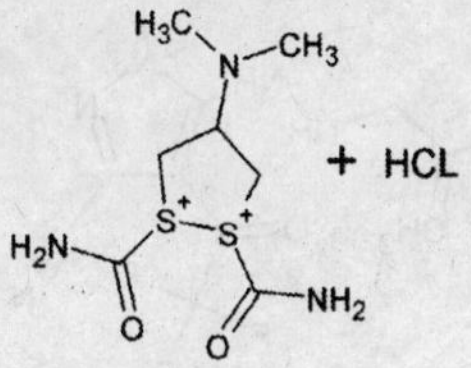 Cartap HCL	SumiTake (1964)	Nereistoxin analog	Prodrug of 2
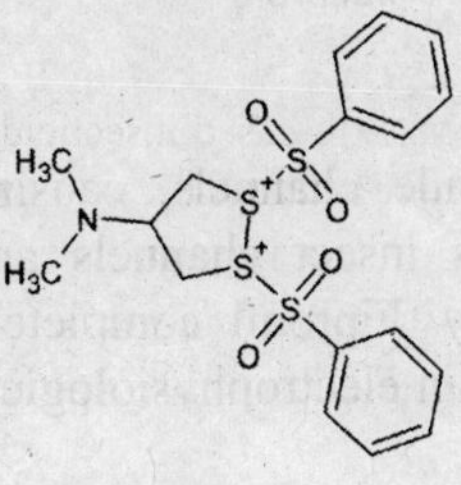 Bensultap	SumiTake (1968)	Nereistoxin analog	Prodrug of 2
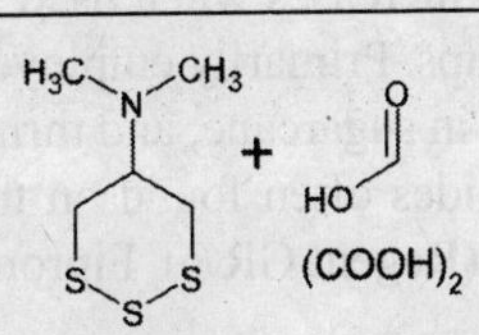 Thiocyclam	Sandoz (1979)	Nereistoxin analog	Prodrug of 2
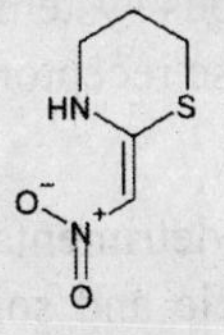 Nithiazine	Shell (1978)	Neonicotinoid	First lead structure for CNIs

10. Phenyl Pyrazole (FIPROLES)

The one insecticide compound in this group is Fipronil. It was discovered and developed by Rhone Poulenc between 1985 and 1987 and released to the market in 1993. Fipronil is a white powder with a moldy odor and however, acts as an insecticide with contact and stomach action. It is the most lipophilic insecticide ever discovered and persistent in soil. It is highly effective against a variety of insect pests (ants, beetles, cockroaches, fleas, ticks, termites, mole crickets, thrips, rootworms, weevils, and other insects).

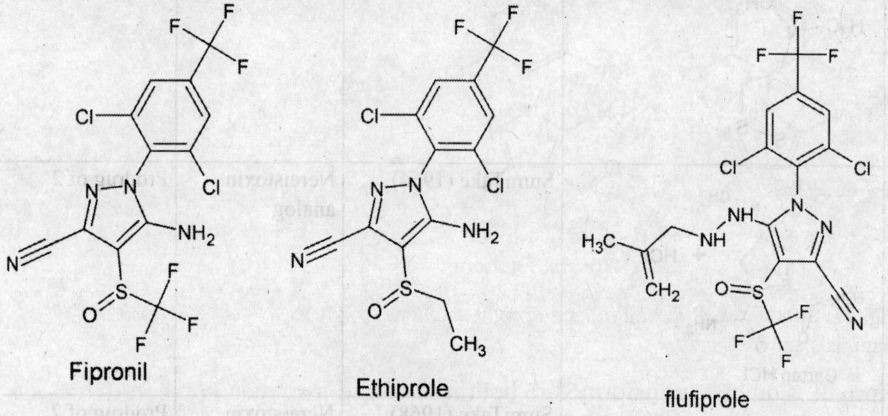

Fipronil

Ethiprole

flufiprole

Fig. 16: Strucutres of Fipronil, Ethiprole, flufiprole

Fipronil inhibits glutamate and GABA-activated chloride channels, causing neuronal excitement and insect death. It also inhibits insect channels and mammalian receptors for GABA although less potently. Fipronil completely inhibits insect chloride channels at values below 100 nM in electrophysiological and biochemical testing.

It has been shown that the effectiveness of this treatment increases when used in smaller quantities, particularly against stem borers and thrips. Primarily employed for the control of stem borer and leaf folder in rice, borers in sugarcane, and thrips in chili, fruit crops, and other similar pests. The insecticides often found on the market include Fipronil 5% SC (Regent®), Fipronil 0.3G (Regent GR®), Fipronil 80% WG (Jump®), and Fipronil 18.87% SC (Johnny®).

Fipronil is a broad-spectrum pesticide. It affects insects' central nervous systems by inhibiting chloride channels regulated by GABA or glutamate. These receptors are weaker or absent in mammals.

It is extremely poisonous to termites and has severe and long-term detrimental effects on their numbers. It poses a long-term threat to nutrient cycle and soil fertility, as termites are "beneficial" important species in these ecological processes. Its toxicity to termites also puts the ecology of areas where termites are a dominating group at danger, given their importance as a food supply for many higher animals.

Fipronil sulfone, one of its principal breakdown products, is more hazardous and long-lasting than the parent chemical. There is evidence that fipronil and some of its metabolites can bioaccumulate, especially in fish.

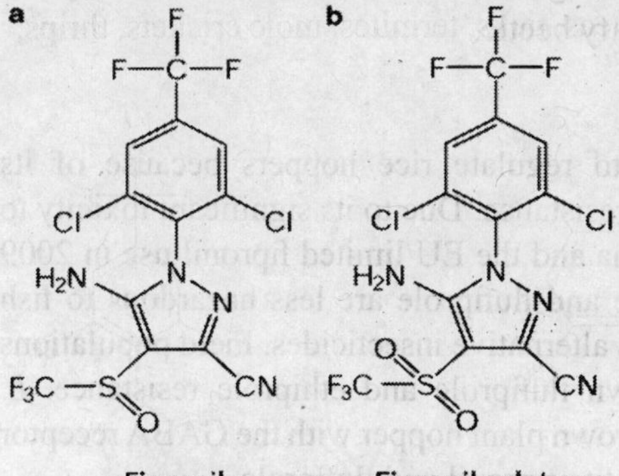

a Fipronil b Fipronil sulphone

Fig. 17: Structures of fipronil (a) and fipronil sulphone (b) (major metabolite of fipronil in mammals and insects).

Fipronil is quickly transformed in both insects and mammals to fipronil sulfone. Human fipronil exposure causes symptoms (headache, nausea, convulsions) usually connected with the antagonism of GABA receptors in the brain.

Fipronil is highly poisonous to bees (LD_{50} = 0.004 µg/bee) and gallinaceous birds (LD_{50} = 11.3 mg/kg for Northern bobwhite quail), but has modest toxicity to ducks (LD_{50} > 2150 mg/kg for mallard duck). It is mildly hazardous to laboratory mammals when administered orally (LD_{50} = 97 mg/kg for rats, 91 mg/kg for mice).

Fipronil breaks down slowly on plants and even more slowly in soil and water. Its half-life can be anywhere from 36 hours to 7.3 months, based on the substrate and the conditions. It doesn't move around much in dirt and doesn't have much of a chance of getting into groundwater.

Plant metabolism: The molecule may be metabolized in plants (rice, cotton, sweet peppers) by oxidizing sulfoxide to a sulfone form.

Bayer Crop Science Japan (formerly Rhone-Poulenc Agrochemicals) discovered ethiprole, a phenylpyrazole insecticide, in 1994. Ethiprole with a relative molecular mass of 397.2 affects the γ-aminobutyric acid-dependent neurotransmission system in insects and used in paddy in India and Indonesia.

Fate in soil: It took 5–71 days for Ethiprole and 535 days for metabolite B to break down in soil. It took 1.3 to 2 days for Ethiprole to break down in natural sunlight in the spring at 35 degrees north latitude.

Animal metabolism: In rats, the main metabolic routes for Ethiprole could be sulfonyl group oxidation or reduction, as well as alkyl group oxidation. Ethiprole's LD_{50} was >7080 mg/kg bw when taken orally, >2000 mg/kg bw when applied topically, and >5.2 mg/L when inhaled.

Flufiprole- A Phenylpyrazole pesticide with an equimolar racemate of (R)- and (S)-flufiprole showing of molecular mass (g mol-1) 491.24 and recommended to manage rice brown plant hopper and leafy beetle.

Resistance to fiprole group

China commonly employed fipronil to regulate rice hoppers because of its diverse modes of action and low cross-resistance. Due to its significant toxicity to honeybees and aquatic organisms, China and the EU limited fipronil use in 2009 and 2013, respectively. Since ethiprole and flufiprole are less hazardous to fish and bees, they are commonly utilized as alternative insecticides. Field populations of *Nilaparvata lugens* have also shown flufiprole and ethiprole resistance. *L. striatellus*, an ethiprole-resistant little brown plant hopper with the GABA receptor A2'N mutation, was also found resistant to fipronil and flufiprole.

Table 15: Toxicity values

Insecticides	Rat/Mouse LD_{50}(mg/kg)		LD_{50} (mg/kg)	LC_{50} (ppm)		LD_{50} (µg/ bee)
	Rat	Dermal	Avian	Fish	Daphnia	Honey bees
Fipronil	100	>2000	31-2150	0.25-0.43	0.19	0.004
Ethiprole	>2000	>2000	--	--	--	

11. Spinosyns (Macrocyclic lactones)

Spinosyns are the newest class of insecticides, represented by spinosad. Spinosad is a new insecticide containing a structurally unique glycosylated macrolactone with selective activity against a wide variety of insect pest species. Spinosyns are produced by fermentation of the actinomycete *Saccharopolyspora spinosa* Mertz and Yao isolated from a Caribbean soil sample (Bret *et al.* 1997). Spinosad consists of about 80% spinosyn A and 20% spinosyn D (Fig. 18) with traces of other structurally related macrolides as minor components. Spinosyn A (Relative molecular mass- 732.0) is also slightly more biologically active than spinosyn D (Relative molecular mass- 746.0)

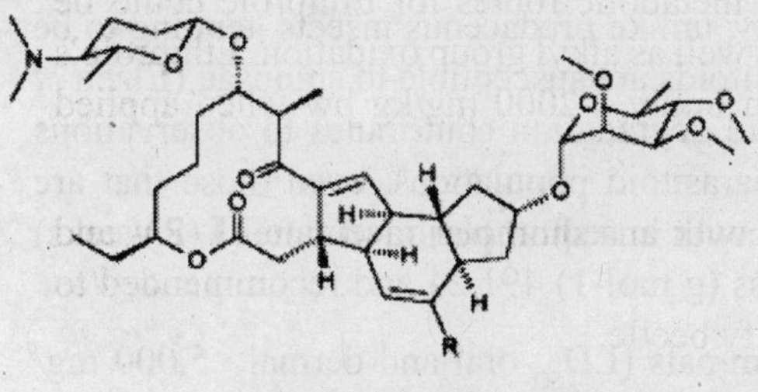

spinosyn A R = H
spinosyn D R = CH_3

Fig. 18: Structures of the components of spinosad insecticide.

Spinosad is both a nerve poison and a stomach poison, so it kills pests that it contacts and those that consume it on the foliage they eat. In insects, the poisoning symptoms of either spinosad or spinosyn. After exposing spinosad leads to muscle contractions due to excitation of the central nervous system, leading initially to postural changes, typically elevation of the body and straightening of the legs (Salgado, 1997, 1998; Salgado *et al.,* 1998). After many hours of excitation, the movements and fine tremors finally cease and the ensuing paralysis is apparently due to neuromuscular fatigue. Spinosyn A has no direct depressant effect on the neuromuscular system and at high concentrations neuromuscular transmission is actually enhanced. In electrophysiological studies with cockroach neurons, 20 nM spinosyn A was shown to activate nAChRs and this action could be blocked by the selective nicotinic receptor antagonist α-bungarotoxin. In addition, spinosyn A also prolonged the action of ACh. The compound also affected GABA receptors in isolated insect neurons

The commercial compound is sold as a water-based suspension concentrate or as water-dispersible granules. It has a novel molecular structure and mode of action that provides excellent crop protection. Spinosyns A and D are highly toxic to Diptera, Lepidoptera, Thysanoptera, and some species of Coleoptera, but they have extremely low toxicity for mammals; therefore, spinosad is classified by the U.S. Environmental Protection Agency as a reduced-risk material (Thompson et al. 2000). It is particularly more effective as a broad-spectrum molecule against *Helicoverpa armigera*, Diamond back moth (DBM) and chilli thrips. It has both contact and stomach activity with long residual activity and the formulation available is Spinosad 45%SC (Tracer®) and Spinosad 2.5 % SC (Success®).

Beneficial insects

Laboratory bioassays have shown some spinosad toxicity to predators (e.g., *Orius insidiosus*), corresponding greenhouse and field tests have generally shown spinosad to have little effect (Studebaker and Kring, 2003), principally due to the rapid environmental degradation of spinosad (Saunders and Bret, 1997; Crouse *et al.*, 2001; Williams *et al.*, 2003). Hymenopterous parasitoids to have varying levels of susceptibility to spinosad, with many, unlike predaceous insects, tending to be relatively sensitive. While not all parasitoids are susceptible to spinosad (Elzen *et al.*, 2000; Papa *et al.*, 2002). This rapid degradation contributes to observations in greenhouse and field studies that parasitoid populations, even those that are relatively sensitive to spinosad, can recover in a short period of time (1–2 weeks) (Williams *et al.*, 2003).

Spinosad has very low toxicity to mammals (LD_{50} oral and dermal> 5,000 mg/kg), birds, and many aquatic invertebrates, is moderately to slightly toxic to fish, but is highly toxic to marine mollusks (shellfish). Spinosad undergoes extensive metabolism and is largely excreted in the feces. In the environment, its solubility is low (above pH 5), tends to bind to soil particles/organic matter, does not persist in the soil, and ultimately breaks down to CO_2 and H_20, so it is unlikely to leach to groundwater.

12. Avermectins

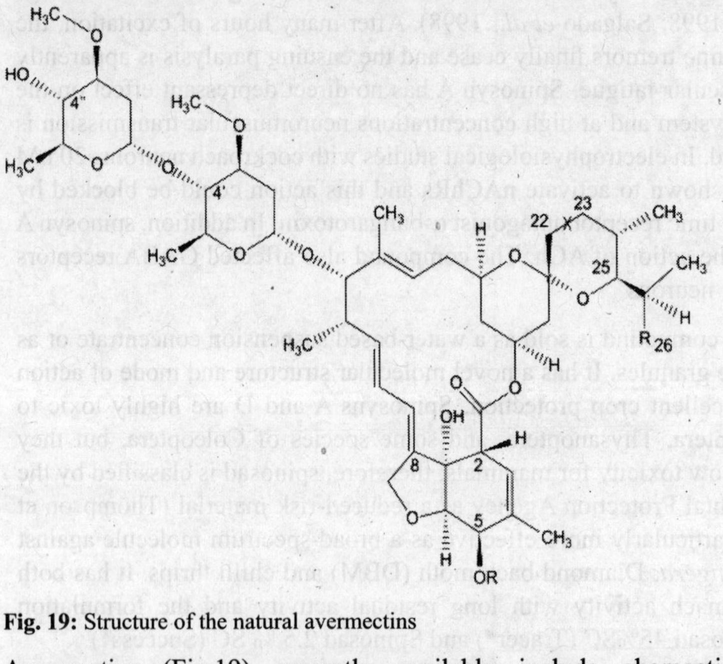

Fig. 19: Structure of the natural avermectins

Avermectins (Fig.19) currently available include abamectin, selamectin, eprinomectin, doramectin, and ivermectin, with emamectin benzoate (EMB) (macrocyclic lactone) being a second-generation derivative of avermectins generated from *Streptomyces* bacteria fermentation products. Abamectin, launched by Merck Sharp & Dohme Agvet in 1997, and milbemectin, introduced by Sankyo in 1990, are 16-membered macrocyclic lactones generated by the fermentation of *Streptomyces avermitilis* and *Streptomyces hygroscopicus*, respectively.

Emamectin Benzoate

Merck scientists discovered emamectin benzoate (homologous semi synthetic macrolides), and Novartis (now Syngenta) introduced to the market in 1997. Emamectin benzoate is a mixture of at least 90% 4”-epi-methylamino-4”-deoxyavermectin B1a (MAB1a) and up to 10% 4”-epi-methylamino-4”-deoxyavernectin Blb benzoate (MAB1b), $C_{56}H_{81}NO_{15}$ (Fig.20). Emamectin benzoate inherits many of the advantages of its progenitor drug, avermectin, but its insecticidal activity has grown by several orders of magnitude.

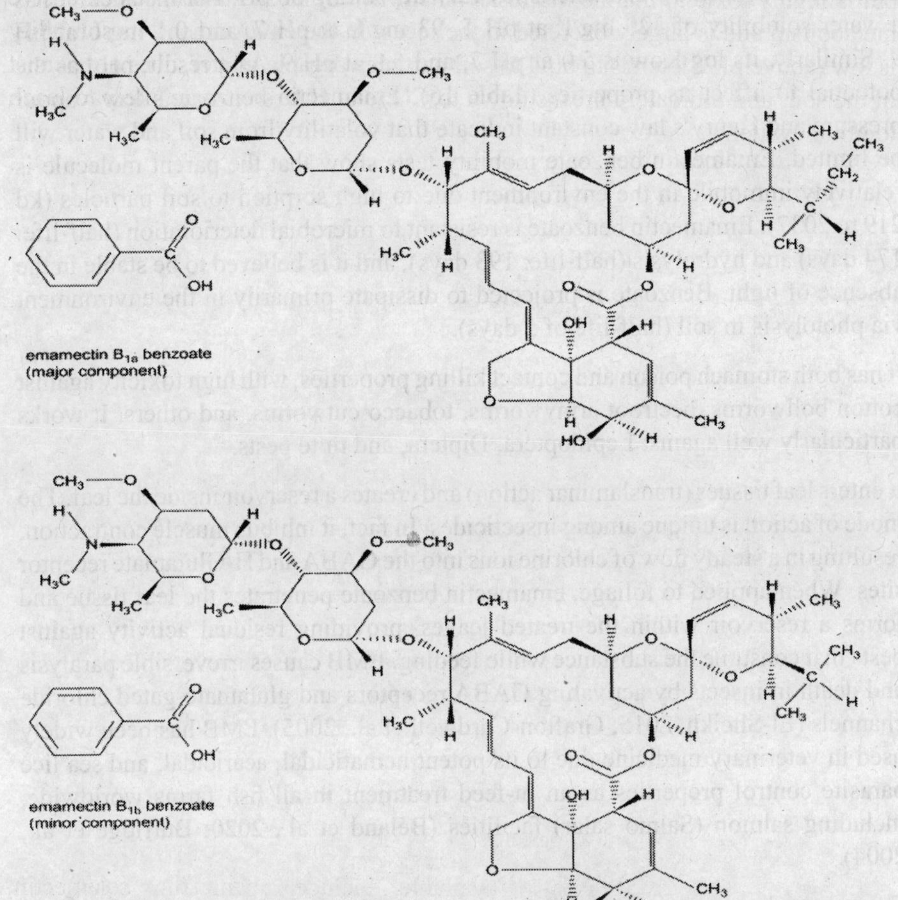

emamectin B_{1a} benzoate (major component)

emamectin B_{1b} benzoate (minor component)

Fig. 20: Structures of emamectin B_{1a} benzoate (major component) and emamectin benzoate B_{1b} benzoate (minor component)

Table 16: Physical and Chemical Properties of Emamectin benzoate

Property	Value	Reference
Water solubility; (pH 7)	93 mg/L	Product Chemistry; MRID 44883704;
Log K_{ow}	5.0 (pH 7)	New Chemical Review (D226628,2000)
Vapor pressure	$3x10^{-8}$ Torr	New Chemical Review @226628,2000); (25OC)
Henry's law constant	3.8 x 10^{-10} atm m^3/mol	Product Chemistry; MRID 44883705
pKa	6.8	http:1/www.aoac.orglpubs/Journal/2001/ ab8403.htm

Emamectin benzoate's characteristics vary depending on pH. For instance, it has a water solubility of 320 mg/L at pH 5, 93 mg/L at pH 7, and 0.1 mg/L at pH 9. Similarly, its log Kow is 5.0 at pH 7, and 5.9 at pH 9. As a result, pH has the potential to affect its properties (Table 16). Emamectin benzoate's low vapour pressure and Henry's law constant indicate that volatility from soil and water will be limited. Emamectin benzoate mobility tests show that the parent molecule is relatively immobile in the environment due to high sorption to soil particles (kd 219 to 2037). Emamectin benzoate is resistant to microbial deterioration (half-life: 174 days) and hydrolysis (half-life: 193 days), and it is believed to be stable in the absence of light. Benzoate is projected to dissipate primarily in the environment via photolysis in soil (half-life of 5 days).

It has both stomach poison and contact killing properties, with high toxicity against cotton bollworms, beetroot armyworms, tobacco cutworms, and others. It works particularly well against Lepidoptera, Diptera, and mite pests.

It enters leaf tissues (translaminar action) and creates a reservoir inside the leaf. The mode of action is unique among insecticides. In fact, it inhibits muscle contraction, resulting in a steady flow of chlorine ions into the GABA and H-Glutamate receptor sites. When applied to foliage, emamectin benzoate penetrates the leaf tissue and forms a reservoir within the treated leaves, providing residual activity against pests that consume the substance while feeding. EMB causes irreversible paralysis and death in insects by activating GABA receptors and glutamate-gated chloride channels (El-Sheikh, 2015; Grafton-Cardwell et al., 2005). EMB has been widely used in veterinary medicine due to its potent nematicidal, acaricidal, and sea lice parasite control properties as an in-feed treatment in all fish farms worldwide, including salmon (Salmo salar) facilities (Béland et al., 2020; Burridge et al., 2004).

Emamectin benzoate was initially registered in Japan in 1998, under the brand name Affirm®. It was used to suppress lepidopteran pests on leafy vegetables and brassicas, as well as to inject the trunk of pine trees to control pine sawfly populations. Proclaim®, an emamectin benzoate-based pesticide, was granted an emergency exemption in Hawaii for diamondback moth control on horticulture crops in 1996 and 1997 and full registration for use was later approved in 1999.

Safety and Toxicity

Table 17: Acute toxicity/irritation and eco toxicity studies of emamectin benzoate 5%SG

Acute toxicity/irritation studies	Result
Ingestion {Oral (LD_{50} Rat)}	1516 mg/kg body weight
Dermal (LD_{50} Rat)	> 2000 mg/kg body weight
Inhalation (LC_{50} Rat)	> 6.28 mg/l air - 4 hours
Eye contact	Mildly Irritating
Skin Contact	Slightly Irritating
Skin Sensitization	Not a Sensitizer (Guinea Pig)
Eco-Acute Toxicity	
Fish (Rainbow Trout) 96-hour LC_{50}	174 ppb
Green Algae 5-day EC_{50}	> 3.9 ppb
Invertebrate (Water Flea) Daphnia Magna 48-hour EC_{50}	1.0 ppb
Bird (Mallard Duck) 14-day LD_{50}	46 mg/kg

Reproductive/Developmental Effects: Developmental and reproductive toxicity observed in dosages that are toxic to mature animals

Chronic/Subchronic Toxicity Studies:

Emamectin Benzoate: Tremors and nerve lesions observed at lowest dose tested in rabbits. Bladder changes reported in rats.

Carcinogenicity: Emamectin Benzoate: None observed.

At first, emamectin was thought to be safe for humans to use because GABA-reactive neurones are only found in the human central nervous system (Roberts, 1984). However, emamectin's lipophilicity makes it easier for it to cross celluar and nuclear membranes, which means it is also harmful to humans (Wolterink, 2012). EMB can damage cells and genes (Ahmed, 2014; Zhang et al., 2016; Wu et al., 2016) because it impacts the reproductive system, liver, and kidneys (El-Sheikh and Galal, 2015; Gamal A. Gabr, 2015; Khaldoun-Oularbi et al., 2013, 2015,2017).

Abamectin

Abamectin was first used in 1985. It is in a group of closely similar macrocyclic lactones that are either produced directly by the actinomycete, *Streptomyces avermitilis* or prepared through semisynthetic modifications (Fisher and Mrozik, 1989). The main part of the insecticide abamectin is a natural product called avermectin B1a, which shows the general structure motif of the avermectins. Abamectin is a pesticide that is made up of 80% or more avermectin B1a and 20% or less avermectin B1b. It is also known as avermectin B1 (Fisher and Mrozik, 1989) (Fig.21).

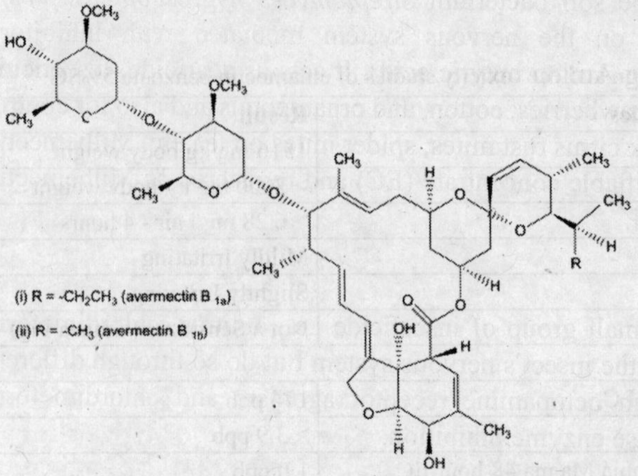

(i) R = -CH_2CH_3 (avermectin B_{1a})

(ii) R = -CH_3 (avermectin B_{1b})

Fig. 21: Structures of avermectin B_{1a} and avermectin B_{1b}

Physico chemical properties

Methanol gives abamectin its off-white to yellow crystal form. No smell. The molecular weight is 887.11, specific gravity is 1.158 at 21 °C, and the freezing and melting points are 150 and 155 °C. Vapour pressure is very low, 1.5 x 10^{-9} mmHg; flash point is between 135 and 150 °C. It's almost impossible for water to dissolve it (solubility = 3.5×10^{-4} g/100 mL).

Abamectin is registered worldwide for controlling mites and certain insect pests on ornamental plants, citrus, cotton, pears, and vegetable crops as a foliar spray at 5 to 27 g/hectare. Abamectin controls agromyzid leafminers, *Liriomyza* spp., Colorado potato beetle, diamond back moth, tomato pinworm, citrus leafminer (*Phyllocnistis citrella*) and pear psylla. It can pierce leaves and enter the plant vascular system.

Abamectin is highly photosensitive and has been proven to degrade rapidly on plant and soil surfaces, as well as in water after agricultural applications. Abamectin was also discovered to be quickly destroyed by soil microorganisms. Abamectin residues in or on crops are extremely low, often less than 0.025 ppm, resulting in negligible human exposure from harvesting or ingestion of treated crops. In addition, abamectin does not linger or accumulate in the environment. Abamectin's bioavailability in non-target organisms is limited by its instability, low water solubility, and tight binding to soil, which also prevents it from leaking into groundwater or entering the aquatic environment.

13. Milbemycins

Milbmectin: Milbemectin is an active ingredient of insecticide created and developed by Mitsui Chemicals Crop & Life Solutions. This active ingredient is found in natural products produced by actinomycetes in soil i.e. a macrolide antibiotic derived from the soil bacterium *Streptomyces hygroscopicus subsp. aureolacrimosus*. It acts on the nervous system mediated with inhibitory neurotransmitter, GABA (g-Amino butyric acid). It acts as acaricide, insecticide for use on apples, pears, strawberries, cotton, and ornamentals and also for control of citrus red mites and pink citrus rust mites, spider mites on tea etc. Milbemectin is formulated as an emulsifiable concentrate (EC) and registered as Milbemectin 01 % EC on rose and chilli.

14. Formamidines

This group comprises a small group of insecticide i.e., amitraz, chlordimeform as both compounds affect the insect's nervous system but do so through different pathways: amitraz through octopamine receptor agonism and chlordimeform through monoamine oxidase enzyme inhibition.

15. Diamide Groups

DIAMIDES- Diamides are a novel class of insecticides that exhibit exceptional insecticidal activity against a variety of lepidopterans and hemipterans. Nihon Nohyaku discovered flubendiamide, the first diamide insecticidal compound, and collaborated with Bayer to develop it. Subsequently, DuPont introduced chlorantraniliprole and cyantraniliprole, which are commercialized by DuPont and Syngenta. The diamide insecticide molecules are classified under group 28, "ryanodine receptor modulators," of the Insecticide Resistance Action Committee (IRAC) mode of action classification, as they share the same target site, the ryanodine receptor. The Diamide group is mainly divided into phthalic diamides and anthranilic diamides (Fig.22).

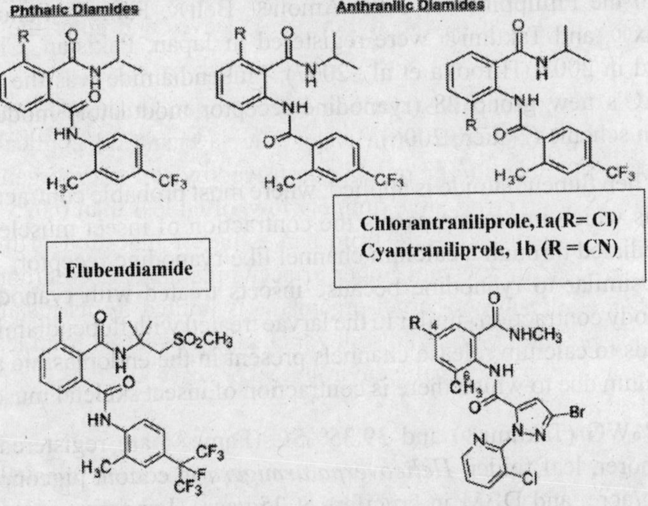

Fig. 22: Structure of Phthalic and anthranilic diamides

Flubendiamide- Phthalic acid diamide or Phthalic diamide

The parent compound structure of flubendiamide was found by Nihon Nohyaku Co., Ltd. during their early 1990s pyrazine dicarboxamide herbicide development effort. Flubendiamide, a phthalic acid diamide insecticide co-developed by Nihon Nohyaku and Bayer Crop Science AG, was synthesized in 1998 after the discovery of more powerful substituents. Flubendiamide is characterized by a three part chemical structure as shown in the general formula a) phthaloyl moiety b) An aromatic amide moiety c) an aliphatic amide moiety (Fig.23).

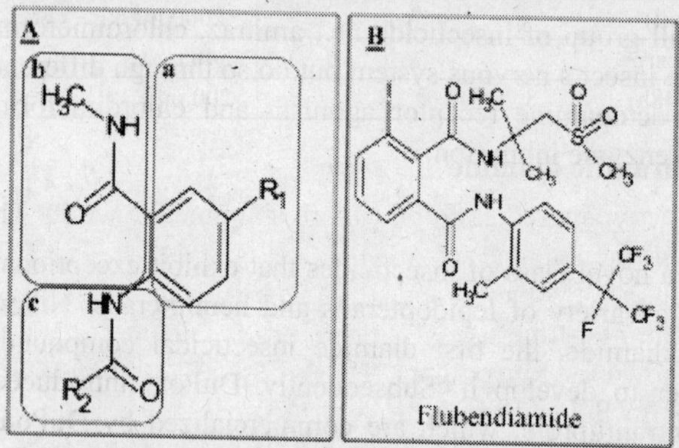

Fig. 23: A. Flubendiamide is characterized by a three part chemical structure as shown in the general formula a) phthaloyl moiety b) An aromatic amide moiety c) an aliphatic amide moiety; B. Flubendiamide structure.

Flubendiamide possesses good biological, ecological, and toxicological properties (Hilder and Boulter, 1999; Hall, 2007; Ebbinghaus-Kintscher at al., 2006). After registering in the Philippines in 2006, Amoli®, Belt®, Fame®, Fenos®, Synapse®, Phoenix®, and Takumi® were registered in Japan, Pakistan, Chile, India, and Thailand in 2007 (Hirooka et al., 2007). Flubendiamide was the first insecticide in IRAC's new group 28 (ryanodine receptor modulator) mode of action classification scheme (Nauen, 2006).

Mode of action: When flubendiamide is applied, where most probable contraction of an insect body is which inturn represents the contraction of insect muscle by calcium release mediated through a calcium channel like ryanodine receptor. The mode of action is similar to ryanodine because insects treated with ryanodine showed sustained body contraction similar to the larvae treated with flubendiamide. i.e., Ryanodine binds to calcium release channels present in the endoplasmic and sarcoplasmic reticulum due to which there is contraction of insect skeletal muscle.

Flubendiamide 20%WG (Takumi®) and 39.35%SC (Fame®) are registered to manage rice stem borer, leaf folder, *Helicoverpa armigera* in cotton, pigeonpea, and tomato at 100g/acre, and DBM in crucifers at 25g/acre. Lead generation to anthranilic diamides from phthalic diamide insecticides

Table 18: Toxicological profile of flubendiamide.

Acute toxicity	**Result**
Acute Oral LD_{50} (rat- male and female)	>2000 mg/kg
Acute Dermal LD_{50} (rat- male and female)	>2000 mg/kg
Eye Irritation- Rabbit	Slight irritant
Skin Irritation-Rabbit	Non irritant
Mutagenicity- Ames test	Negative
Ecological toxicity	
Avian oral LD_{50} (Bob white quail)	> 2000 mg/kg
Aquatic organisms LC_{50} (96h) carp	>546 µg/l
Honeybee (Oral/contact LD_{50} (48h)	>200 µg/l

Chlorantraniliprole- Anthranilic diamide

In 2007, products with this active ingredient hit the market around the world. The product is available in various brand names as Acelepryn®, Altacor®, Coragen®, Dermacor® X-100, Prevathon®, Voliam Flexi®, Voliam® Xpress, Durivo®, and Virtako® right now. The first-time use of chlorantraniliprole was in 2008. Chlorantraniliprole, also known as Rynaxypyr® (Dupont, USA), was the first anthranilic diamide, similar to flubendiamide, it is very good at getting rid of insects in the order Lepidoptera (Temple et al., 2009).

Insecticidal Potency of Chlorantraniliprole (Rynaxypyr®)

A wide variety of Lepidoptera are effectively controlled by chlorantraniliprole, which is used to combat economically significant parasites. Through the rapid cessation of insect feeding, chlorantraniliprole treatments provide plant protection. Desiccation and susceptibility to predatory and environmental hazards are the consequences of the rapid paralysis of the affected insects. Chlorantraniliprole is also exceedingly potent against neonates as they hatch from the eggs. The young larvae are influenced by either consuming the treated chorion or eating small bites from the treated foliage or fruit. Compared to applications made post-oviposition, the ovicidal effects are intensified on the eggs that are deposited on chlorantraniliprole-treated surfaces.

The lepidopteran and dipteran adults consuming treated exudates or water droplets, chlorantraniliprole controls them well. Other research has demonstrated that when these insects come into contact with spray droplets, the effect is amplified by the application of adjuvants. In addition to direct toxicity, study on C. pomonella adults found that sub-lethal exposure to chlorantraniliprole significantly reduced egg laying and subsequent egg hatch.

The spray intervals will fluctuate in accordance with the growth rate and the production of new leaves, as chlorantraniliprole does not protect new growth when administered through foliar applications. In the presence of intense insect pressure, a spray interval against lepidopteran species may be less than seven days on a canopy that is rapidly expanding and growing. Nevertheless, residual control may persist for a period of 14-21 days in mature plants that exhibit minimal or no new growth. Chlorantraniliprole is effective in the control of foliar-feeding lepidopteran pests when used as a soil treatment, as it is able to migrate through the xylem. In a three-year study conducted on bell peppers with *Ostrinia nubilalis* (European maize borer), chlorantraniliprole, when applied via drip irrigation, demonstrated an efficacy that was equivalent to or greater than that of conventional grower programs that involved foliar sprays.

Chlorantraniliprole is both persistent and mobile in the environment. It is difficult to dissolve in water, but it is capable of moving through the environment due to its sluggish degradation in soil and its poor binding to soil (half-lives: 1,130 days). (Chlorantraniliprole Pesticide Fact Sheet; US Environmental Protection Agency, Office of Prevention, Pesticides and Toxic Substances, U.S. Government Printing Office, 2008).

Table 19: Toxicology and Ecotoxicology- Chlorantraniliprole

Acute toxicity	**Result**
Inhalation LC_{50}	> 2.0 mg/L
Dermal LD_{50}	>5000 mg/kg
Primary Eye Irritation	Minimal effects clearing in less than 24 hours
Primary Skin Irritation	Mild or slight irritation at 72 hours (no irritation or erythema)
Ecological toxicity	
Fish LC_{50}	> 10–100 mg/L (low toxicity)
Honeybee (direct contact or oral exposure) Acute LD_{50}	> 11 µg/bee (Low toxicity)

In oral, dermal, inhalation, and eye examinations, chlorantraniliprole exhibits minimal toxicity to mammals. Chlorantraniliprole does not cause cutaneous irritation. It may cause a minor irritation to the eyes. It is not a skin sensitizer (Chlorantraniliprole Pesticide Fact Sheet; US Environmental Protection Agency, Office of Prevention, Pesticides and Toxic Substances, U.S. Government Printing Office, 2008). The results were also favorable in a variety of experiments designed to monitor reproductive toxicity, neurotoxicity, carcinogenicity, and mutagenicity.

The results of numerous laboratory and field trials have demonstrated that chlorantraniliprole has a negligible effect on the populations of a variety of parasitic wasps (e.g., *Aphelinus*, *Trichogramma*), predatory mites (e.g., *Amblyseius*, *Typhlodromus*), predatory insects (e.g., *Coccinella, Orius*, *Nabis*, and *Chrysoperla*), and spiders. In both laboratory toxicity investigations and field tests, formulated chlorantraniliprole products have been demonstrated to have negligible effects on honeybees. In a direct contact investigation with the bumble bee, *Bombus impatiens*, no mortality was observed 72 hours after exposure to chlorantraniliprole at $1.0 gL^{-1}$.

When birds are exposed to chlorantraniliprole for a short time, it is practically nontoxic to them. Chlorantraniliprole is not highly toxic to the majority of soil invertebrates, including earthworms (Final Human Health and Ecological Risk Assessment for Chlorantraniliprole Rangeland Grasshopper and Mormon Cricket Suppression Applications; US Department of Agriculture, Marketing and Regulatory Programs, Animal and Plant Health Inspection Service, 2019). Chlorantraniliprole is regarded as exceedingly toxic to aquatic invertebrates. The U.S. Environmental Protection Agency (USEPA) has stated that there is "no evidence of chlorantraniliprole toxicity" to mammals (Chlorantraniliprole: Problem Formulation for Registration Review; US Environmental Protection Agency, Office of Chemical Safety and Pollution Prevention, U.S. Government Printing Office, 2020).

Safety to mammals: Chlorantraniliprole demonstrates exceptional differential selectivity for insect ryanodine receptors in comparison to mammalian ryanodine receptors. This selectivity is likely a significant contributing factor to the mammalian safety observed with Chlorantraniliprole. It does not have any significant acute toxicity through the oral, dermal, or inhalation routes of exposure, and it does not induce skin sensitization, eye irritation, or skin irritation.

Cyantraniliprole-Anthranilic diamide

Cyantraniliprole (Cyazypyr™), a second anthranilic diamide discovered by DuPont and co-developed with Syngenta (Wiles et al., 2011), is chemically similar to chlorantraniliprole. The cyantraniliprole discovery was achieved by modifying the physical properties, with a specific focus on reducing logP in order to increase water solubility. Chlorantraniliprole (pH 7, 20°C) exhibited a logP of 2.86; however, the introduction of a cyano-group at the 4-position of the anthranilic core resulted in a logP of 1.91 (pH 7, 20°C) for cyantraniliprole. This resulted in a rise in water solubility from approximately 1 ppm for chlorantraniliprole to approximately 15 ppm for cyantraniliprole. The biological attributes were indicative of this physical property change. In particular, cyantraniliprole demonstrates a broader spectrum of insecticidal activity and effectively manages both sucking and piercing insects, including whiteflies and aphids (Foster et al., 2012; Gravalos et al., 2015) and lepidopterans also, i.e called as cross spectrum insecticide while due to high logP value, chlorantraniliprole action was restricted to lepidopterans only.

The improved plant systemicity of cyantraniliprole is the most plausible explanation for any improved control of sucking insects, as Chlorantraniliprole and cyantraniliprole insecticides have comparable potencies against hemipteran RyRs. In comparison to other diamide insecticides, the broad spectrum of this anthranilic diamide is believed to be a result of its physical properties, a higher water solubility and a lower log P. Consequently, it is more suitable for systemic applications (Selby et al., 2013). In 2012, Benevia®, Spinner®, and Exirel® were introduced as products that contained cyantraniliprole.

Biological Profile of Cyantraniliprole (Cyazypyr™)

Cyantraniliprole is capable of controlling a wide range of piercing sucking and gnawing pests, including leafminers, fruit flies, psyllids, caterpillars, and whiteflies. Despite the fact that cyantraniliprole is notably effective against the immature stages of pests, it also exhibits exceptional activity on the adult stage for specific species. Cyantraniliprole is generally effective at utilization rates ranging from 10 to 200 g a.i. ha^{-1}, subject to the pest, crop, application method and timing, and the use of adjuvants.

The cyantraniliprole protects plants by stopping pests from eating quickly; most of the time, the insects are unable to feed within minutes to hours of being exposed to cyantraniliprole. Although insects can die within hours, it usually takes one to three days for biting-chewing pests to be controlled and up to five days for piercing-sucking pests to be seen to have worked in the field. Because cyantraniliprole can penetrate into the leaf cells, it protects against rain wash-off, increasing its effectiveness.

When applied to plant foliage, cyantraniliprole exhibits translaminar activity, resulting in the suppression of significant aphid, whitefly, and caterpillar populations. The use of oil adjuvants promotes leaf penetration, hence increasing the level and uniformity of pest control. The physicochemical characteristics of cyantraniliprole enable xylem mobility and consequently upward translocation in plants. When administered to the root zone, cyantraniliprole is translocated to the above-ground sections of the plant, making it suitable for soil application methods such as drenching, furrow application, drip chemigation, and seed treatments. However, phloem mobility from foliar cyantraniliprole treatments has not been widely documented.

In India, the product has label claim to control thrips, flea beetles, cabbage aphids, mustard aphids, DBM, and tobacco caterpillars in pomegranates; thrips, *Helicoverpa armigera*, and Spodoptera spp. in cabbage; leaf miner, *Aphis gossypi*, *Thrips tabaci*, white fly, and *Helicoverpa armigera* in tomatoes; and leaf miner, red pumpkin beetles, aphids, thrips, white fly, pumpkin caterpillars, and fruit flies in gherkins @ 240–360 ml per acre.

Table 20: Physico-chemical properties of chlorantraniliprole and cyantraniliprole

Property	**Chlorantraniliprole**	**Cyantraniliprole**
Chemical class	Anthranilic diamide	Anthranilic diamide
Molecular weight (g mol^{-1})	483.15	473.72
Physical form	Off-white crystalline powder	Solid, powder
Melting point (°C)	208–210	224
Water solubility (20°C;mg/l)	0.9–1.0	14.2–17.2
Log Pow	**2.86 at 20◦C**	**1.9 at 20◦C**

Table 21: Toxicological profile of Chlorantraniliprole and Cyantraniliprole

Species	Chlorantraniliprole	Cyantraniliprole
Acute oral LC_{50}, rat (mg kg^{-1})	>5000	>5000
Avian, acute oral LD_{50}, bobwhite quail (mg kg^{-1})	>2250	>2250
Aquatic. rainbow trout LC_{50} (mgl^{-1})	>13.8 (solubility limit)	>12.6
Crayfish LC_{50} (mgl^{-1})	>1.42	4.0
Earthworm, acute LD_{50} ($mgkg^{-1}$ soil)	>1000	>1000

Chlorantraniliprole and cyantraniliprole have low mammalian toxicity and good nontarget toxicology. Receptor-based studies of natively produced mammalian RyR isoforms show that chlorantraniliprole selects insect RyRs 300-fold to over 2000-fold over mammalian RyRs. Cyantraniliprole has considerably lower mammalian RyR activity. This chemical is about 500-fold less potent than insect RyRs against mouse RyR1, the most sensitive mammalian receptor isoform examined. Target site selectivity appears to be a fundamental role in the low mammalian toxicity of anthranilic diamide insecticides.

Tetraniliprole-Anthranilamide

Tetraniliprole, is a ryanodine receptor (RyR) modulator and works by keeping the ryanodine receptors open, which allows calcium to be released in an uncontrolled way.

Mode of action

Insect calcium channels might provide an ideal pesticide target for commercial use. Calcium homeostasis is essential in many biological processes, including cell communication, muscular contraction, neurotransmitter release, and fertilization. Muscle contraction includes the control of two types of channels: voltage-gated channels, which govern external calcium entry, and ryanodine receptor channels (RyRs), which regulate the release of internal calcium storage. (RyRs are named after the natural substance ryanodine, which is found in trees and shrubs of the *Ryania* genus). Ryanodine and similar extracts from these bushes' bark interfere with muscle contraction by altering the channel's conductance state. Ryanodine, either alone or in combination with rotenone and pyrethrum, has been marketed as a "organic alternative" to synthetic insecticides. Attempts to improve the insecticidal effectiveness and selectivity of ryanodine by structural modification have had limited success.

Anthranilamides selectively activate ryanodine receptors using in situ muscle preparations and calcium mobilization tests on native and recombinant cells. Radioligand-binding experiments reveal these chemicals bind to a receptor location different from ryanodine. Calcium mobilization studies in mammalian cell lines show that anthranilic diamides choose insect receptors over mammalian receptors by >500-fold. Thus, anthranilic diamides provide a novel pest-management option and a pharmacological tool for calcium signaling study.

Chlorantraniliprole and cyantraniliprole are anthranilic diamides, while flubendiamide is phthalic. Both classes bind to the receptor at the same location despite their structural differences. Anthranilic diamides (antranilamides) are the first safe and effective RyR insecticides.

Flubendiamide and the anthranilic diamide insecticides, chlorantraniliprole and cyantraniliprole target the RyR. These novel pesticides are in Group 28, Ryanodine Receptor Modulators, according to IRAC. Insects and vertebrates have two types of calcium channels, voltage-gated, which govern Ca^{2+} entry, and RyR, which regulate calcium release. RyRs are named after the plant alkaloid ryanodine, which locks them open. Ryanodine's insecticidal characteristics were first identified in the 1940s, but before diamide insecticides, there were no synthesized RyR molecules for commercial insect control. RyR's four identical subunits create a tetramer with a central pore that regulates Ca^{2+} flow. Anthranilic diamides directly activate the receptor channel, unlike ryanodine, which locks it open after RyR activation.

Chlorantraniliprole stimulates receptor-mediated Ca^{2+} release in cockroach, *Drosophila, Heliothis*, and hemipteran chimeric insect RyRs at 40-50nM, while cyantraniliprole is more potent against sucking insects, it is equivalent to chlorantraniliprole against lepidopteran and hemipteran RyRs. As with flubendiamide, anthranilic diamides bind to a different location than ryanodine.

The potential ryanodine binding site is the pore area of the RyR (amino acid residues). Amino acid residues are essential for sensitivity to phthalic and anthranilic diamide insecticides, as demonstrated by photoaffinity experiments and replacement of certain areas of recombinant insect RyRs [Kato et al. 2009]. Compared to anthranilic diamide, these amino acid residues seemed to be more important for flubendiamide sensitivity.

Diamide resistance development

Diamide resistance is a rising global issue. *P. xylostella* has developed resistance to 93 pesticides and has become one of the most difficult pests to manage in cruciferous vegetables. Diamondback moth larvae have long been known for their rapid resistance to novel products. This is most likely due to their genetic adaptability, rapid generation time, high fecundity, and the fact that new chemistry is frequently utilized, resulting in intense selection pressure in the field.

Mutations at the RyR target site are clearly involved in conferring diamide resistance, and there is additional evidence that a metabolic component contributes to the resistant phenotypes and following this report, control failures with diamides have been recorded in various lepidopteran pests. The tomato leaf miner *Tuta absoluta* (Roditakis et al., 2015) collected in Sicily (Resistance factor >2000 fold) and the smaller tea tortrix *Adoxophyes honmai* (Resistance factor 77-105 fold) collected in Shizuoka Prefecture, Japan (Uchiyama and Ozawa, 2014) both exhibit high levels of diamide resistance. A field population collected from Lat Lum Kaew, Pathum Thani Province, demonstrated an exceptionally high resistance to flubendiamide (RF = 26,602) and high resistance (RF = 775) to chlorantraniliprole. Similarly, a very high increase in resistance to flubendiamide (RF = 4817.4) and high resistance to chlorantraniliprole (RF = 87.7) was noted.

Just 18 months after flubendiamide's introduction, the *P. xylostella* population in Thailand began exhibiting signs of resistance to the drug and cross-resistance to chlorantraniliprole. Larvae grown from a field population collected in Sai Noi, Nonthaburi Province (a vegetable growing area near Bangkok) in 2010 had resistance factors for both flubendiamide and chlorantraniliprole of 66.3 and 35.4, respectively. Over-reliance on a single mode of action, minimal crop rotation (caused by continuous cruciferous plantings), under-dosing with insecticide (to save costs), irrigation practices that resulted in excessive product wash-off (which allowed insects to be exposed to sub-lethal levels), and a lack of cogent insecticide resistance management (IRM) strategies were some of the major factors linked to diamide resistance in Thailand. In the absence of effective and long-lasting integrated pest management strategies, selection pressure for resistance development will only rise when more active pesticides, including tetraniliprole and cyclaniliprole (Sparks and Nauen, 2015), hit the market.

16. Meta Diamide

The *meta*-diamide structure was generated *via* drastic structural modification of a lead compound, flubendiamide, and the subsequent structural optimization of *meta*-diamides on each of its three benzene rings led to the discovery of broflanilide.

Diamide insecticides: Diamide insecticides, such as chlorantraniliprole and flubendiamide, are characterized by the presence of a single amide group (-CONH-) within their chemical structure. They typically contain a single aromatic ring. They specifically target lepidopteran pests (moths and butterflies), as well as other chewing insects.

Meta-diamide insecticides: Meta-diamide insecticides, broflanilide, have two amide groups (-CONH-) attached to a meta position on an aromatic ring. This structural difference distinguishes them from diamide insecticides. Meta-diamides are effective against a broader range of pests, including sucking insects, aphids and whiteflies, as well as certain beetle species.

Broflanilide insecticide has strong broad-spectrum efficacy and can be applied in a range of ways, including: foliar, in-furrow and seed treatment for the control of tough chewing insects and soildwelling pests, like caterpillars (Lepidoptera), beetles (Coleoptera) and certain thrips in specialty and row crops, as well as urban and rural pests such as termites, ants, cockroaches and flies.

Unique MOA

With its unique MOA, Broflanilide insecticide is among the first compounds in the market introduced under the new IRAC group 30. Plus, there's no known cross-resistance with existing products in the market, making it a superior insecticide resistance management tool.

Fast Acting, residual control

It is a fast acting compound that quickly works at all insect life stages to provide excellent, long-lasting control of pests and protects initial growth stages of seeds and young plants. It offers 14-21 days of residual control for foliar applications, a month or longer as a seed treatment, as well as up to five years of termite control in residential applications

New Mode of Action

The Insecticide Resistance Action Committee (IRAC) approved the unique mode of action of Broflanilide insecticide and designated this insecticide among the first members of its new IRAC Group 30. There is no known cross resistance with currently available products in the market. In the insect's nervous system there are two types of signals transferred between cells-excitatory signals and inhibitory signals. The balance between these signals is crucial for normal insect behavior. Neurotransmitters, such as GABA, communicate these signals between cells. In a normally functioning nervous system, GABA released by one cell activates GABA receptors of another cell, resulting in inhibition. Broflanilide insecticide prevents GABA from transmitting inhibitory signals, which causes overexcitation of the nervous system and leads to incapacitation of the insect. It does this by binding to a novel insecticide site on the GABA receptor that locks it closed. Ultimately, Broflanilide insecticide reduces the pest population and insects are prevented from causing damage. As a top-performing product with broad-spectrum control of Lepidoptera pests and key Coleoptera and thrips species.

Fast Action

Broflanilide insecticide is a fast acting compound and at the recommended dose (12.5 g a. i/ha), reduced coordination and appetite of *Spodoptera eridania* (Southern armyworm) were recorded within the first hour after direct exposure. 100% mortality of the pest was achieved at 2 hours after exposure to Broflanilide insecticide. The current chlorantraniliprole insecticide, applied at the labeled rates, took 5x longer to reach the same result, enabling pests to further damage crops. Several factors might affect speed of kill, such as insect species, its life stage and water volume.

Translaminar movement- A valuable and important aspect of Broflanilide insecticide is its ability to move across leaf tissue after a foliar application (from one side of the leaf to the other), otherwise known as translaminar movement, ensuring that the active ingredient moves throughout insect-feeding areas. Translaminar movement offers growers the ability to control targeted insect populations, such as thrips that may feed on the underside of the leaf, even if the product was applied on the upper side of the leaf and may not have come into direct contact with the target pest during the foliar application. To demonstrate the translaminar movement of Broflanilide insecticide, droplets of the product solution were placed on the upper surface of the leaf. For this bioassay, Broflanilide insecticide was applied at approximately half the recommended dose for thrips control to better exhibit the differences between the treatments applied. Thrips were then placed on the underside of the leaf, and control was evaluated for each treatment. The testing included the use of penetrating adjuvants, which enhance translaminar movement resulting in greater leaf protection through increased control. Because of its very low water solubility, Broflanilide insecticide stays where its applied

17. Pyrimidines

1. **Benzpyrimoxan 10% SC (Orchestra®)** is a member of the class of pyrimidines. It is an insecticide discovered and developed by Nihon Nohyaku, inhibits molting of nymphs of plant and leaf hoppers and reduces the pest population in paddy fields. BPX® technology is registered by Nichino India Pvt. Ltd. (See chapter 13)

Benzpyrimoxan 10% SC

Crop	Common Name of the pest	Dosage/acre			Waiting period (days)
		a.i	Formulation	Dilution in Water (Liter)	
Rice	BPH, WBPH	30-40g	300-400ml	200	31

18. Pyrazoles

Tolfenpyrad belongs to new class of chemistry known as "Pyrazoles" group. It has unique mode of action i.e. mitochondrial electron transport inhibitor. It is registered as Tolfenpyrad 15 EC and used @ 400ml/acre (60 g a.i./acre) in cabbage (DBM), okra to control major sucking insect pests viz., aphids, jassids, thrips and whitefly, cotton (aphids, jassids, thrips, whitefly), chilli (aphids, thrips), mango (hoppers, thrips), onion (thrips) and cumin(aphids, thrips)

Fenpyroximate is an acaricide and a tert-butyl ester. It acts on mitochondrial NADH: ubiquinone reductase inhibitor. Fenpyroximate 05 % EC is effective against spider mite/pink mite/purple mite/yellow mite/eriophyid mites. This acaricide is particularly recommended for root feeding in coconut @ 10ml/litre of water (0.50 g a.i./tree) against eriophyid mites.

19. Thiourea Compounds/Derivatives

Diafenthiuron

Diafenthiuron a thiourea acaricide, insecticide and an aromatic ether was developed by Ciba-Geigy AG (now Syngenta Crop Protection AG) in 1988 and released in 1991. According to the investigations, diafenthiuron is a proinsecticide that converts to the insecticidal carbodiimide (pro-insecticide) by oxidative desulfurization. Diafenthiuron failed to show any impact on biochemical and neurophysiological assay systems, including mitochondria, cuticle development, axonal sodium channel, and key neuronal receptors. However, it was found to be a substrate for cytochrome P-450.

Diafenthiruon is converted into carbodiimide in field crops (cotton, cabbage), microsomes, and entire insects and animals. Diafenthiuron's conversion to carbodiimide and its precursor, S-monoxide, competes with metabolic deactivation, which largely creates hydroxylated derivatives and ureas (Fig.24).

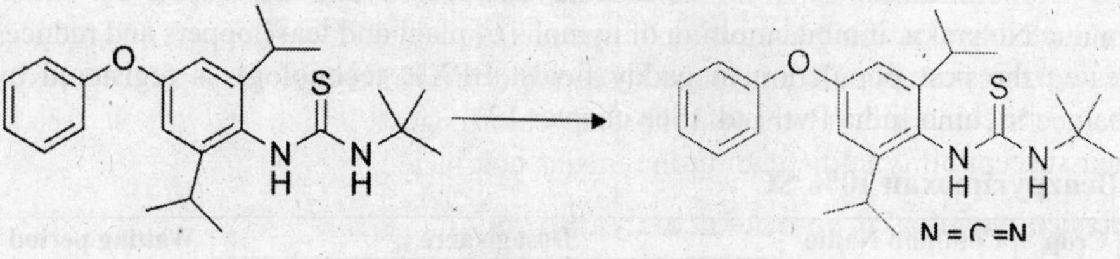

Fig. 24: A. Diafenthriuron ------Oxidative desulfurization-------B.Diafenthriuron carbodiimide (Pro-insecticide)

Diafenthiuron was initially developed at Ciba Giegy (now Syngenta Crop Protection Ag) based on its strong performance in field testing against spider mites in citrus and cotton, but it also showed surprising promise against *B. tabaci,* a type of whitefly.

In order to manage sucking insect pests, whiteflies, aphids, leafhoppers, and mites in cotton, diafenthiuron was first applied commercially in 1991. Despite not being systemic, the pesticide affects the insects present on the ventral side of leaves translaminarly, even if the spray droplets do not come into contact with them and this characteristic, is particularly helpful in crops viz. cotton with dense canopies, which may be either due to efficient uptake into the leaf cuticle or vapor-phase activity. Nymphs and adults were shown to be the most susceptible to the effects of diafenthiuron on different stages of whiteflies in cotton, however the decrease in egg hatching was not as notable.

It was proven in the lab with mites that diafenthiuron has strong translaminar activity, which is good for performance in the field. Field conditions make it easy to see, and it has been said that the vapor phase action may be stronger in the field because more vapor is made when spraying in the field. It worked well against the diamondback moth, *Spodoptera litura* as well. In India, commercially available as an insecticide and acaricide, sold under the brand names Pegasus and Polo.

Diafenthiuron does no harm to the majority of beneficial arthropods in the field, including cotton-specific ones. Nonetheless, considerable toxicity was noted against predatory bugs, with beneficial insects (predatory mites) (*Aphidius rhopalosiphi*) being somewhat detrimental at 500 g ha[1] and beneficial insects (parasitic wasps) (*Typhodromus pyri*) being harmless at 500 g ha^{-1}.

There have been no reports of cross resistance to other insecticides or acaricides due to the distinct mode of action of diafenthiuron. Most importantly, despite field cage selection studies in Malaysia and Thailand showing no observable resistance after 25 and 55 generations, respectively, the whiteflies, *Bemisia tabaci*, *Trialeurodes vaporiorum*, and the aphid, *Aphis gossypii*, which have developed strains resistant to all major insecticides (OPs, pyrethroids, growth regulators, and neonicotinoids), remained fully susceptible. The resistance monitoring trials conducted in Taiwan have also confirmed that *P. xylostella* strains in heavily treated areas are significantly less susceptible to modern insecticides, including abamectin, emamectin benzoate, fipronil, chlorfenapyr, and Spinosad. However, their susceptibility to diafenthiuron remains unaffected.

Physico-chemical properties of diafenthiuron- It doesn't dissolve well in water, but it dissolves very well in a number of organic solvents. Because of how it is chemically made, it is not likely to leak into groundwater. Most of the time, it doesn't stay in the soil, but it can stay in water for a long time.

Table 22: Physico-chemical properties of diafenthiuron

Solubility	In water: 0.06 mg/l (20- 25 °C). Organic solvents: acetone (320), n hexane (9.6), methanol (47), n-octanol (26), toluene (330) g/l (20- 25 °C).
Melting point (°C)	146
Octanol-water partition coefficient at pH 7, 20 °C	Log P=5.76
Vapour pressure (mPa)	<0.002 mPa (25 °C)

Table 23: Toxicity of diafenthiuron

Acute studies-Rat	**LD_{50} (mg/kg)**
Oral	2068
Dermal LD_{50} (mg kg^{-1} body weight)	2000
Inhalation LC_{50} (mg l^{-1})	0.558
Percutaneous LD_{50}	>2000 mg kg^{-1}
Birds-Acute LD_{50} (mg kg^{-1})	1500
Earthworms - Acute 14 day LC_{50} (mg kg^{-1})	1000
Honeybees (*Apis* spp.) (worst case from 24, 48 and 72 hour values)	

Acute studies-Rat	LD_{50} (mg/kg)
Contact acute	1.5 µg bee^{-1}
Oral acute	2.1 µg bee^{-1}
Aquatic ecotoxicology	
Temperate freshwater fish – Acute 96 hours LC_{50}	0.0007 mgl^{-1}
Temperate freshwater aquatic invertebrate *(Daphnia magna)* - Acute 48 hour EC_{50}	> 0.5 mgl^{-1}

20. Sulfite Ester Group- Propargite is a sulfite ester and a terminal acetylenic compound. It has a role as a sulfite ester acaricide. It is a dark colored liquid and formulated as a wettable powder or water emulsifiable liquid. Practically insoluble in water (10.5 mg/L) and as an ester slowly reacts with water to form sulfur dioxide and alcohols; reaction is more rapid under basic or acidic conditions.

It is used for control of phytophagous mites (particularly motile stages) on vines, fruit trees, tomatoes, vegetables, ornamentals, cotton, maize, groundnut and sorghum and effective against spider mites. eriophyid mites, purple mites, pink mites, scarlet mites in tea, chilli, apple and brinjal and Propargite is formulated as an emulsifiable concentrate liquid as 57EC in India and available in market as Omite®.

Propargite exhibits low acute toxicity via oral and dermal exposures. The oral LD_{50} value in rats is about 1-4 g kg^{-1}. With dermal exposure, the LD_{50} values are 250 mg kg^{-1} (male) and 680 mg kg^{-1} (female) in rats and >3 g kg^{-1} in rabbits. Propargite is, however, a strong eye and skin irritant that causes erythema, edema, and eschar formation in rabbits.It can cause illness by inhalation, skin absorption and/or ingestion. In Humans, Dermatitis is the major form of toxicity following dermal propargite exposure. Signs include erythema, burning, itching, exfoliation, and hyperpigmentation. Ocular exposure produces irritation. Changes in the chemical formulation have alleviated many of the acute irritant effects associated with propargite use.

The primary hazard is the threat to the environment. Since it is a liquid it can easily penetrate the soil and contaminate groundwater and nearby streams.

21. Thiazolidine Group- Hexythiazox working as acaricide comes under this group and recommended against spider mites in tea, chilli, apple, brinjal, rose apple, grapes and okra @200ml/acre and marketed under the trade name Maiden®.

22. Quinazoline Group- A acaricide, Fenazaquin belongs to quinazoline group. It is registered as Fenazaquin 10 % EC and sold as Magister® that was recommended to manage mite complex in tea, spider mite in vegetables and apple. Another important insecticide from this group is pyrifluquinazone and popular tarade name is Clasto®.

23. Quinoline Group- Acequinocyl is a natural product found in ***Apis cerana*** and comes under class quinoline. Acequinocyl is an acetate ester consisting of 1,4-naphthoquinone bearing acetoxy and dodecyl substituents at positions 2 and 3 respectively. It has a role as a mitochondrial cytochrome-bc1 complex inhibitor and an acaricide. It is an acetate ester and a member of 1,4-naphthoquinones. Eg: Acequinocyl 15 % SC

24. Halogenated Pyrroles

Chlorfenapyr-Dioxapyrrolomycin is the fundamental structure employed in the development of chlorfenapyr. It was suggested that the insecticidal activity of these compounds was a result of the uncoupling of oxidative phosphorylation. The ability to uncouple depends on physical and chemical factors i.e. the molecule's lipophilicity (log P), which lets it pass through the mitochondrial membrane and acidity (pKa), which breaks up the proton gradient that is needed to convert ADP into ATP. The studies have shown that for the best insecticidal action, the pKa level needs to be between 7.0 and 7.9 and the log P level needs to be 6.0±1.

The chemical structure of chlorfenapyr is lipophilic, but it doesn't have the acidic proton that is required for strong uncoupling action. It is a unique compound that works as an insecticide when it is released by metabolic dealkylation of the parent molecule, partly with the help of mixed function oxidases (MFOs) (Fig.25).

This insecticide works on larvae and adults of many pest species, such as crop mites and insects from the orders Lepidoptera, Coleoptera, Isoptera, Thysanoptera, Orthoptera, Hymenoptera, and Acarina present in vegetables, fruits, cotton, and flowers.

Chlorfenapyr was first registered for use on crops in Israel in 1995. It was then made available in South Africa, Japan, Chile, and Taiwan, where it was tested and found to be effective against Lepidoptera, Coleoptera, and mites. In the US, it was registered in 2001 for more than just controlling field insects. It was also registered to kill termites. In the same way, South Korea, Mexico, Australia, Indonesia, Thailand, and France all registered.

Chlorfenapyr is absorbed mostly through feeding, followed by contact, because of its unique mechanism of action, it controls pests that are resistant to other pesticide classes, and no cases of target site cross resistance have been reported. Chlorfenapyr demonstrates good translaminar mobility in plants. It has very little systemic and/or ovicidal action. Because chlorfenapyr is non-repellent, it has found special application in non-crop control, especially termites, for example, when applied as a barrier treatment around buildings, termites are unable to detect its presence in the soil and hence walk across the treated zone, contacting a deadly amount. In reality, termite control professionals have discovered chlorfenapyr to be a vital tool for removing termites from homes while also offering effective residual protection against future termite attack. Chlorfenapyr is particularly efficient as a spray and in baits for controlling cockroaches, ants, and other home pests that resist irritating pesticides due to non-repellent nature possessed by the compound.

Chlorfenapyr has been shown to be rather benign to natural enemies because to its pro insecticidal qualities, that gets converted to an active form in the insect body by the action of MFOs. For example, research conducted in the United States and Australia found that chlorfenapyr had a considerably lower impact on predatory bugs, parasitic wasps, and spiders in the cotton habitat.

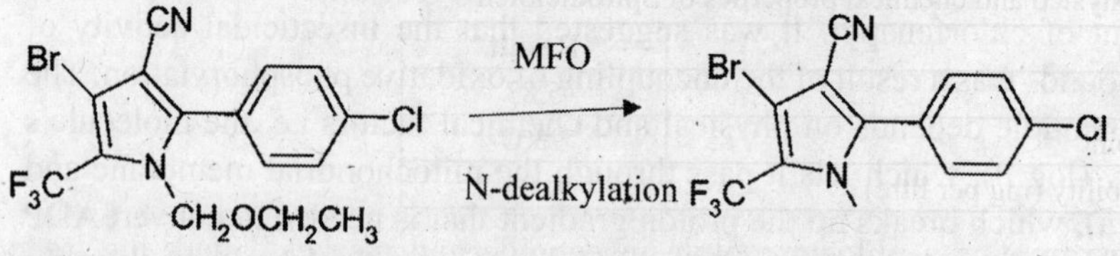

CN Br Cl N F_3C $CH_2OCH_2CH_3$ —MFO→ N-dealkylation CN Br Cl N F_3C

Chlorfenapyr Pyrrole

Fig. 25: Chemical structures of chlorfenapyr and pyrrole

Since chlorfenapyr's commercial use, resistance has been reported in a few cases, primarily metabolic cross-resistance to pyrethroid esterases in Australian *Helicoverpa armigera* bollworms and increased esterase and/or glutathione-S-transferase activity in Australian and Japanese *Tetranychus urticae* mites. Because MFOs bioactivate chlorfenapyr, some pyrethroid-resistant species may develop cross-resistance to chlorfenapyr that has been hypothesized for *Haematobia irritans* hornflies in the USA.

Chlorfenapyr 10 % SC is registered in India to manage DBM (*Plutella xylostella*) and mites in chilli @ 300-400ml/acre under brand name as Intrepid.

25. Tetranic Acid Derivatives

There are three insecticides in this group namely, Spirodiclofen, Spirotetramet & Spiromesifen

1. Spirodiclofen was effective against all developmental stages of mites, including eggs, but did not kill adult mite. Spirodiclofen significantly lowered female adults' fertility, resulting in a significant reduction in the number of eggs laid. Furthermore, female mites exposed to sublethal amounts produced sterile eggs. Notably, the lipid content in treated female adults of *T. urticae* was drastically reduced, indicating that the chemical had interfered with lipid production.

It exhibited superior activity against numerous strains of red spider mite that exhibited a high level of resistance to commercial acaricides, and it has a long-lasting effect and excellent compatibility in field conditions. It is a novel non-systemic foliar acaricide that exhibits long-lasting activity. A new non-systemic foliar acaricide, Envidor, is effective in early to late-season applications and provides outstanding long-lasting activity and its active ingredient is spirodiclofen.

It was first launched in 2002 in Korea under trade name Envidor® (widely used tradename) and later on available in Germnay, USA, Turkey, Japan, China, Brazil and Spain. It is used on citrus, grapes, stone fruits, pome, almonds and nuts. The other trade names of spirodiclofen marketed worldwide are in Japan it is sold as Ecomite on pome fruit, Daniemon in Japan on citrus and Sinwai in Korea.

Table 24: Physico and chemical properties of Spirodiclofen

Properties	**Result**
Log *Pow*	5.83
Melting point	95°C
Water solubility (mg per litre)	0.05

2. Spirotetramet is a phloem mobile insecticide and marketed with the trade name "Movento®". Spirotetramat 240g/L is recommended for use on fruits, vegetables, cotton showing control against sucking pests viz. aphids, psylla, scales, mealybugs, whiteflies and mites with the recommended dose of 0.0048-0.0144% a.i. per litre of water. Spirotetramat has been shown to greatly reduce the lipid content of aphids feeding on leaves treated with this compound. It has also demonstrated good efficacy against neonicotinoid resistant whiteflies. It shows broader spectrum of activity against different aphid species (*M. persicae, A.gossyii*) The juvenile stage of aphids are particularly affected by spirotetramat whereas adults are greatly affected in terms of their fecundity which lead to drastic reduction in population development in fields. It also showed systemic efficacy against aphids and whiteflies when applied through foliar route.

Spirotetramat is a pro insecticide as it is readily transformed into enol form in leaves after penetration into the plant showing better efficacy against aphids. It also protects young leaves due to its systemic properties and well distribution with in the plant

Spirotetramat enol, a pro insecticide, exhibits a significant water solubility and is a weak acid (pKa 4.9). Consequently, the compound is mobile within the plant's symplast (phloem) in accordance with the weak acid hypothesis. Therefore, it has the potential to move both acropetally and basipetally, and it may even provide protection to the plant roots when administered in a foliar manner. Movento is the trade name for the active constituent spirotetramat, the first broad-spectrum phloem mobile insecticide that has been approved for use on cotton, vegetables, fruits, and tea. It has been shown to be effective against sucking pest complex, aphids, scales, mealybugs, and whiteflies.

Though the action at initial is slow, but long lasting and good larvicidal activity is noticed duly protecting new roots and shoots of the plant. The product has a favorable toxicological profile and can be used in IPM. It was first launched in 2007 in Indonesia and later on in USA, China, Mexico, Australia and Turkey.

Physico and chemical properties of Spirotetramat

Properties	Result
Log *Pow*	2.50
Melting point	142°C
Water solubility (mg per litre)	30

3. Spiromesifen: Spiromesifen though particularly kills young immature insects, but also has a strong effect on mite and whitefly adults affecting reproduction and this effect depends on dosage.

Spiromesifen has ovicidal effects on mites, however egg hatch in whiteflies was significantly decreased by transovarial effects upon pre-exposure of female adults to the compound. Against whiteflies resistant to pyrethroids, organophosphates, carbamates, cyclodienes and neonicotinoids as well as Tetranychus strains resistant to abamectin, pyridaben, fenpyroximate, hexythiazox and clofentezine, Spiromesifen was also demonstrated to be rather successful. is a foliar contact insecticide indicated against mite and whiteflies extensively used in USA, Brazil, Mexico, India, Indonesia and sold under the trade name Oberon®.

Developed globally for use on vegetables, fruits, cotton, corn, beans, tea and some ornamentals, Oberon comprises the active component spiromesifen and is a novel foliar insecticide -acaricide. Excellent to good control of whiteflies (Bemisia spp., Trialeurodes spp.), spiromesifen is also quite efficient against mites including spider mites (Tetranychid spp., Tarsonomeid mites, wide mites). It works also powerfully against tomato and pepper psyllids. First registered in Indonesia in 2003, it subsequently showed up in Brazil, Mexico, and USA. Towards predatory mites, spiders, and beneficial insects this chemical is safe.

Table 25: Physical and biological properties of Spiromesifen

Properties	Result
Log *Pow*	4.55
Melting point	98°C
Water solubility (mg per litre)	0.05

Properties	Spirodiclofen	Spiromesifen	Spirotetramat
Spider mites	Yes	Yes	Yes
Whiteflies	No	Yes	Yes
Aphids	Yes	No	Yes

Target crops and pests: It is used to control pests infesting vegetables, fruits, cotton, tea and found effective against whiteflies, highly effective against mites. The product is even controls mites population that are resistant to abamectin, pyridaben, fenpyroximate, hexythiazox, and against whiteflies resistant to organophosphates, carbamates, pyrethroids, pyriproxyfen, and neonicotinoids

Table 26: Efficacy of Tetranic acid derivatives against pests

Product available in market	Crops	Target pests
Spiromesifen 22.90 % SC	brinjal, cotton, apple, chilli, tea, okra tomato	mites and whitefly
Spirotetramat 15.31 %OD	chilli, okra, grapes, cotton, citrus	aphids, thrips, whiteflies, mealybugs, psylla, mites

Table 27: Acute toxicity of Tetranic acid derivatives

Acute toxicity	LD_{50} (mg g^{-1})		
	Spirodiclofen	Spiromesifen	Spirotetramat
LD_{50} for rat (oral)	>2500 mg/kg	>2500 mg/kg	>2000 mg/kg

Spirodiclofen Spiromesifen Spirotetramat

Fig. 26: Structures of Spirodiclofen, Spiromesifen, Spirotetramat

26. Fluoraliphatic Sulfonamides

Sulfluramid is a sulfonamide obtained by the formal condensation of perfluorooctane-1-sulfonic acid with ethylamine. It has a role as an environmental contaminant, a xenobiotic, an acaricide and an insecticide. It is functionally related to a perfluorooctane-1-sulfonic acid and an ethylamine.

27. Pyridine Carboxamide

Flonicamid (N-cyanomethyl-4-trifluoromethylnicotinamide) belongs to a new class of chemistry known as "Pyridinecarboxamide", a selective insecticide discovered by Ishihara Sangyo Kaisha, Ltd and developed jointly with FMC Corporation that was launched in 2005 and is currently registered in more than 40 countries for the control of a broad range of aphid pests as well as some other hemipteran and thysanopteran pests. Flonicamid is a proinsecticide whose bioactive metabolite 4-trifluoromethlynicotinamide (TFNA-AM) disrupts chordotonal organ function in a way that is indistinguishable from the Group 9 insecticide, pymetrozine.

It has a strong systemic and translaminar action and registered as solid formulation i.e. **Flonicamid 50WG** that is found to be effective at 30-40 g a.i./acre. It is recommended to control major sucking insect pests in cotton viz., aphids, jassids @ 60g/acre, whiteflies @ 80g/acre, and in rice to control BPH, GLH WBPH @ 60g/acre, in okra to manage aphids, jassids, whiteflies @ 80g/acre and in brinjal to control aphids, jassids, whiteflies @ 80g/acre. It is registered in India by United Phosphorus Limited and marketed as **"ULLALA"**.

28. Pyridine Azomethines

One compound in this group is **Pymetrozine** which is availabe as Pymetrozine 50%WG (Chess®) recommended control brown plant hopper in rice @ 120g/acre.

Crop	Pest	Dosage/acre		Dilution in Water (liter)	Waiting period (days)
		a.i (g)	Formulation		
Paddy	BPH	60	120g	200	19
Mango	Hoppers	60	120g	400	36

29. Pyripyropenes- Pyripyropene is a class of naturally occurring compounds with potent insecticidal properties. These compounds are derived from the fermentation products of certain strains of the fungus *Mycelia sterilia*, particularly species within the Pyripyropene A, B, and C groups. Afidopyropen, a novel insecticide, is a derivative of pyripyropene A, which is produced by the filamentous fungus ***Penicillium coprobium***. Afidopyropen has strong insecticidal activity against aphids and is currently used as a control agent of sucking pests worldwide.

Afidopyropen 50 g/L DC is registered against sucking pests mainly whitefly and jassids in brinjal, cotton and cucumber and marketed under trade name "SEFINA".

Crop	Pest	Dosage/acre			Waiting period (days)
		a.i (g)	Formulation	Dilution in Water (liter)	
Brinjal	whitefly, jassids	20	400 ml	200 – 300	01
Cotton	whitefly, jassids	20	400 ml	200 – 300	25
Cucumber	whitefly	35 – 20	700 – 400ml	200	05

30. Isoxaline Group

Isoxazoline is a 5-membered heterocycle present in the active compounds of few commercial agro products. The molecular target of isoxazolines is the inhibition of GABA-gated chloride channels in insects. These facts have inspired the use of the isoxazoline scaffold in the design of novel insecticide compounds. The main strategies used for isoxazoline synthesis are either the 1,3-dipolar cycloaddition between a nitrile oxide and an alkene or the reaction between hydroxylamine and an α,β-unsaturated carbonyl compound. Isocycloseram is registered as PLINAZOLIN® technology by Syngenta.

Eg : Fluxametamide, Isocycloseram

1. Fluxametamide belongs to the isoxazoline insecticides which is developed and produced as a racemate by Nissan Chemical Industries Ltd. It has broad-spectrum bioactivity to many kinds of pests, including Lepidoptera, Acarina, Thysanoptera, and Diptera. These pests are harmful to vegetables, such as cabbage and Chinese cabbage, cotton, tea plant, and soybean. Fluxametamide is a novel type of ligand-gated chloride channel (LGCC) antagonist, which mainly affects γ-aminobutyric acid gated chloride channel (GABACl) of pests with novel mode of action. Thus, the application of fluxametamide could help delay the development of insecticide resistance, and it has been in the registration process in several countries.

Enantioselectivity in Acute Toxicity to Honeybees: The decline in the number of pollinators, such as honeybees, has become a global concern. So it is necessary and important to evaluated the risk of fluxametamide and its enantiomers to honeybees. The results of acute toxic effects on *Apis mellifera* L.are summarized in Table 29. The result indicated that the acute contact toxicity of *rac*-fluxametamide (LD_{50}=1.50 μg/bee) is far more less than that of other two GABACl antagonists, *rac*-fipronil (LD_{50}=0.0039 μg/bee) and *rac*-ethiprole (LD_{50}=0.0187 μg/bee). Furthermore, the acute toxicity increased with the order of *R*-(-)-fluxametamide, *S*-(+)-fluxametamide and *rac*-fluxametamide. The *rac*-fluxametamide exhibited 4.3 times higher acute toxicity than *S*-(+)-fluxametamide for honeybees, and *S*-(+)-fluxametamide was shown to be over 30 times more toxic than *R*-(-)-fluxametamide to honeybees. *Rac*-fluxametamide showed the highest acute toxicity towards honeybees.

Table 28: Acute contact toxic (LD_{50}, 48 h) of *rac*-fluxametamide and its enantiomers against honeybees

Compound	**LD_{50} (95% CIa) μg a.i./bee**	**R^2**
rac-fluxametamide	1.50 (1.30-1.74)	0.9821
S-(+)-fluxametamide	6.47 (5.94-7.04)	0.9938
R-(-)-fluxametamide	>200	

*95% confidence intervals. Correlation coefficients (R^2)

Fluxametamide 10%EC is registered in India against leaf hopper, thrips, fruit and shoot borer in brinjal; DBM, tobacco caterpillar, semi looper in cabbage; thrips, fruit borer & tobacco caterpillar in chilli, leaf hopper, thrips & fruit borer in Okra; spotted pod borer & pod borer in Redgram and thrips, fruit borer in tomato @ 160ml/acre with a waiting period of 5 days and sold under the brand name "GRACIA".

2. Isocycloseram

Mode of action

Isocycloseram is a novel isoxazoline insecticide and acaricide with activity against lepidopteran, hemipteran, coleopteran, thysanopteran and dipteran pest species. Isocycloseram selectively targets the invertebrate Rdl GABA receptor at a site that is distinct to fiproles and organochlorines. The widely distributed cyclodiene resistance mutation, A301S, does not affect sensitivity to isocycloseram, either in vitro or in vivo, demonstrating the suitability of isocylsoseram to control pest infestations with this resistance mechanism. The detailed studies demonstrated that the binding sites relevant to the insecticidal activity of avermectins and isocycloseram are distinct. Isocycloseram was shown to compete for binding with metadiamide insecticides related to broflanilide. In addition, a G335M mutation in the third transmembrane domain of the Rdl GABA receptor, impaired the ability of both isocycloseram and metadiamides to block the GABA mediated response. As such the Insecticides Resistance Action Committee (IRAC) has classified isocycloseram in Group 30 "GABA-Gated Chloride Channel Allosteric Modulators".

Isocycloseram 9.2% DC (10% w/v)					
Crop	**Pest**	**Dosage/acre**			**Waiting period (days)**
		a.i (g)	**Formulation in ml**	**Dilution in Water (liter)**	
Brinjal	shoot and fruit borer	24	240		
Cabbage	leaf Feeder, DBM	8-12	200-300	200	10
Chilli	yellow mites	8	200	200	7
	thrips and fruit borer	24	240		
Cotton	jassids	8	200	200	37
	thrips, bollworm	24	240		
Red Gram	gram pod borer, spotted pod borer	20-24	200-240	200	58
Groundnut	leaf miner, leaf feeder, thrips, Jassids	20-24	200-240	200	48
Soybean	leaf worm, semi-looper, girdle beetle, stem fly	24	240	200	35
Isocycloseram 18.1% SC (20 % w/v)					
Rice	leaf folder	8	40	200	29
	stem borer	24	120		
Maize	fall armyworm, stem borer	24	120	200	48

31. Mesoionic Group

The term "mesoionic" refers to the chemical nature of these compounds, which possess both positive and negative charges within the same molecule, giving them an overall neutral charge. This unique feature allows mesoionic insecticides to interact effectively with biological targets within insects while minimizing their impact on non-target organisms and the environment.

The origins of the discovery of triflumezopyrim lie in a fungicide discovery program which led to proquinazid (5). The program began with pyridopyrimidinone lead 1 (Figure 2) which morphed into the quinazolinone proquinazid. The chemistry used to propare the initial lead compounds shown in Figure 2 produced the desired O-alkylated products in generally good yield but also produced in many cases a much more polar yellow compound observed as a baseline yellow spot in tlc. Some of the chemists on the program isolated these more polar compounds and characterized them as the N-alkylated mesoionic products.

Pexalon™ insecticide contains Triflumezopyrim active ingredient. It belongs to a novelmesoionic chemical class which is chemically distinct from any other existing class of insecticides. Pexalon™ provides outstanding and long-lasting control of plant hoppers and leafhoppers in rice, providing excellent plant protection from direct pest damage and hopper transmitted virus diseases. Due to its unique mode of action, Pexalon™ is a useful tool in rice hopper insecticide resistance management programs.

Mode of action: Triflumezopyrim, a newly commercialized molecule from DuPont Crop Protection, belongs to the novelclass of mesoionic insecticides. This study characterizes the biochemical and physiological action of this novel insecticide. Using membranes from the aphid, *Myzus persicae*, triflumezopyrim was found to displace 3H-imidacloprid with a Ki value of 43 nM with competitive binding results indicating that Triflumezopyrim binds to the orthosteric site of the nicotinic acetylcholine receptor (nAChR). In voltage clamp studies using dissociated *Periplanetaa mericana* neurons, triflumezopyrim inhibits nAChR currents with an IC50 of 0.6 nM. Activation of nAChR currents was minimal and required concentrations_100 mM. *Xenopus* oocytes expressing chimeric nAChRs (*Drosophila* a2/chick b2) showed similar inhibitory effects from triflumezopyrim. In *P. americana* neurons, co-application experiments with acetylcholine reveal the inhibitory action of triflumezopyrim to be rapid and prolonged in nature. **Such physiological action is distinct from other insecticides in IRAC Group 4 in which the toxicological mode of action is attributed to nAChR agonism**. The inhibitory action of Triflumezopyrim is unlike other product acting at the acetylcholine-binding site and elicits a distinct physiological response at the α2-nAChR subunit. Mesoionic insecticides act via inhibition of the orthosteric binding site of the nAChR despite previous beliefs that such action would translate to poor insect control. Triflumezopyrim is the first commercialized insecticide from this class and provides outstanding control of hoppers, including the brown plant hopper, *Nilaparvata lugens*, which is already displaying strong resistance to neonicotinoids such as imidacloprid.

Movement in Plants- Triflumezopyrim

A. Translaminar movement: After foliar application, Triflumezopyrim active is able to move across the plant cuticle and then penetrates into the leaf tissue and moves locally within the leaf. This allows the product to move and reach the pests where they are located. This bioavailability whether through ingestion or contact activity, ensures excellent and long lasting control of pest populations. In addition, the translaminar movement ensures rain fastness, which helps Pexalon™ insecticide and other Triflumezopyrim based products to record relatively better wash-off. As is typical of xylem mobile compounds, when Triflumezopyrim penetrates the leaf tissue the active ingredient is translocated via the xylem towards the leaf tip.

B. Root-uptake systemic movement: Application of products powered by Triflumezopyrim active to soil media in nursery boxes allows the rice seedling roots to take up the product and redistribute it throughout the plant. The bioavailability of Triflumezopyrim within the plant provides excellent and long-lasting protection of rice crop against damage by pests

Protection by Pexalon™ insecticide against virus diseases vectored by the major rice hoppers in Asia-Pacific, when used as part of a virus management program.

Virus name	***Abbreviation***	***Insect vector***	***Type of transmission***
Rice ragged stunt virus	RRSV	*Nilaparvata lugens*(BPH)	Persistent
Southern rice black streaked dwarf virus	SRBSDV	*Sogatella furcifera*(WBPH)	Persistent
Rice stripe virus	RSV	*Laodelphaxstriatellus*(SBPH)	Persistent
Rice tungrobacillifrom virus	RTBV	*Nephotettix spp.* (GLH)	Non-Persistent

Eg: Triflumezopyrim 10 % SC; Triflumezopyrim 20 % WG

Crop	**Pest**	**Dosage/acre**			**Waiting period (days)**
		a.i (g)	**Formulation**	**Dilution in Water (Liter)**	
Triflumezopyrim 10%SC					
Paddy	BPH, WBPH	10	95ml	200	21
Triflumezopyrim 20% WG					
Paddy	BPH, WBPH	10	50g	200	21

A. Favourable mammalian toxicology profile: In all acute, sub-chronic and chronic toxicology studies conducted, triflumezopyrim active showed low toxicity to test animals which indicates a good fit in crop management by allowing the establishment of a short pre-harvest interval (PHI).

Table 29: Mammalian toxicity profile of Triflumezopyrim active

Acute oral LC_{50}	4930 mg/kg
Dermal LD_{50}	>50 00 mg/kg
Inhalation LC_{50}	>5.0 4 mg/L
Dermal irritation	Not a dermal irritant
Eye irritation	Slight eye irritation
Ames mutagenicity	Negative
In-vitro chromosomal aberration	Negative
In-vivo micronucleus	Negative

B. Product degradation profile: Triflumezopyrim active breaks down under a variety of field and laboratory conditions. Under field conditions, Triflumezopyrim remained mainly in the upper soil horizon. Triflumezopyrim poses low risk for bio-accumulation or bio-magnification in the soil.

C. Low toxicity to non-target organisms: Data from several studies indicate that Triflumezopyrim active has low impact on non-target organisms such as fish, *Daphnia* and earthworms, and beneficial arthropods such as pollinators and arthropod natural enemies of rice hoppers. The low risk to non-target organisms coupled with the degradation in the environment makes Triflumezopyrim and branded products containing Triflumezopyrim ideal for rice production systems.

Environmental toxicology of Triflumezopyrim active	
Species/toxicity test	**End point**
Bobwhite quail -Acute oral LD_{50}	2109 mg/kg
Bobwhite quail - Dietary LC_{50}	>5620 ppm
Rainbow trout -96 hrs. LC_{50}	>107 mg/L
Carp - 96 hrs. LC_{50}	>100 mg/L
Daphnia magna – 48 hrs. EC_{50}	>122 mg/L
Honeybee acute oral (LD_{50})	0.51 μg a.i./bee
Earthworm acute- LC_{50} at 7 d & 14 d	>1000 mg/kg soil

Source: DuPont Crop Protection, Stine-Haskell Research Centre (SHRC), Newark DE, USA

D. Physical and chemical properties: Triflumezopyrim active possesses ideal physical-chemical properties and this coupled show high intrinsic potency on rice plant hoppers and leafhoppers and with a unique mode of action make these products powered by Triflumezopyrim superior additions to rice pest management programs and successful crop production.

Table 30: Physical and chemical properties of Triflumezopyrim technical active ingredient

Chemical Class	**Mesoionic; Pyridopyrimidine-dione compound**
Solubility	Water: 64.6-76.1 g/L at 20 ° C Dichloromethane: 71.89-116.5 g/L Acetone : 65.9-91.5 g/L
Partition coefficient in octanol/water (Log Pow)	1.24
Volatility	non-volatile

Source: DuPont Crop Protection, Stine-Haskell Research Centre (SHRC), Newark DE, USA

32. Pyridazine Pyrazolecarboxamides or Pyridazine Amides

Pyridazine pyrazolecarboxamides (PPCs) are a novel insecticide class discovered and optimized at BASF. Dimpropyridaz is the first PPC to be submitted for registration and controls various aphid species as well as whiteflies and other piercing sucking insects.

Eg: Dimpropyridaz 120 g/l SL (Efficon®)

Mode of action

Dimpropyridaz is being developed by BASF as the first PPC, a novel insecticide class discovered through in vivo screening and optimized at BASF. Picolinic thioamide 2, the original hit active only on aphids, was the starting point for an extensive synthesis program that yielded PPCs with expanded spectrum, including whiteflies and other piercing-sucking insects leading to the selection of dimpropyridaz for commercialization. We show that PPCs are a new class of chordotonal organ modulator insecticide with a mode of action that is different from IRAC groups 9 and 29. PPCs containing a secondary amide linker are significantly more active than their corresponding tertiary amide versions in ex vivo assays of chordotonal organ function. Consistently, we demonstrate that dimpropyridaz is actively metabolized in green peach aphid through N-de-ethylation and hydroxylation at the 3-methylbutan-2-yl side chain. Stretch activation of chordotonal organs involves mechanotransduction and amplification processes, with Ca^{2+} influx through Nan–Iav TRPV channels leading to action potential generation. We show that dimpropyridaz acts upstream of TRPV channels, independently of the target of flonicamid, decreasing calcium levels in chordotonal organs and, thus, disrupting chordotonal organ function.

Crop	Target pest	Dosage/acre			Waiting period (days)
		a.i (g)	Formulation in ml	Dilution in Water (Liter)	
Dimpropyridaz 120g/l SL					
Brinjal	whitefly	32 – 40	360 – 400	200	03
	jassids, aphids	33.60	280	200	
Cotton	whitefly	108 – 120	360 – 400	200	31
	jassids, aphids	33.60	280	200	
Cucumber	whitefly	32 – 40	360 – 400	200	3
	jassids, aphids	33.60	280	200	
Tomato	whitefly, jassids, aphids	32 – 40	360 -- 400	200	3
Chilli	whitefly, aphids	32 – 40	360 – 400	200	3

33. Unclassified Compound

The insecticide in this group is Pyridalyl and marketed as Pyridalyl 10EC under the brand name of Sumipleo® and found to be effective against bollworms in cotton, fruit and shoot borer in okra, diamond back moth in cabbage etc.

Pyridalyl 10 % EC

Crop	Target pest	Dosage/acre			Waiting period (days)
		30 – 40	300 – 400	.200 – 300	07
Okra	Fruit & shoot borer	20 – 30	200 – 300	200 – 300	03
Cabbage	Diamond back moth	20 – 30	200 – 300	200 – 300	03

34. Insect Growth Regulators (IGRs)

Insect Growth Regulators, or IGRs, attack the insect's endocrine system, which produces the hormones needed for growth and for development into an adult form. IGRs interfere with development and disrupt metamorphosis and reproduction.

Hormones involved in growth and reproduction of insects are

1. **Juvenile hormones**: These hormones prevent insect from moulting to next instar
2. **Moulting hormones or ecdysones**: These hormones are involved in re-absorption of old cuticle and also in deposition, hardening and tanning of new cuticle

Insects poisoned with IGRs cannot molt or reproduce, and eventually they die. Many of the currently available IGRs mimic a special protein called juvenile hormone. In a normal insect, juvenile hormone is circulated throughout the insect's body and "tells" the insect to stay in its current stage. After a certain amount of time, the insect stops producing juvenile hormone, and the insect metamorphoses, or changes, into its next life stage. When an insect is poisoned by an IGR that mimics juvenile hormone, the insect doesn't receive the signal to metamorphose because, even though the insect may have stopped producing juvenile hormone, the IGR is still circulating throughout its body and sending the signal to stay in the current stage i.e. juvenile hormones keep insects at specific instars when present and allow maturation when absent.

The derivatives of juvenile hormones have been synthesized and used for chemical control of insects, which are commonly referred as juvenile hormones analogues or juvenile hormones mimics, or juvenoids. These juvenile hormone mimics are compounds bearing a structural resemblance to the juvenile hormones of insects. These are lipophilic sesquiterpenoids containing an epoxide and methyl ester groups. Eg: Methoprene, Kinoprene, Hydroprene, Fenoxycarb and Pyriproxifen.

The juvenile hormone, methoprene bears a close structural resemblance to juvenile hormones, and fenoxycarb, which possesses a phenoxybenzyl group instead of a carbon chain with an epoxide. Both these compounds are soluble in organic solvents and have extremely low toxicity to mammals. The timing of application is important for successful control.

Another hormone important in metamorphosis is ecdysone. These interferes with the production of ecdysone, causing the insect to be unable to molt. They are steroid-like molecules but too expensive to be synthesized and used as control agents. However, some new insect growth regulators which are not based on steroids have proven effective against Lepidoptera. Eg: Chromafenozide, Halofenozide,Tebufenozide, Methoxyfenozide.

35. Chitin Synthesis Inhibitors

Chitin synthesis inhibitors (CSI), which interfere with chitin biosynthesis in insects and are safe for most nontarget organisms, represent one major class of compounds in integrated pest management (IPM). Benzoyl phenyl ureas (BPUs, IRAC group 15) are the most commonly used chitin synthesis inhibitors. The selectivity of benzoylureas is achieved by their interference with chitin deposition. To date, 15 benzoÿlurea chitin synthesis inhibitors have been commercialized after more than four decades of research and development.

I). Benzyl phenyl ureas (BPUs): Benzoyl phenyl ureas possess a number of halogen substituents. Diflubenzuron (Dimilin®) is the prototypical compound in this group though in this series, second generation compounds also exist. Water solubility of these compounds is extremely low (<1ppm). The insects exposed to these compounds are unable to form normal cuticle because the ability to synthesize chitin is lost. These benzoylureas are taken up more by ingestion than by contact. The research on the analogues of DU19111, the first commercial product diflubenzuron was discovered by Philip-Duphar B.V. introduced to the market with the trade name Dimilin® in 1975.

Diflubenzuron indeed inhibits the process of incorporation of N-acetylglucosamine into insect chitin in vivo, but there are no direct biochemical effects on enzymes, receptors, or intracellular organelles under a cell-free system.

By virtue of the unique action mechanism and low toxicity for bees, birds, and humans along with their high biological activity against pests, dichlorbenzuron and penfluoron were developed commercially as insecticides at Philip-Duphar B.V. Co. and Thompson-Hayward Chemical Co. Penfluoron was found to possess a high larvicidal activity, against *P. brassicae* in laboratory tests. At the same time, chlorbenzuron was discovered at JiangSu Institute of Ecomoes and Soochow University, China. Teflubenzuron is a multihalo substituted BPU and was introduced in Thailand by Celamerck GmbH & Co. KG (now BASF). Teflubenzuron exhibited better ovicidal and larvicidal activities against the African armyworm (*Spodoptera exempta*) than diflubenzuron. The most successful benzoylurea compound after diflubenzuron is triflumuron, discovered by Bayer and it is the first commercial product of the second generation BPUs.

Triflumuron, a multihaloalkoxy substituted BPUs at the aniline moiety (B ring) belonging to the first generation of benzoylphenylureas displayed both larvicidal and ovicidal activities and showed broad-spectrum larvicidal activity against cotton bollworm (*Helicoverpa armigera*), in comparison teflubenzuron and diflubenzuron (first-generation BPUs) where were inactive.

Hexaflumuron was intended to be used in products to control termites by Dow AgroSciences. In the following years, novaluron, lufenuron and noviflumuron were developed successively by Makhteshim-Agan, Ciba-Geigy (now Novartis), and Dow AgroScience, respectively. Bistrifluoron was obtained from more than 2000 BPUs at Dongbu Hannong Chemical Co., Korea. It is active against lepidopteran pests and especially whitefly as an insect growth regulator. The third-generation BPUs exhibit stronger topical insecticidal activity and broader spectrum larvicidal activity (high acaricidal activity). Flufenoxuron was developed at Shell Research Ltd. (now BASF AG), a broad-spectrum acaricide (*Tetranychus urticae*) and insecticide that is effective against Lepidoptera, Homoptera, Diptera and Hemiptera pests. Flucycloxuron combines a high insecticidal activity with a high acaricidal activity against both tetranychid and eriophyid mites. It is the result of an optimization program with benzoylphenylureas on acaricidal activity at Philip-Duphar B.V. (now Platform Specialty Products Co.). Similarly, Chlorfluazuron was launched in 1989, which was discovered and developed by Ishihara Sangyo Kaisha Ltd and showed more larvicidal activity against specific target insects viz. *Spodoptera litura*, and diamondback moth (*Plutella xylostella*) larvae and was 10-fold more inhibitory than lead compound, diflubenzuron. Fluazuron was discovered and developed at Ciba-Geigy AG (now Novartis). The chemical structure of fluazuron is very similar to that of chlorfluazuron, but the insecticidal spectrum is different controlling cattle tick (*Boophilus microplus*).

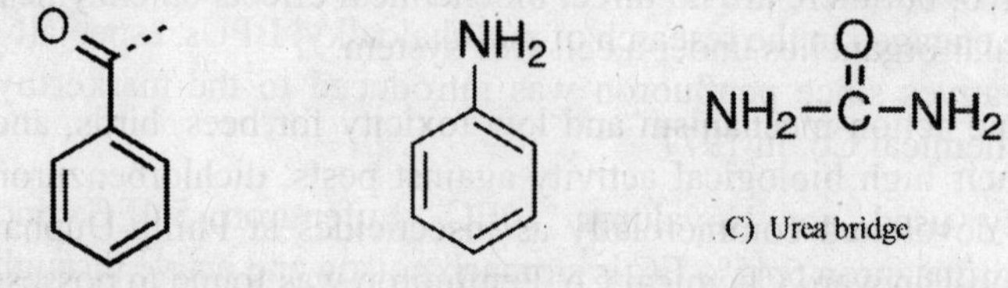

A) Benzoyl ring B) Aniline ring C) Urea bridge

Fig. 27: Benzoylphenylureas can be separated into three parts namely, the benzoyl ring (A), aniline ring (B), and urea bridge (C)

Structure–Activity Relationships (SARs): The benzoylphenylureas can be separated into three parts namely, the benzoyl ring (A), aniline ring (B), and urea bridge (C) (Figure 27). At the beginning of BPU research, it was established that a 2-Cl, 2, 6 -diCl, or 2, 6-diF substitution pattern on the benzoyl moiety (A) was required for optimal activity, with the 2, 6-diF substitution pattern being most effective. The larvicidal activities against *P. brassicae* of the 2, 6 -difluorobenzoylphenylureas were about 20 times higher than the activities of the 2, 6-dichlorobenzoylphenylureas. These results suggested that the conformation having an angle of 30° between the –CO (NH)– moiety and the benzoyl ring (A) has optimum larvicidal activity. As per the bioisosterism principles, the benzoyl ring (A) of BPUs could be replaced by the aromatic heterocycles or cyclic alkyl substituents (such as pyridine, pyrazole, furan, and cyclopropyl groups, but the larvicidal activities decreased significantly. In terms of the urea bridge (C) moiety, the larvicidal activity of thiourea derivatives with the same substructure was worse than that of the urea derivatives. Research has shown that N′-substituted benzoylureas can improve the poor solubility of BPUs in organic solvents. Generally, N′-alkyl products displayed better solubility but worse larvicidal activity in comparison with their parent structure.

The structural optimization of the aniline moiety (B ring) is the focus for most researchers, and various functional groups on the aniline (B ring) produced three generations of commercial benzoylphenylureas. Research on the QSARs of BPUs has shown that different functional groups and pests have different SARs. In general, bulky groups at the 2,6-position of the aniline moiety (B ring) are adverse to the larvicidal activities, and electron-withdrawing groups at the 4-positon of the aniline moiety (B ring) are helpful to the insecticidal effects. The replacement of aniline moiety with aromatic heterocycles keeping the insecticidal activities same, or reduced in most conditions.

The earliest research was halo-substitution at the aniline moiety (B ring after the discovery of DU19111, it was found that the larvicidal activity of alkyl-substituted BPUs was worse than that of halo-substituted compounds, and 4-alkyl-substituted BPUs displayed better larvicidal activity than multialkyl-substituted derivatives. Many companies were engaged in the research of multihaloalkyl BPUs, especially multifluoroalkyl derivatives since penfluoron was introduced to the market by ThompsonHayward Chemical Co. in 1977

The IGRs commonly used are Novaluron 10EC, Lufenuron 5.4EC and Buprofezin25SC. Chlorfluazuron 5.45% EC is stomach poison and active through ingestion, available as liquid formulation, It is used in cabbage to manage diamond back moth @ 600ml/acre and in cotton to manage *Spodoptera, Helicoverpa* @ 600-800 ml/acre. It is marketed as Atabron® by United Phosphorus Limited.

Lufenuron 5.4%EC (Match®) is generally recommended to manage DBM in crucifers, Pod borer, pod fly in pulses, *Helicoverpa* in cotton, fruit borer in chilli @ 320 ml/acre etc.

Novaluron 10%EC (Rimon®) is used to manage fruit borers in tomato, pod borers in bengalgram, DBM in cabbage, bollworm in cotton @ 300-400 ml per acre, whereas in chilli @ 150ml/acre against fruit borers.

II). Thiadiazinones

Buprofezin 25%SC (Applaud®) is used to manage hoppers in paddy @ 330 ml/ acre, mango hoppers @ 1-2ml/litre and in grapes to control mealybug @ 400-600ml/acre.

Environment fate and Ecotoxicology

Most BPUs suffer a rapid degradation in both soil and water by photolysis and hydrolysis (Table 32). The half-life of benzoylureas in the environment varies by type of medium such as soil, water, or sunlight. The half-life of diflubenzuron ranges from hours to months, that of novaluron in tropical soils ranges from 11 to 59 days and noviflumuron and chlorfluazuron do not readily break down in the environment. The metabolites of BPUs may have adverse effects to humans. For example, diflubenzuron is quickly degraded in the environment mainly by hydrolysis and photodegradation,

Table 31: Physicochemical Properties of Commercial Benzoylphenylureas

Compound	solubility in water ($mg\ L^{-1}$) at 20 °C	melting point (°C)	log *P*[a]	Soil degradation DT_{50}[b] (days)	Aqueous photolysis DT_{50}[c] (days)	Aqueous hydrolysis DT_{50}[d](days)
diflubenzuron	0.08[e]	227.6	3.89[f]	3	80	96
teflubenzuron	0.01	225.0	4.30	92	10	stable
triflumuron	0.04	194.0	4.90	22	32.8	stable
hexaflumuron	0.027	203.5	5.68	57	6.3	stable
novaluron	0.003	177	4.30	72	Stable	stable
lufenuron	0.046	169.1	5.12	16.3	0.75	stable
flucycloxuron	0.001	143.6	6.97	208	18	28
chlorfluazuron	0.016	222.6	5.80	90	Stable	stable
bistrifluoron	0.03	171.5	5.74	80.5	10	stable
noviflumuron	0.194	156.2	4.94	250	-	stable
flufenoxuron	0.0043	167.0	5.11	42	6	267
fluazuron	0.02	219.0	5.10	-	-	-

[a]Octanol–water partition coefficient(P) at pH 7, 20°C.; [b]Aerobic. [c]At pH 7. [d]At 20°C and pH7. [e]≤50-low; 50–500= moderate;>500-high. [f]2.7-low bioaccumulation,2.7-3.0=moderate; <30- non persistent high. [g]365, very persistent. [h]30, stable. [i]365, very persistent. [j]–,no data. K Calculated Clog P value by Chem Bio Draw Ultra 12.0. l pH sensitive: stable at pH 5;8.7 days at pH9; at pH 7 95%of substance remained after 30days. [m]Stable at pH 5 and 7, DT50=57 days at pH 9, all at 25°C. [n]Stable at pH5. [o]DT50=139days, natural light,40 deg N. [p]pH sensitive: stable pH5–7, DT50=101days at pH 9,25°C. [q]Stable pH 5–9. [r]Stable pH 5–9, 20°C. [s]pH sensitive: stable in acid and neutral media, DT50=19 days at pH 9. [t]Not sensitive to pH.

Benzoylurea insecticides possess favorable environmental properties and low acute toxicology to mammals and birds as well as crop pollinators such as bees (Tables 33). These specific features usually provide a large margin of safety with regard to ecological and toxicological effects. Bioconcentration factor (BCF) can be expressed as the ratio of the concentration of a chemical in an organism to the concentration of the chemical in the surrounding environment. The BCF is a measure of the extent of chemical sharing between an organism and the surrounding environment. Flufenoxuron (BCF = 33856) possessing high bioconcentration potentials was banned in the European Union in 2011 due to its high potential for bioaccumulation in the food chain and high risk to aquatic organisms. Flufenoxuron is marketed as having "high persistence" in the environment, and it does not biodegrade easily,

Most BPUs have low to moderate toxicity to fish; however, BPUs exhibit high toxicity to aquatic invertebrates and crustaceans because these chitin synthesis inhibitors may adversely affect nontarget organisms including beneficial insect species and crustaceans which possess chitin as a component of the exoskeleton. For example, BPU viz. teflubenzuron show have very high toxicity against *Daphnia magna* (EC_{50}=0.0028). (Table 2). Hexaflumuron application had significant effects on *C. septempunctata* in cotton cultivation, suggesting potential risks to beneficial arthropods.

Table 32: Ecotoxicology of Commercial Benzcyl phenyl ureas

Compound	Acute toxicity						
	BCF[a]	Mammals oral LD_{50}	Birds LD_{50} (mg kg^{-1}) (*Colinus virginianus*)	Fish LC_{50}(96h) (mgL^{-1}) (*Oncorhynchus mykiss*)	aquatic invertebrates EC_{50}(48h) (mgL^{-1}) (*Daphnia magna*)	earthworms LC_{50}(14days) (mg kg^{-1}) (*Eisenia fetida*)	honeybees LD_{50}(48h) ($\mu gbee^{-1}$)
diflubenzuron	320[b]	>4640[c]	>5000[d]	>0.13[e]	0.0026[f]	>500[h]	>25(oral)[i]
teflubenzuron	640	>5038	>2250	>0.0065[l]	0.0028	>500	>72(oral)
triflumuron	612	>5000	561	>0.021[l]	0.0016	>500	>200(contact)
hexaflumuron	4700	>5000	2000	100[l]	0.0001	880	0.1(oral)
novaluron	2091	>5000	>2000	>1.0	0.058	>1000	>100(oral)
lufenuron	5300	>2000	2000	>29[l]	0.0013	>500	>197(oral)
flucycloxuron	-	>5000	>2000[n]	>100	>0.00027	1000	>100(contact)
chlorfluazuron	-	>8500	>2510[n]	>300[o]	0.000908	>1000	>100(oral)
bistrifluoron	2414	>5000	>2250	>0.5[o]	>1.2	>1000	>100(contact)
noviflumuron	3000	>5000	>2000	1.8	3.11	>10000	>100(contact)
flufenoxuron	33856	>3000	>2000	>0.0049	0.000043	>500	>100(contact)
fluazuron	-	>2000	>2000[n]	15	0.0006	1000	

[a]Bio concentration factor. [b]<100-low potential; 100-5000= threshold for concern ;>5000- High potential. [c]>2000-low;100-2000=moderate;<100-high. [d]>2000-low ;100-2000=moderate;< 100-high; [e]>100, low;0.1-100=moderate;<0.1-high; [f]>100-low;0.1-100=moderate;<0.1-high; h>1000-low;10-1000=moderate; <10-high. [i]>100-low;1-100=moderate;<1-high. [l]Lepomis macrochirus. [n]*Anas platyrhynchos*. [O]Cyprinidae.

36. Fumigants

The fumigants are small, volatile, organic molecules that become gases at temperatures above 40°F. They are usually heavier than air and commonly contain one or more of the halogens. Most are highly penetrating, reaching through large masses of material. Fumigation at lower temperatures requires a higher dosage rate for a longer exposure period than fumigation at higher temperatures.

The common fumigants are Hydrogen phosphide, Methyl bromide, Ethylene dichloride, Hydrogen cyanide, Sulfuryl fluoride, and the familiar home-use moth repellents, Napthalene crystals and Paradichlorobenzene crystals. The use of phosphine gas (PH_3) has also replaced methyl bromide in a few applications, primarily for insect pests of grain and food commodities. The use of aluminum or magnesium phosphide pellets, reacts with atmospheric moisture to produce the gas, phosphine but the phosphine gas is very damaging to fresh commodities and is highly adsorbed onto oil, thus does not perform as a soil fumigant.

A fumigant can be called an ideal fumigant if it meets the following conditions

a) It should be highly toxic to target pest
b) Properties of fumigant should not change under the action of environmental factors viz. temperature, humidity, light etc.
c) It should not change or affect the nutritive properties of grain and its products
d) It should not leave any harmful residues after fumigation
e) No unpleasant odour should be left after fumigation
f) It should not corrode metals and damage the storage container material
g) It should not cause fire or explosion
h) It should have good diffusion properties
i) It should not get absorbed by the product being disinfested and gets quickly dissipated

Classification of fumigants

The fumigants are classified broadly into three categories

a) Solid fumigants: Fumigants in the form of tablets/pellets/powder form react with atmospheric water vapor to release toxic gas eg. Aluminium phosphide tablets
b) Liquid fumigants: Fumigants having high boiling point and exist in liquid form at normal room temperature. Eg. EDCT mixture, Ethylene dibromide (EDB), Carbon disulphide (CS_2)
c) Gaseous fumigants : Fumigants having low boiling point and highly volatile at normal room temperature exist in gaseous form. Such chemicals are stored in metallic cylinders strong enough to withstand pressure exerted by the gas. Eg. Methyl bromide.

Phosphine/Hydrogen Phosphide (Ph_3)/Phostoxin

Phosphine contains aluminium phosphine powder which reacts slowly with moisture in the air. It has approximately the same gas weight as air (1.18). It is generally regarded as non-flammable (Spark Point 100°C) and toxic (Permissible limit for man, 0.05ppm). Because of its slow speed of activation (it generally takes 1-3 hour to become effective) and the ease of handling as tablets, it is considered a fumigant.

Chemical formula	**HP_3**
Odour	Garlic
Boiling Point	-87.4°C
Freezing point	-133.5°C
Specific gravity	Gas 1.214
Lowest Explosion Point	1.79% by Volume in air

Phosphine is an effective fumigant used to control stored grain insect pests and prepared foods. Aluminium phosphide liberates phosphine on reaction in atmosphere. The aluminium phosphide tablet also contains ammonium carbamate which releases CO_2 and ammonia which reduces the risk of explosion at the time of diffusion of gas from the tablet.

$$Alp + 2NH_4OC(O)\,NH_2 + 3\,H_2O = PH_3 + Al(OH)_3 + NH_3 + CO_2$$

Aluminium phosphide tablets are of 3 and 1 gram which are capable of liberating 1/3 in weight as phosphine gas i.e. a 3gram tablet releases 1 gram of gas. It is sold under the brand names such as Celphos, Quickphos etc. It is formulated as tablets, pellets and sachets of powder with materials such as ammonium carbamate, urea and paraffin to regulate release of fumigant and suppress flammability. It is ineffective at low temperature, below 12°C it cannot be used. The threshold limit value time weighted average for an eight-hour daily exposure in a five-day week is set at 0.3ppm. under no circumstances, phosphine be used in vacuum fumigation as the compound is unstable at reduced pressures. High moisture seed should not be fumigated as the viability germination capacity is insignificant of moisture content is >12%.

Status of Aluminium phosphide: Restricted fumigant

Details of Restrictions: The Pest Control Operations with Aluminium Phosphide may be undertaken only by Govt./Govt. undertakings / Govt. Organizations / pest control operators under the strict supervision of Govt. Experts or experts whose expertise is approved by the Plant Protection Advisor to Govt. of India

The production, marketing and use of Aluminium Phosphide tube packs with a capacity of 10 and 20 tablets of 3 g each of Aluminium Phosphide are banned completely. (www.cibrc.nic.in)

Table 33: Aluminum Phosphide 56 % (3g Tablet, 10g Pouch) - Recommendations for use

Commodity type	Pests	Dose	Exposure period	Aeration Waiting period
Stored Whole Cereals and Seed Grains Millet, Pulses Dry Fruits, Nuts Spices & Oil Seeds	Rice Weevil Lesser Grain Borer, Khapra Beetle, Rust Red Flour Beetle, Saw Toothed Grain Beetle, Caddle Beetle, Drug Store Beetle, Cigarette Beetle, Pulse Beetle	3 tablets /ton or 150 gm/100 m^3 or 10 gm Pouch/ton of commodity	Minimum 05 Days (*Sitophilus oryzae*) or 07 Days (*Trogoderma granarium*)	One hour of partial aeration in case non- polyethylene packed commodities allowed by 6-8 hrs of full aeration. For polyethylene packed commodities minimum aeration period is 48 hrs. The waiting period for the release of stock is 48hrs in both the cases. Recommendation for bag stock 15 days.
Milled Products: De-oiled Cakes, Rice Bran Flour, Grain Animal & Poultry Food Split Pulses (Dal) & other Processed Food	Long Headed Floor Beetle, Coffee Borer, Dried Fruit Beetle, Flat Grain Beetle, Carpet Beetle	03 tablets/10 gm per ton or 225 gm/100 m^3	05 days	Aeration is waiting Period 07 days to be checked PH3 detector strips.
Empty Godowns & Sheds	Rice Moth, Almond Moth, Mites, Fruit Fly, Granary Weevil, Caddle or Flour worm, Red Flour Beetle, Indian Meal Moth, Larger cabinet Moth, Wheat Kernel Damage in the field Cockroach.	14 tablets /1000 m^3 or 150 gm/100 m^3 or 4 pouch 10 gms each/1000 CFT or 150 gm/100 m^3	72 hrs.	Aeration Period 24 hrs detectors trips or 4hosphine detect tubes should be used in the premises to signal safety of atmosphere.
Rodents Burrows	Rodents	01 Tablet / Burrow	-	-

Source: www.cibrc.nic.in

Methyl bromide (CH_3Br): Methyl Bromide (MB) has been commercially used as a fumigant since the 1930s (MBTOC, 1994). Methyl bromide has a boiling point of 3.6°C and freezing point -93°C and is non-flammable.

Known for its versatility, MB is widely applied to control soilborne pests, including nematodes, fungi, weeds, and insects in high-value crops. Additionally, it has been extensively employed for managing pests such as insects and rodents in structures, transportation, and stored commodities. First proposed for postharvest pest control by Mr. Le Goupil in 1932, MB quickly gained popularity, particularly for whole-site fumigation of grain mills and bagged grain stacks, especially in the U.S. and Africa.

What makes MB particularly effective is its gaseous form, which allows it to penetrate effectively across a broad temperature range. Its quick action and ability to dissipate rapidly from treated environments result in minimal disruption to both crop production and commercial operations. This combination of features has made MB a widely adopted and essential fumigant across various industries. MB was classified as a "controlled substance" under the Montreal Protocol in 1992 (Article 1 and Annex E). This decision arose from growing concerns not only about its contribution to ozone depletion but also its toxicity, safety risks, negative impacts on biodiversity, and potential for water pollution. Consequently, strict regulations were imposed on MB use. Methyl bromide (MB) fumigation remains a preferred treatment for certain perishable and durable commodities in international trade due to its well-established reputation for preventing the spread of quarantine pests.

It is the only fumigant available in India for quarantine treatment of commodities where quick disinfection is desired. It has also the ability to penetrate quickly and deeply into sorptive material at normal atmospheric pressure. CH_3Br is extensive. It used in quarantine stations for plants, vegetables and some fruits, stored products, warehouses, containers, ships, railway wagons. MB at low temperature is more suitable than other fumigants because of its low boiling point. It is stored in metallic cylinders due to volatile nature at low temperature. The continuous exposure and higher dosages of methyl bromide adversely affects the seed viability and germination. This chemical has been listed as an ozone depleting substance by the parties to montreal protocol with agreement from 1995 to limit future product. Unlike phosphine, MB itself has no odour, hence it is mixed with chloropicrin @ 2 percent as a warning gas.

Status of Methyl Bromide: Restricted fumigant

Details of Restrictions: Methyl Bromide may be used only by Govt./Govt. undertakings/Govt. Organizations / Pest control operators under the strict supervision of Govt. Experts or Experts whose expertise is approved by the Plant Protection Advisor [G.S.R.371 (E) dated 20th May, 1999 and earlier RC decision]

Methyl Bromide 98 % w/w- Recommendations for use

Commodity	Pests	Fumigation method	Dose	Exposure period	Residues
Stored Whole Cereals and Seed, Millet, Pulses	Rice Weevil, Lesser Grain Bore, Khapra Beetle, Rust Red Flour Beetle, Saw Drug Store Beetle	Air tight cover	24 gm/m^3	6-8 hours waiting Period 24 hrs.	As when residues not to exceed 25 ppm
Milled Products: Flour	Khapra Beetle, Rust Red Flour Beetle, Lesser grain borer	Air tight cover	24 -32 gm/m^3	12-24 hrs waiting Period 72 hrs	As when residues not to exceed 25 ppm
Dry Fruits, Nuts Spices & Oil Seeds	Rust Red Flour Beetle	Air tight cover	24 -32 gm/m^3	24 hrs waiting Period 72 hrs	As when residues not to exceed 25 ppm

Source: www.cibrc.nic.in

Acrylonitrile: It gives off mustard like odour and has a boiling point of 77°C and freezing point of -82°C. The compound is highly inflammable, hence never used alone and always mixed with another suitable material such as carbon tetrachloride to reduce the risk of fire or explosion. It is highly toxic to human beings.

Carbon disulphide (CS_2): Carbon disulphide reacts with peptides and aminoacids of the foodstuffs and is also toxic to man

Carbon tetrachloride: carbon tetrachloride has a molecular formula of CCl_4. It is normally used in mixture with highly inflammable compounds viz. ethylene dichloride to overcome the problem of fire hazards. It is extremely poisonous to humans and causes liver damage.

Chloropicrin ($CCl_3.NO_2$): Chloropicrin is used as a warning gas with odourless compounds viz., hydrocyanic acid and methyl bromide. It has strong irritating tear gas and corrosive with the metals.

Hydrogen cyanide (HCN): HCN is largely used in vacuum and because of high degree of sorption at atmospheric pressure, it does not penetrate well into small packing materials. It does not affect the viability of seeds.

Ethylene dibromide (1,2- dibromoethane): It is commonly abbreviated as EDB. It has odour similar to chloroform. EDB has boiling point and is sorbed by many materials, so it does not penetrate deeper into the fumigated material. It is mainly used for soil fumigation when food stuffs are fumigated with EDB ampules, they lead to formation of inorganic bromides.

Ethylene dichloride (1, 2 dichloroethane): It is highly inflammable, hence EDC is mixed with carbon tetrachloride in the ratio of 3:1 to overcome the problems of fire hazard. The mixture does not affect the seed germination.

Dichlorvos: (Dimethyl 2,2-dichlorovinyl phosphate): DDVP is used as contact insecticide and applied as fog or spray and is used to control insects in open space of the structure. It doesn't leave behind any residues and does not affect the viability of seeds.

Sulfur-containing compounds

Sulfur-containing compounds viz. carbonyl sulfide and sulfuryl fluoride are gas fumigants used as insecticides. They were developed to replace the ozone-depleting insecticide methyl bromide.

Sulfuryl fluoride (SO_2F_2) was first registered as an insecticide and rodenticide in 1959. It was developed by Dow AgroSciences, LLC (former Dow Chemical) in the 1950s to control drywood termites typically found in warm climates (Derrick et al., 1990) under the trade names Vikane gas fumigant and ProFume gas fumigant. Sulfuryl fluoride is prepared by the direct reaction of fluorine gas (F_2) with sulfur dioxide gas (SO_2). Sulfuryl fluoride is an odorless, colorless gas with a melting point of −135.82°C and a boiling point of −55.38°C. The vapor pressure is 1.3 × 10^4 torr at 25°C. The solubility in water is 0.75 g/kg at 25°C. Upon contact with water, it hydrolyzes to form fluorosulfonic and fluoride ions, and further to sulfuric acid and hydrogen fluoride, according to the reaction (National Center for Biotechnology Information, 2022; Cady and Misra, 1974):

$SO_2F_2 + 2OH^- \rightarrow SO_3F^- + F^- + H_2O$

Sulfuryl fluoride is of low solubility in most organic solvents but is miscible with methyl bromide. Sulfuryl fluoride is not hydrolyzed by water, but is hydrolyzed by NaOH solution. It is used on structures, vehicles, and wood products to control dry termites and wood-infesting beetles. Initial concentrations in fumigated structures are typically 2000–4000 ppm, although other concentrations may be used depending upon the target pest to be controlled, temperature, and the length of the exposure period. Sulfuryl fluoride is restricted to use only by certified applicators. Because sulfuryl fluoride is a gas, structures are completely sealed prior to fumigation and a small amount of chloropicrin is introduced into the structure to warn people and animals that the structure is being fumigated. If it comes into direct contact with skin, it can result in frostbite. It does not remain inside the room after fumigation and completely dissipates without leaving any surface residues (National Center for Biotechnology Information, 2022).

It is known that sulfuryl fluoride is an inhibitor of lipoprotein lipase, an enzyme responsible for the hydrolysis of triacylglycerols of lipoproteins (Kokotos et al., 2000). Degradation of sulfuryl fluoride is via two-step hydrolysis and requires a high pH. The half-life of sulfuryl fluoride in basic water (pH 9.2 at 20°C) is 3.3 min (Cady and Misra, 1974), while the half-life of fluorosulfate (FSO_3) in basic water is 20 h (Jones and Lockhart, 1968).

Animals exposed to lethal doses of sulfuryl fluoride had tremors and convulsions. The dermal route does not appear to play a significant role in the toxicity of sulfuryl fluoride. Treatment for sulfuryl fluoride exposure is similar to other means of intoxication with systemic fluoride: (1) oral administration of dilute calcium hydroxide or calcium chloride to prevent further absorption and (2) injection of calcium gluconate to increase the blood calcium concentration.

Recent developments in fluorine-containing pesticides represent a significant advancement over previous generation of pesticides

1. **Enhanced Efficacy:** Fluorine atoms can increase the bioactivity of pesticides, making them more effective at lower doses. This is due to fluorine's ability to create stronger bonds within molecular structures, enhancing the pesticide's affinity for its target site in pests.

2. **Increased Stability:** Fluorine-containing compounds are often more stable in environmental conditions such as heat, light, and microbial activity. This stability helps pesticides last longer on crops, reducing the need for frequent reapplication and improving overall efficiency.
3. **Target Specificity:** Advances in fluorinated pesticides often allow for better targeting of specific pests without affecting non-target organisms. By focusing on specific metabolic pathways unique to pests, fluorinated compounds can reduce off-target toxicity, making them safer for beneficial insects, wildlife, and humans.
4. **Environmental Persistence and Biodegradability:** While many fluorine-containing pesticides are designed for stability, recent research has focused on balancing environmental persistence with biodegradability. Efforts are being made to develop compounds that are potent against pests but break down more safely over time, reducing long-term environmental impact.
5. **Resistance Management:** Pests can develop resistance to traditional pesticides over time. The unique mechanisms and improved efficacy of fluorine-containing pesticides can help manage resistance, providing a new mode of action that may work where other pesticides fail.

Overall, these developments make fluorine-containing pesticides powerful tools in agriculture with better performance, lower environmental risks, and higher safety for non-target organisms compared to earlier pesticide formulations.

Table 34: Pesticides containing Fluorine

Year of launch & manufacturer	**Common Name (Trade name)**	**MoA (IRAC)**	**Target**	**Moieties (Total number of Hal: type of Hal atoms)**
2018 and Dupont	Triflumezopyrim (Pexalon™)	4E	nAChR	3-(F3 C)-Phenyl (3: 3 x **Fluorine**)
2019 and Nissan Chemical Industries	Fluxametamid (Gracia®)	30	GABA	5-F3 C-4,5-dihydro-3-isoxazole, 3,5-Cl2 -Phenyl (5: 3 x **Fluorine**, 2 x Chlorine)
2020 and Bayer Crop Science	Tetraniliprole (Vayego®)	28	RyR	3-F3 C-Tetrazole, 3-Cl-Pyridine (4: 3 x **Fluorine**, 1 x Chlorine)
2021 and Mitsui Chemicals, BASF	Broflanilide (Exponus®)	30	GABA	2-F3 C, 4-[CF(CF 3)2],6-Br-Phenyl, 2-F-Phenyl (12: 11 x **Fluorine**, 1 x Bromine)

Year of launch & manufacturer	Common Name (Trade name)	MoA (IRAC)	Target	Moieties (Total number of Hal: type of Hal atoms)
2021 and Nihon Nohyaku	Benzpyrimoxan (Orchestra®)	UN	UN	4-F3 C-Benzyl (3: 3 x **Fluorine**)
2021 and Syngenta	Isocycloseram (Plinazolin®)	30	GABA	5-F3 C-4,5-dihydro-3-isoxazole, 3,5-Cl2 , 4-F-Phenyl, (6: 4 x **Fluorine**, 2 x Chlorine)
2022 and Meiji Seika	Flupyrimin (Kevuka®)	4F	nAChR	F3 C-CO-N=, 6-Cl-Pyridine (4: 3 x **Fluorine**, 1 x Chlorine)

Metaldehyde ($C_8H_{16}O_4$)

Metaldehyde a tetramer of acetaldehyde and is used primarily to kill snails and slugs.was discovered by von Liebig in 1835, and a century later its use as a molluscicide was proposed by Gimingham and Newton in 1937. It is manufactured by reacting acetaldehyde with various acids at a low temperature. It is a white crystalline powder, flammable solid with mild menthol odor. The molecular weight is 176.24, boiling point@112–116 °C and freezing/melting point at 47 °C showing flash point of 36 °C. It is moderately soluble in water (60 mg/L at 30 °C), highly volatile, and non-persistent with a soil degradation DT_{50} of 5.1 days. Furthermore, it is moderately mobile in soil (logKoc = 2.38) and has low potential for bioaccumulation (BCF = 11 L/kg) and is therefore a moderate alert for both environmental fate and ecotoxicology (*University Hertfordshire*, 2007; Williams et al., 2017). It is applied in the form of liquid, granules, sprays, dusts, or pelleted/ grain bait to kill slugs, snails.

As a molluscicide, metaldehyde is used for controlling slugs and snails in different crops. It is sold as Metaldehyde 2.5% Pellet and to be used @ 15-25 kg /Acre and available in market as SNAILKILL. Its pesticidal action is due to contact with the foot of the mollusc, making it torpid and increasing the secretion of mucus leading to dehydration (RSC, 1987).

Metaldehyde is toxic to all domestic animal species, with reports of intoxication of dogs, cats, birds, horses, sheep, swine, goats, and cattle. Acute poisoning is common in pets, birds, domestic, and wild animals. Dogs are the domestic animal species most often intoxicated (Andreasen, 1993; Talcott, 2004; Yas-Natan *et al.*, 2007). The LD_{50} in rats is 227–690 mg/kg b.w., in rabbits 290–1250 mg/kg, and in cats 100–300 mg/kg. A very small amount of metaldehyde is required to cause poisoning or death. The major target organs for metaldehyde toxicity include central nervous system (CNS), liver, kidney, and lung. Animals that ingest metaldehyde may exhibit a variety of toxicological signs including vomiting, tachycardia, ataxia, tremors, seizures, and death (Dobler, 2003). Upon dermal contact, metaldehyde may cause irritation.

Metaldehyde 2.5% Pellets

Crops	Target pest	Dosage	Method of application	Compatibility
Citrus, Rubber, Paddy (Rice), Tea, Vegetables	Snails, Slugs, Giant, African snails	15-25 kg / Acre	Find out affected areas in the evening, place small quantity of baits (about 50-80 g per 100 sq. ft. area) near crawling path, near plant base, between the rows of crop plants or any other place where the snails and slugs are a nuisance.	Should be used in solitary

B. Classification based on mode of entry in the insect

Insecticides play a crucial role in managing pest populations, safeguarding crops, and ensuring food security. Understanding their modes of action is key to their effective utilization. Let's delve into the modes of action based on mode of entry:

1. **Contact Insecticides:** These insecticides act upon pests upon direct contact (Fig. 28). They form a protective barrier on the surface of plantspinos, disrupting the insect's physiology upon contact. Contact insecticides are particularly effective against pests that feed on exposed plant parts such as leaves and stems. They can cause paralysis, desiccation, or interference with the insect's nervous system.
2. **Stomach Insecticides:** Once ingested by the insect, stomach insecticides disrupt vital physiological processes within the pest's digestive system (Fig. 28). They can interfere with enzyme systems, leading to metabolic dysfunction or block essential nutrient absorption, ultimately resulting in the insect's demise. Stomach insecticides are effective against chewing insects such as caterpillars, beetles, and grasshoppers.
3. **Systemic Insecticides:** These insecticides are absorbed by the plant and transported through its vascular system, making them accessible to pests that feed on different plant parts, including those not directly sprayed (Fig. 28). Systemic insecticides provide continuous protection against pests, even after application. Upon ingestion by the insect, they disrupt physiological processes, often targeting the nervous system or interfering with vital metabolic pathways
4. **Fumigant Insecticides:** Fumigants are gaseous insecticides that penetrate insect respiratory systems, disrupting cellular respiration and causing asphyxiation. They are highly effective against pests in enclosed spaces

such as storage facilities, warehouses, and shipping containers. Fumigants have the advantage of reaching inaccessible areas where other insecticides might not penetrate effectively.

5. **Translaminar Insecticides or local systemic:** Translaminar insecticides are absorbed by the plant's foliage and then redistributed within the leaf tissue, forming a reservoir of active ingredient from upper surface to lower surface of leaf (Fig. 28). When pests feed on treated foliage, they ingest the insecticide along with plant tissues, leading to their demise. Translaminar insecticides provide targeted protection against foliar-feeding pests while minimizing environmental exposure. They are particularly effective against piercing-sucking insects such as aphids, whiteflies, and leafhoppers. Examples include certain neonicotinoids and diamides.

Each mode of action offers unique advantages and applications in integrated pest management strategies, contributing to sustainable agriculture practices while minimizing environmental impact. Understanding these modes of action empowers growers to select the most appropriate insecticide formulations for effective pest control tailored to their specific needs and cropping systems. The majority of contact insecticides are older chemistry, and they mostly have multisite modes of action (rather than a specific mode of action like modern systemic insecticides), which provides them with resistance management benefits.

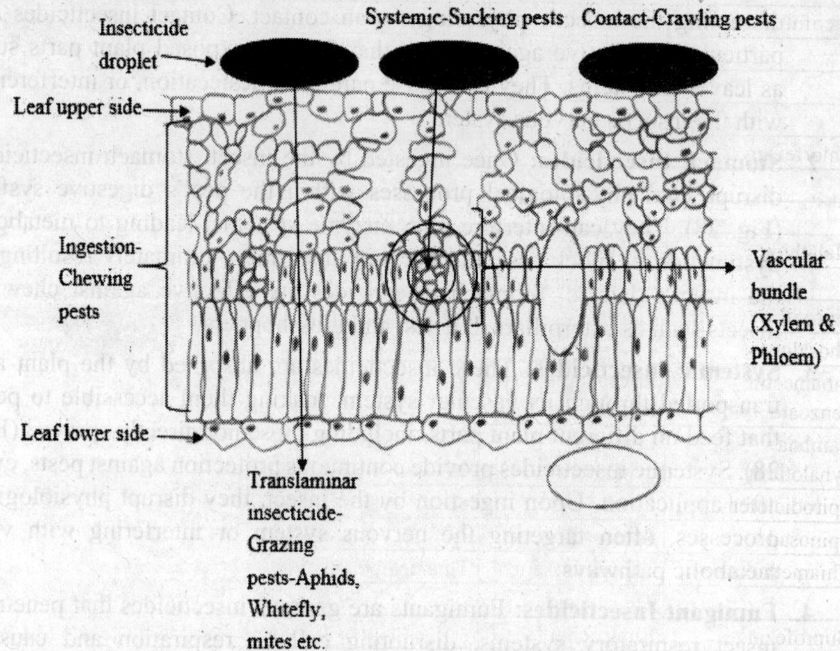

Fig. 28: Classification based on mode of entry in the insect

Table 35: Categorisation of Insecticides as Contact/Stomach or Ingestion/Systemic/Translaminar or local systemic /Fumigant

Contact	Stomach/ Ingestion	Systemic	Translaminar/ Local systemic	Fumigant
Abamectin	Abamectin	-	Abamectin	-
Alphamethrin	-	-	-	-
Acetamiprid	Acetamiprid	Acetamiprid	Acetamiprid	-
Amitraz	-	-	-	-
-	-	Acephate	-	-
-	-	Aldicarb	-	-
-	-	-	-	Aluminium phosphide
-	Allethrin	-	-	-
Bifenthrin	Bifenthrin	-	-	-
Cyfluthrin	-	-	-	-
-	-	Carbofuran	-	-
-	-	Clothianidin	Clothianidin	-
Chlorantraniliprole	-	Cyantraniliprole	-	-
Chlorpyriphos	-	-	-	-
-	Cypermethrin	-	-	-
-	-	Cartap hydrochloride	-	-
Dicofol	-	-	-	-
-	-	Dimethoate	-	-
-	Deltamethrin	-	-	-
-	-	-	Diafenthiuron	-
Chlorfenapyr	-	-	Chlorfenapyr	-
-	Fluvalinate	-	-	-
-	-	Chlordimeform	Chlordimeform	-
Malathion	Malathion	-	-	-
-	Fenitrothion	-	-	-
Fipronil	Fipronil	Fipronil	-	-
Phosalone	-	-	-	-
Emamectin benzoate	Emamectin benzoate	-	Emamectin benzoate	-
Lambda cyhalothrin	Lambda cyhalothrin	-	-	-
Spirodiclofen	-	-	-	-
Spinosad	Spinosad	-	Spinosad	-
Thiamethoxam	Thiamethoxam	Thiamethoxam	Thiamethoxam	-
-	Thiacloprid	-	-	-
Buprofezin	Buprofezin	-	-	-
Flufenoxuron	Flufenoxuron	-	-	-

Contact	Stomach/ Ingestion	Systemic	Translaminar/ Local systemic	Fumigant
Indoxacarb	Indoxacarb	-	-	-
Isocycloseram	Isocycloseram	-	-	-
Broflanilide	Broflanilide	-	Broflanilide	-
-	Pymetrozine	Pymetrozine	Pymetrozine	-
Spirotetramat	Spirotetramat	Spirotetramat	Spirotetramat	-
-	Imidacloprid	Imidacloprid	Imidacloprid	-
Carbofuran	Carbofuran	-	Carbofuran	-
Diflubenzuron	Diflubenzuron	-	Diflubenzuron	-
Abamectin	Abamectin	-	Abamectin	-
	-	-		
-	-	Sulfoxaflor	Sulfoxaflor	-
Propargite	-	-	-	-
Profenophos	-	-	-	-
-	Propoxur	Propoxur	-	-
-	-	Phosphomidon	-	-
-	Zinc phosphide	-	-	-
Novaluron	-	-	-	-
-	-	Methomyl	-	-
-	-	Monocrotophos	-	-

C. Toxicity class

Classification of the Insecticides	LD_{50} (Oral)	LD_{50} (Dermal)	Colour of the triangle
Extremely toxic	1-50	1-200	Bright red
Highly toxic	51-500	201-2000	Bright yellow
Moderately toxic	501-5000	2001-20000	Bright blue
Slightly toxic	> 5000	> 20000	Bright green

The warning colors furnished on a pesticide container implies-Safety towards mammalian animals

6

Movement of Insecticide in Plants

The way a chemical moves, persists, and behaves in its surroundings is known as its behaviour. A chemical's behaviour will dictate both its residual properties and field efficacy. The behaviour of insecticides is determined by their chemical and physical characteristics as well as the interaction of environmental elements. Our capacity to comprehend and forecast how insecticides will behave in the environment will determine our ability to keep utilizing insecticides.

Important features of systemic insecticides

1. They are not susceptible to wash off by rain
2. By virtue of their movement inside plants, they can afford to protection to regions where they have not been sprayed
3. They are often been transported towards region of growth
4. Since they have very weak contact action, safe to beneficial organism.

Pesticide Redistribution and Its Implications on Pesticide Efficacy

Agrochemical companies try to make pesticides that kill insects effectively while causing as little harm as possible to people using them, the environment, and their health. An important thing about insecticides that makes them work better is that the active ingredient can move from one part of the plant to another easily. When insecticides are applied, they rarely cover all the vulnerable tissues evenly. Redistributing the pesticides helps them work better and be more reliable.

Redistributing pesticides means moving them from where they were first applied to a different spot on or inside a plant, where they can kill the pest or disease

Adsorption versus absorption

The active components of most foliar insecticides remain on plant tissue after the water as a carrier dries. Compounds adsorb to plant tissue via intermolecular interactions. Pesticide is magnetically held to plant tissue by this molecular force. Because rain can dislodge these pesticides, more applications may be needed to maintain protection. They must be applied to cover new growth as plants grow.

Absorption is the process of a pesticide passing through the cuticle and epidermis, the two outermost layers of the plant tissue and into the interior of the plant

(Figure 1). Once inside the plant tissues, a pesticide will be less susceptible to being washed off by rain. Many synthetic pesticides attach to the plant surface through **adsorption** and can also be **absorbed** into plant tissues.

Pesticides that can be absorbed by plants often have a longer residual efficacy due to less wash-off and greater resistance to UV destruction. Adjuvants in spray formulations can improve the adhesion and absorption of synthetic insecticides into plant tissues. Following absorption, some pesticides can travel locally and over large distances via the xylem or phloem. Rain can remove contact insecticides from plant surfaces. The factors which affect the distribution of insecticide after spraying are

Pesticide formulation - Dusts and wettable powders, for example, are easier to dislodge because they contain the active ingredient as well as a carrier material, such as clay/talc, which can be physically washed around or off plant surfaces by rain, as opposed to oil-based formulations (such as emulsifiable concentrates), which dry after application because oily residues adhere to plant cuticles and waxy outer layers.

- The amount of rain received has a greater impact on the dispersion of water, with half an inch or less causing residues to move across the canopy and more rain washing them away from plant tissues also.
- High-toxicity pesticides are more likely to kill the insect even presence of small quantity of residues on the target pest is more effective than lower-toxicity pesticides.

Soil-applied systemic pesticides

Systemic pesticides can be applied to the soil by drenches, drips, or injections. Once within the rhizosphere, plant roots have the ability to assimilate pesticides in a manner analogous to their absorption of nutrients. After being absorbed by the roots, pesticides can be transported systematically by either the xylem or phloem. In certain instances, soil-applied insecticides can retain their effectiveness for a period of up to 12 weeks. Nevertheless, the process of applying pesticides to the soil can result in a delayed attainment of a sufficient concentration within the plant. As an illustration, it took a period of six to eight days for the insecticide imidacloprid to reach the necessary concentration to effectively suppress the glassy-winged sharpshooter (*Homalodisca coagulata*) in grapevines that were 20 years old.

The use of soil-applied systemic pesticides has disadvantages, as their extended persistence might lead to higher amounts of residues on or in harvested products. Systemic insecticides can have detrimental effects on pollinators when they are found in flower structures. Acephate, a type of systemic insecticide, does not move to floral parts when it is used on soil. However, certain substances, such as imidacloprid, have the ability to go to other parts of a flower. Thats way imidacloprid is not safe to honeybees.

An insecticide efficacy can be improved via application if three factors are considered i.e. crop developmental stage/ phenology stage, target pest and weather conditions.

Possession of redistribution attributes among systemic and contact pesticides

Character	Systemic	Contact
Absorbs into plants	Yes	No
Rain fast	Yes	No
Redistributes to improve pesticide coverage	Yes	Moderate
Some can be applied to soil for plant uptake	Yes	No

- Many crops have plant tissues that are difficult to penetrate or cover with pesticides. For example, grape clusters have complex shapes that are difficult to effectively cover with pesticides. Yet, they are the most important plant part to protect from diseases and pests.
- Incomplete coverage can occur for orchard trees or field crops due to reduced spray penetration caused by a dense canopy. Instead of using a contact pesticide, a pesticide with movement or redistribution capabilities might be more effective in some cases. The reason behind this is that the product has the ability to redistribute and offer more uniform protection to all parts of the plant.
- Redistributing pesticides can be beneficial when plants are quickly producing new tissue. When xylem and phloem-systemic compounds are applied to actively growing plants, the active ingredient migrates to unprotected tissues as the plants expand. This can lead to improved protection and longer spray intervals when compared to contact materials that must be applied as new tissue develops. Redistributing insecticides can also successfully control the flying insects *viz.*, aphids, psyllids, whiteflies, and other piercing-sucking insects that can hide, relocate, or resist direct spraying. The xylem and phloem-systemic pesticides can reach any plant vasculature-feeding insect. Piercing-sucking insects eat the insecticide wherever they feed on the plant. Systemic pesticides are useful when pest pressure is considerable.
- Keep in mind that weather might affect pesticide effectiveness during and after applications. Avoid applying pesticides if rain is expected within 24 hours.
- Adjuvants and spreaders accelerate pesticide penetration into leaf tissue, aiding synthetic and contact pesticides during rainy weather. Sticker adjuvants improve contact material retention during rain or irrigation.

- Pesticides can also be affected by heat. A common contact acaricide, sulfur, can produce phytotoxicity if the temperature (90°F/32°C) and relative humidity are high (70%) during or shortly after an application (less than two hours). Younger tissues also suffer sulfur damage more easily. Use pesticides carefully during hot weather. Oils and organosilicone surfactants can cause phytotoxicity, as can pesticide combinations. Spray plants in the morning when temperatures are moderate and tissues dry quickly.

Nevertheless, the level of toxicity of the pesticide has a significant impact on its effectiveness after rains.

Translaminar

Translaminar redistribution refers to the movement of a substance from the side of the leaf where it was sprayed to the other side, providing protection to both sides of the leaf. In order for translaminar redistribution to take place, the pesticide must initially be assimilated by the plant. Subsequently, the pesticide has the ability to migrate from its initial deposition site to the opposite side of the leaf.

Xylem systemic

Xylem systemic refers to the procedure by which a pesticide is absorbed and moves systematically through the xylem vessels of a plant. Xylem vessels facilitate the transport of water and minerals in a unidirectional manner, from the roots to the rest of the plant, a process referred to as acropetal movement. Water and nutrients do not travel significantly downwards or backwards along branches or leaves in xylem channels. The redistribution of xylem can provide protection to actively growing tissues by facilitating the movement of pesticides to those tissues, hence preventing pest damage. The majority of systemic insecticides are transported throughout the plant via the xylem.

Phloem systemic

Phloem systemic, also known as real systemic, refers to the bidirectional transport of pesticides within a plant, moving both upwards to leaves and other structures and downwards to roots (also termed basipetal movement) following absorption. The phloem vessels facilitate the transportation of sugars and other essential nutrients throughout the plant, with a primary focus on directing them towards carbohydrate sinks, such as fruits and roots.

When pesticides are described as "locally systemic," it typically means that they have the ability to move within a plant's tissues, either through diffusion across leaf layers or as a gas, rather than being transported through the plant's xylem or phloem.

Pesticide redistribution (excluding water splash redistribution) often occurs by more than one mechanism.

Insecticides with redistribution processes

Type of redistribution	active ingredients
Xylem translocated	Neonicotinoids- imidacloprid, acetamiprid, clothianidin, dinotefuran, thiamethoxam
	Pymetrozine
	Acephate
	Fipronil
	Flonicamid
Phloem translocated	Spirotetramat
Translaminar	Spinosad
	Abamectin, Emamectin benzoate
	Cyantraniliprole, Chlorantraniliprole
	Chlorfenapyr
Non-translocated	Bifenthrin, Cypermethrin, Permethrin
	Carbaryl
	Spinosad

7

Pro-Insesticides

Pesticide use in agriculture has evolved at several stages in the last few decades. Pesticides, as defined by the FAO, are any substance or mixture of substances intended to prevent, destroy, or control any pest, including vectors of human or animal disease, unwanted species of plants or animals causing harm during or otherwise interfering with the production, processing, storage, transport, or marketing of food, agricultural commodities, wood and wood products, or animal feedstuffs, or substances that may be administered to animals. A plant growth regulator, defoliant, desiccant, or agent used to thin or prevent premature fruit fall, as well as substances applied to crops before or after harvest to prevent them from deteriorating during storage and transportation, are included in this category.

Today's agrochemistry has a massive challenge: ensuring the availability of high-quality active ingredients to enable long-term control of pest species, weeds, and agricultural diseases. The differential uptake and transfer of the active ingredient by target and non-target organisms can contribute to diverse biological activity.

A proinsecticide is an inert molecule that must undergo some structural change(s) before it may be used as a insecticide. Proinsecticide can undergo activation by (i) a chemical (non-enzymatic) process, (ii) a biochemical (enzymatic) process, or (iii) a physical (e.g. photochemical) process. Despite the fact that proinsecticidal chemicals are occasionally active without chemical modification when tested in vitro, their metabolites considerably contribute to the overall biological activity of the applied material.

The structure of pesticides is highly optimized, with multiple biodegradable chemical groups controlling selectivity and metabolic activation in vivo, releasing naturally active biological compounds that can then exert the desired biological effects.

Since scientists are learning more about the biochemistry and genetics of major pest insects, weeds, and crop pathogens, the search for selectivity is an important in discovery of new pesticides. This can be done by changing the structure of the active ingredient in the right way.

Proinsecticide possess a modified chemical structure that can affect the parent molecule performance and biological activity as an active ingredient; for example:

1. Chemical stability, solubility, and lipophilicity, as well as other physical properties, play a crucial role in determining how a substance is distributed within an organism; also called systemic distribution.
2. Pharmacokinetic behavior, including processes *viz.*, adsorption, distribution, translocation, bioavailability and target selectivity which further influence the substance behavior within the organism.
3. Additionally, pharmacodynamic properties, such as toxicity, can be altered to reduce harm towards non-target species and minimize side effects through the differential metabolism of the propesticide and its active metabolite.
4. The slow release of an active metabolite from a pesticide structure can delay or sustain the biological activity of a substance.

Activation principle of Proinsecticide

Active ingredient derivatives are converted to the parent chemical for activity. There are many processes required to convert a pesticide to a pesticidally active one. These activation steps should ideally occur only in the target organism. Despite the fact that it can occur in the plant, environment, including soil and atmosphere. Proinsecticide need to have one or more than one mechanism or process.

1. **Activation by primary biochemical target:** At the target tissue, the activating enzyme is disrupted during the enzymatic conversion of a proinsecticide to an active toxophore. When active toxophore is released from the proinsecticide, the target organism regular processes are disrupted, resulting in their death.
2. **Activation by detoxification system:** Detoxification occurs in the organism body at the target site by degradation mechanisms such as oxidation, reduction, hydrolysis, or conjugation reactions. These reactions act on xenobiotic substances, such as pesticides, resulting in the creation of a more harmful material than the original.
3. **Activation by symbiont metabolism:** Most insects have mutual microorganisms living in their guts and these microorganisms have enzymes, which are not present in insects. So, they are involved in xenobiotics (pesticides) to proinsecticide.
4. **Activation by symbiotic routes:** In this process, there are no enzymes involved. Here the process of activation is not metabolic, but it leads to toxicity due to poison changes in a biological setting.

Classification of Proinsecticides

There are two ways to classify pro-pesticides: One-by the number of activation steps required and the two by the type of pest to be controlled. On the basis of the number of activation stages, pro-pesticides are classified as single-step activation and multiple-step activation.

I. Single step activation

1. **Juvenogens:** Juvenogens are a class of complex chemical compounds that, when exposed to biotic or environmental factors, produce products with juvenile hormone properties. When juvenogen esters are applied topically, the wax-like ester enters the insect body, where the carboxylesterase enzymes break it down into an alcohol. In this case, the juvenogen has higher juvenile hormone activity than the alcohol.
2. **Procarbamates and proformamidines:** Toxophore is activated either enzymatically or non-enzymatically by the insecticide derivatives of toxic methyl carbamates. There are also two other putative activation mechanisms for this insecticide groups.

These are:

- Acid catalyst hydrolysis of the N-S bond
- Thiol induced thiolysis to form a mixed disulfide and toxic methyl carbamate.

II. Multistep activation

These Proinsecticides are pesticides that require more than one metabolic pathway to activate.

1. **Flourocitrate precursors-** The earliest poison where biochemical mechanism of action was precisely elucidated. Nissol and Fluemethyl are two commercially available acaricides with relatively low toxicity to mammals.
2. **Cycloprate-**The activation process consists of two stages. The cycloprate is first hydrolyzed to free acid, and then carnitine ester is formed. As a result, it decreases carnitine's function in fatty acid transport. The amino acid carnitine aids in the transfer of fatty acids across mitochondrial membranes in the liver and skeletal muscles.
3. **Flouromevalenate:** These are strong inhibitors of insect juvenile hormone synthesis. The steps involved in C-oxidation include dealkylation (chlorfenapyr), ester hydrolysis, decarboxylation (indoxacarb), demethylation, and N-diacylation (acephate).

Based on the insecticide group

Pro-Pesticides are classified based on the type of pest they are intended to control namely, proinsecticides, proherbicides, profungicides, and prorodenticides. According to the Insecticide Resistance Action Committee's (IRAC) categorization, a professional committee of Crop Life International; see http://www.irac.org), about 40% of insecticidal mode of action classes contains compounds that use a proinsecticide mechanism.

1. **Pro-Quinazolines:** By opening TRPV channels in insect chordotonal organs, pyrifluquinazon causes desired effects. It is proven that pyrifluquinazon is a proinsecticide since it spontaneously deacetylates and is 100 times more effective than pyrifluquinazon alone at opening Drosophila TRPV channels.
2. **Pro- Pyridine carboxamide:** The metabolite of flonicamid, TFNA-AM, is 100 times more effective than flonicamid, demonstrating that flonicamid is a proinsecticide.
3. **Pro- Pyridazine amides:** Dimpropyridaz is a proinsecticide that are converted into secondary amide active form pyridazine pyrazolecarboxamide by N-dealkylation. Active secondary amide metabolite potently inhibit the function of insect chordotonal neurons.
4. **Pro-Synthetic pyrethroids:** The primary mechanism of action of tralomethrin is likely attributed to its prompt conversion to deltamethrin through the process of dehalogenation. The comparative potency of tralomethrin was found to be 100 times lower than that of deltamethrin. Tralomethrin is the sole pyrethroid compound that has been identified as a proinsecticide.
5. **Pro-oxadiazon:** The compound indoxacarb undergoes a chemical transformation known as N-decarbomethoxylation, resulting in the formation of N-decarbomethoxylated JW062. DCJW is a metabolite that exhibits a significantly higher potency and toxicity compared to its parent compound (indoxacarb). It acts on voltage-dependent blocker of Na^+ dependent compound action potentials. Indoxacarb exhibits bioactivation in mammals as well, although the process of N-decarbomethoxylation leading to the formation of DCJW is comparatively less efficient in mammals than in comparison to insects.
6. **Pro-Organochlorines:** Dieldrin, a proinsecticide derived from aldrin via epoxidation of the non-chlorinated double bond was extensively used as an insecticide prior to its prohibition in the 1970s due to persistence organic pollutant.
7. **Pro-Organophosphates:** A desulfuration reaction by cytochrome P450 (CYP) can turn chlorpyrifos into chlorpyrifos-oxon (CPO), an effective anticholinesterase agent. Similarly, insects metabolize acephate into methamidophos by hydrolysis and it is active acetylcholinesterase inhibitor, whereas mammals metabolize acephate more readily into des-O-methylacephate (major metabolite) and methamidophos was found at a low level. In this way acephate shows relatively high selectivity against insects (Mahajna *et al.*, 1997 and Roberts, 1999).

   The selective toxicity of malathion can be explained by the balance between bioactivation (oxidative desulfuration) and the active acetylcholinesterase inhibitor, malaoxon. Similarly, through oxidative desulfuration, parathion is transformed into paraoxon, where P=S form are turned into P=O form. Likewise, hydrolysis converts formothion to dimethoate and dimethoate carboxylic acid. Although schradan alone is ineffective against AChE, studies have shown that it is transformed into a phosphoramide oxide, a powerful AChE inhibitor, by an oxidase.

8. **Pro-Carbamates:** In Fukuto's lab, pro-carbamates were prepared systematically changing them to N-phosphoryl, N-sulfenyl, and similar carbamates. The biological and toxicological properties of these carbamates can be altered by changing the derivatizing moiety, affecting the final product's physicochemical properties, such as lipophilicity (log P).

   Metam-sodium, a pro insecticide is converted rapidly into methyl isothiocyanate by dehydrosulfuration. In soils, plants, and animals, thiodicarb is metabolized by hydrolysis or thiolysis of the N-S bond to methomyl. Carbosulfan, furathiocarb and benfuracarb are pro-pesticides that degrade to the active substance carbofuran by hydrolysis. Aldicarb is broken down into sulfoxide though sulfoxide and aldicarb were equally effective at killing insects, the sulfoxide is more powerful inhibitor of acetylcholinesterase (AChE).

9. **Pro-Phenyl pyrazoles (Fiproles) :** In insects, the phenylpyrazole, fipronil possesses a sulfoxide group that can be oxidized by cytochrome P450 to produce a more potent, fipronil sulfone metabolite. Further, sulfone is 300 times more effective than fipronil at desensitizing glutamate gated chloride channels. Fipronil may be intrinsically active, but it converted into another active form (fipronil sulfone) before reaching its target.

10. **Pro-Nereistoxin:** Nereistoxin is a cyclic disulfide isolated from a marine annelid. It was the starting point for the invention of the proinsecticides *viz.*, cartap, bensultap, and thiocyclam, where all these insecticides were transformed into nereistoxin, that acts on the insect nicotinic acetylcholine receptor.

11. **Pro-Neonicotinoids**: Neonicotinoids are pesticides that act as agonist at the insect nicotinic acetylcholine receptor (nAChR). They are commercially available, that are effective against a wide variety of nuisance insects, including hemipterans i.e. aphids, whiteflies, and planthoppers as well as coleopterans and even some lepidopterans.

    Transformation of ring systems into non-cyclic structures

    Non-cyclic insecticidally active neonicotinoid structures might be converted into six-membered ring proinsecticides via acetal, thioacetal, and aminal synthesis. For example, thiamethoxam (X=O) has a six-membered ring, which cleaves (e.g., by bioactivation or oxidation) into clothianidin without a six-membered ring in insect and plant tissues when hydroxylation of the X-CH_2 moiety occurs by cytochrome P450 monooxygenases during oxidative metabolism.

12. **Pro-Thio-Urea Derivatives:** An oxidative phosphorylation reaction results in the release of an electrochemical potential gradient that is generated by protons passing across the inner mitochondrial membrane by the mitochondrial ATP synthase, and it catalyzes ADP phosphorylation, which releases ATP to the cell. An insecticide/acaricide diafenthiuron (thiourea derivative), is a biologically inactive proinsecticide that is oxidatively desulfurated into the insecticidally active carbodiimide derivative using sunlight. These insecticidally active carbodiimide metabolites are prepared from the precursor, diafenthiuron S-monoxide.

    In general, the presence of NADPH and active microsomes are required for the synthesis of all metabolites. Carbodiimide inhibits mitochondrial respiration through selectively adhering to inner membrane protolipids (8 kDa) and outer membrane porins (30 kDa). A light and cytochrome-P450-dependent monooxygenases in insects can both activate diafenthiuron.

13. **Pro- Halogenated pyrroles:** Chlorfenapyr a broad-spectrum proinsecticide that kills mites and insects, including termites. In plants, it has good translaminar property but not much systemic activity. When the pest species consumes or gets contact with it, metabolic N-dealkylation, which is just the oxidative removal of the N-ethoxymethyl group of its pyrrole ring by mixed function oxidases, creates an active metabolite called CL 303268 that is lipophilic, weakly acidic (it has an acidic proton at the pyrrole). This is a powerful uncoupler of oxidative phosphorylation. It does this by disrupting the proton gradient in insect mitochondria, which is done partly by mixed-function oxidases (MFOs) but through N-dealkylation, the insecticidal activity of the proinsecticide was improved. As a result, this active pyrrole molecule possess a good balance between metabolic activation and phytotoxicity.

14. **Pro-fluoraliphatic sulfonamides:** Sulfluramid is breaks down into perfluorooctane sulfonate (PFOS) by N-de-ethylation. PFOS is a toxic, extremely persistent and bioaccumulative pollutant.

15. **Pro-Tetranic acid and Pro-Tetramic acid derivatives:** The acetyl CoA carboxylase (ACCase; EC 6.4.1.2), which is important for the biosynthesis of fatty acids, is a biotin-dependent carboxylase that turns bicarbonate into malonyl-CoA and ATP. After foliar application, Spirodiclofen, Spirotetramate, Spiropidion and Spiromesifen moves into the leaf, but they can also be spread acropetally. After getting through the plant's the epidermis these proinsecticides are broken down in the plant's tissues to make enol metabolites that kill insects and stop the Acetyl CoA carboxylase from working.

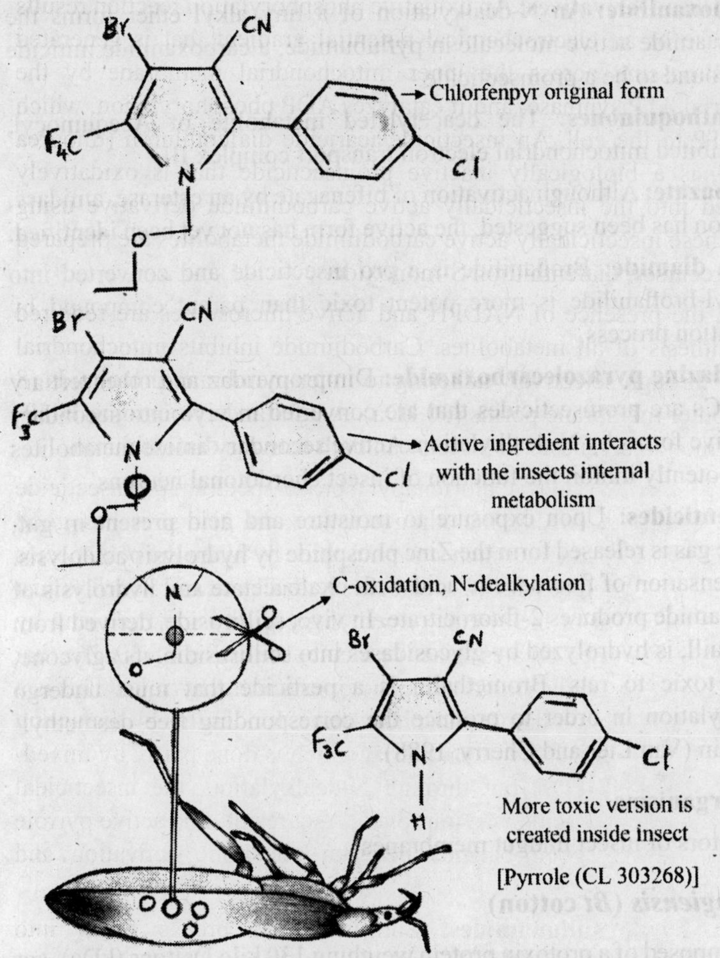

Br
CN
F4C
N
Cl
O
Chlorfenpyr original form
Br
CN
F3C
N
Cl
O
Active ingredient interacts
with the insects internal
metabolism
C-oxidation, N-dealkylation
N
O
Br
CN
F3C
N
H
Cl
More toxic version is
created inside insect
[Pyrrole (CL 303268)]

16. **Pro-formamidines:** N-demethylation of chlordimeform creates a highly toxic compound, demethyl chlordimeform (DCDM).

17. **Pro-Amidines:** Amitraz is a proinsecticide and converted into DPMF (Dimethylphenyl N methylformamidin) by hydrolysis process.

18. **Pro-Anti juvenile hormone or Proallatotoxins**: The precocenes viz., precocene 2 isolated from *Ageratum* plants and converted to dihydrodiol (DHD), which was then discovered to impede the terminal (oxidative) phase of JH production in the corpora allata, leading to premature growth of the larva.

19. **Pro-Pyrazole and Pro-Beta-Ketonitrile:** Mites are more easily penetrated by the cyenopyrafen and cyflumetofen forms, which are converted into the active keto-enol forms.

20. **Pro-Carboxanilide:** An N-dealkylation of a tiny alkyl ether forms the secondary amide active molecule in pyflubumide, a carboxanilide miticide that was found to be a proinsecticide.
21. **Pro-Naphthoquinones:** The deacetylated metabolite of Acequinocyl, DHN, inhibited mitochondrial electron transport complex III.
22. **Pro-Carbazate:** Although activation of bifenazate by an esterase, amidase, or oxidation has been suggested, the active form has not yet been identified.
23. **Pro-meta diamide:** Broflanilide is a pro insecticide and converted into des-methyl-broflanilide is more potent toxic than parent compound by dimethylation process.
24. **Pro-pyridazine pyrazolecarboxamide:** Dimpropyridaz and other tertiary amide PPCs are proinsecticides that are converted in vivo into secondary amide active forms by N-dealkylation. Active secondary amide metabolites of PPCs potently inhibit the function of insect chordotonal neurons
25. **Pro-Rodenticides:** Upon exposure to moisture and acid present in gut, phosphine gas is released form the Zinc phosphide by hydrolysis/acidolysis. The condensation of fluoroacetic acid with oxaloacetate and hydrolysis of fluoroacetamide produces 2-fluorocitrate. In vivo, scilliroside, derived from the red squill, is hydrolyzed by glycosidases into scillirosidin, its aglycone, which is toxic to rats. Bromethalin is a pesticide that must undergo N-demethylation in order to produce the corresponding free desmethyl-bromethalin (Van Lier and Cherry, 1988).

26. Pro-Microorganisms

Microbial disruptors of insect midgut membranes

Bacillus thuringiensis (*Bt* cotton)

The crystal is composed of a protoxin protein weighing 130 kilo Daltons (kDa), get soluble in the larval midgut's alkaline pH and further enzymatically cleaved into an active toxin weighing 60-70 kDa. The toxin exhibits an affinity for receptors located on the midgut epithelium, leading to the subsequent development of pores. The presence of a pore disrupts the inward potassium gradient, leading to the enlargement of microvilli and subsequent lysis of cells. The occurrence of ion leakage from the gastrointestinal tract into the haemolymph leads to an imbalance of ions within the haemolymph. The insects that have been exposed to toxins experience rapid mortality due to the direct action of the toxin, or alternatively, they may cease to feed and succumb to septicemia, resulting in mortality within a span of approximately 48 to 72 hours.

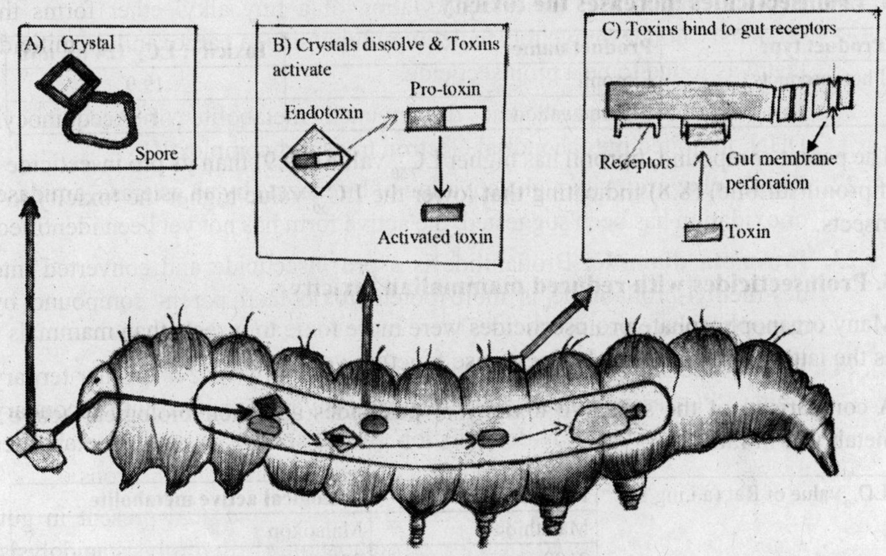

Potential advantages of a proinsecticide can be summarized as follows:

1. Alteration of physicochemical properties:

An increase in stability, solubility, or lipophilicity that influences distribution within an organism (systemicity).

Lipophilicity

Product type	Product name	Lipophilicity: log Kow
Photoproducts	Fipronil Fipronil sulfone	2.74 2.81

The above table depicts that parent compound, fipronil showed less lipophilicity than its pro insecticide fipronil sulfone.

2. Sustaining activity

Slow release of the active ingredient from its derivatives causes delayed or prolonged activity.

3. Increases selectivity

Due to the distinct metabolism of the parent compound and its derivatives, there is an increase in selectivity, i.e. a decrease in toxicity towards non-target species.

1.1 Proacaricides show increased in selectivity to mites

Cyflumetofen is a proacaricide that inhibits mitochondrial complex II at very low concentrations and shows high selectivity against spider mites *viz.*, two-spotted spider mite (*Tetranychus urticae*) and the citrus red mite (*Panonychus citri*).

4. Proinsecticides increases the toxicity

Product type	Product name	Toxicity: LC_{50} (24 h), nM
Photoproducts	Fipronil	19.9
	Fipronil sulfone	8.8

The parent compound, fipronil has higher LC_{50} value (19.9) than its pro insecticide (fipronil sulfone) (8.8) indicating that lower the LC_{50} value higher the toxicity to insects.

5. Proinsecticides with reduced mammalian toxicity

Many organophosphate proinsecticides were more toxic to insects than mammals as the latter stopped acetylcholinesterase function very quickly.

A comparison of the selective toxicity of pesticides and their biological active metabolite in rats.

LD_{50} value of Rat (a.i.mg/kg)	Propesticide	Biological active metabolite
	Malathion	Malaoxon
	2600	308
	Acephate	Methamidophos
	866-945	20

The above table reveals biological active metabolites, malaoxon and methamidophos showed less LD_{50} value than malathion and acephate respectively. In mammals, malathion gets hydrolysed through an enzyme carboxylesterases, which is not observed in insects. Similarly, methamidophos was extremely toxic to mammals, but its acetylated form i.e. acephate reduces that toxicity by 45-folds while retaining its insecticidal properties.

6. Increases systemicity

Unlike the proinsecticide spirotetramate, its insecticidally active enol metabolite possess physicochemical properties that allow it to enter the plant's vascular system for translocation upwards and downwards via the plant's long-distance transport systems namely xylem and phloem.

Biochemical and chemical activation processes of proinsecticides

S.No.	Insecticide group	Pro-insecticide	Biological active metabolite	Biological activation Process
1.	Pro-Quinazolines	Pyrifluquinazon	Deceacetylated Pyrifluquinazon metabolite B	Deacetylation
2.	Pro- Pyridine carboxamide	Flonicamid	TFNA-AM (4-Trifluoromethylnicotinamide)	N-dealkylation
3.	Pro- Pyridazine amides	Dimpropyridaz	Secondary amide metabolite (pyridazine pyrazolecarboxamide)	N-dealkylation
4.	Pro-Synthetic pyrethroids	Tralomethrin	Deltamethrin	Dehalogenation (Debromination)
5.	Pro-Oxadiazines	Indoxacarb	DCJW (N-decarbomethoxyllated JW062)	Ester hydrolysis, N-decarbomethoxylation
6.	Pro-Organochlorines	Aldrin	Dieldrin	Epoxidation
7.	Pro-Organophosphates	Acephate	Methamidophos	Demethylation, N-Deacylation
8.		Chlorpyrifos	Chlorpyrifos-oxon	CYP(Cytochrome P450enzymes)-dependent desulfuration
9.		Malathion	Malaoxon	Oxidative desulfuration
10.		Disulfoton	Disulfoton oxon	Oxidative desulfuration
11.		Trichlorfon	Dichlorvos (DDVP)	Dehydrochlorination
12.		Parathion	Paraoxon	Oxidative desulfuration
13.		Formothion	Dimethoate and Dimethoate carboxylic acid	Hydrolysis
14.		Schradan	Phosphoramide N-oxide	N-oxidation
15.		Profenofos	Profenofos Phosphorothiolate	Phosphorothiolate oxidation
16.		Dimethoate	Omethoate (Dimethoate oxon),	Oxidation
17.	Pro-Carbamates	Metam-sodium	Methyl isothiocyanate (MIC)	Dehydrosulfuration
18.		Carbosulfan	Carbofuran	Hydrolysis
19.		Benfuracarb	Carbofuran	Hydrolysis
20.		Furathiocarb	Carbofuran	Hydrolysis
21.		Thiodicarb	Methomyl	Hydrolysis
22.		Aldicarb	aldicarb sulfoxide	S-oxidation

S.No.	Insecticide group	Pro-insecticide	Biological active metabolite	Biological activation Process
23.	Pro-Phenyl pyrazoles (Fiproles)	Fipronil	Fipronil-sulfone	S-oxidation
24.	Pro-Nereistoxin	Cartap	Nereistoxin	Hydrolysis
25.		Bensultap	Nereistoxin	Hydrolysis
26.		Thiocyclam	Nereistoxin	Hydrolysis
27.	Pro-Neonicotinoids	Thiamethoxam	Clothianidin	N-oealkylation
28.	Pro-Thio-Urea Derivatives	Diafenthiuron	Diafenthiuron carbodiimide	Oxidative desulfurization
29.	Pro-Halogenated pyrroles	Chlorfenapyr	Pyrrole (CL 303268)	C-oxidation, N-dealkylation
30.	Pro-Fluoraliphatic sulfonamides	Sulfluramide	Perfluorooctane sulfonate (PFOS)	N-de-ethylation
31.	Pro-Tetramic acid derivatives	Spirotetramat	Spirotetramate-enol	Hydrolysis
32.		Spiropidion	Spiropidion-enol	Hydrolysis
33.	Pro-Tetranic acid derivatives	Spiromesifen	Spiromesifen-enol	Hydrolysis
34		Spirodiclofen	Spirodiclofen-enol	Hydrolysis
35.	Pro-Formamidines	Chlordimeform	Demethyl chlordimeform (DCDM)	N-demethylation
36.	Pro-Amidines	Amitraz	DPMF (Dimethylphenyl N methylformamidin)	Hydrolysis
37.	Pro-Anti juvenile hormone	Precocene	Dihydrodiol (DHD)	Oxidation
38.	Pro-Pyrazole	Cyenopyrafen	keto-enol forms	ester hydrolysis
39.	Pro-Beta-Ketonitrile	Cyflumetofen	De-esterified Cyflumetofen (AB-1) keto-enol forms	ester hydrolysis
40.	Pro-Carboxanilide	Pyflubumide	NH pyflubumide	N-dealkylation
41.	Pro-Naphthoquinones	Acequinocyl	DHN	Deacetylation
42.	Pro-Carbazate	Bifenazate	Bifenazate-Carbamic acid (diazene)	Oxidation

S.No.	Insecticide group	Pro-insecticide	Biological active metabolite	Biological activation Process
43.	Pro-meta diamide	Broflanilide	des-methyl-broflanilide	Dimethylation
44.	Pro-pyridazine pyrazolecarboxamide	Dimpropyridaz	Pyridazine pyrazolecarboxamide (secondary amide active form)	N-dealkylation
	Rodenticide			
45.	Pro-Inorganic	Zinc phosphide	Phosphine	Hydrolysis/acidolysis
46.	Pro-Organic	Flouroacetic acid or flouroacetamide	Flourocitric acid	Condensation via hydrolysis
47.	Pro-Plant origin	Scilliroside (Red squill)	Scillirosidin	Hydrolysis
48.	Pro-Non-anticoagulant	Bromethalin	Desmethyl-bromethalin	N-Dimethylation
	Microorganism			
49.	*Bacillus thuringiensis* (Dipel)	Delta-endotoxins	Proto-toxin	Proteolytic cleavage

8

Insecticides Mode of Action

Insecticide mode of action is divided in to four major chapters 1. Nerve & Muscle 2. Growth & development 3. Respiration & Energy production 4. Metabolic process 5. Unknown or Non-Specific (multisite inhibitors) (Source: IRAC). To understand the mode of action of insecticides, it is necessary to understand the function of nervous system and how it works in insect. The nervous system functions transmits information throughout the body. This system has three components: 1) **Peripheral nervous system (PNS)** to receive and transmit incoming signals (taste, smell, sight, sound, and touch) and to transmit outgoing signals to the muscles and other organs, effectively communicating them how to respond, and 2) **Central nervous system (CNS)** to interpret the signals and coordinate the body's responses and movements. 3) **Visceral or sympathetic nervous system (VNS)** directly connected with the brain that supplies nerves to the anterior part of the alimentary canal (foregut and midgut), heart, and spiracles. Some additional nerves arise from posterior compound ganglion of ventral nerve cord (VNC) which supply nerves to the posterior part of gut and reproductive system.

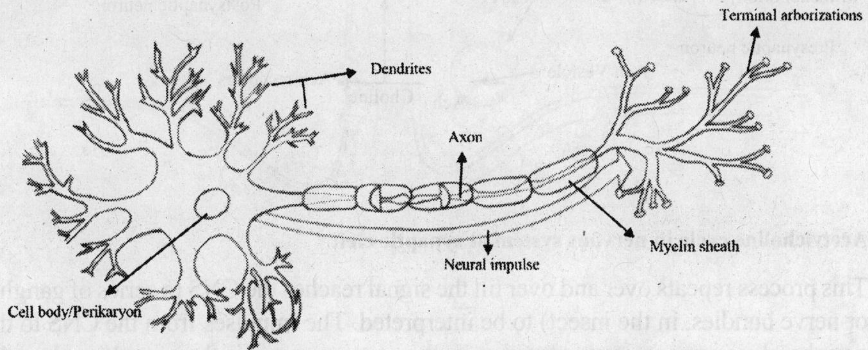

Basic structure of neuron

A neuron is a single nerve cell. It connects with other neurons and with muscle fibers (the basic units of muscles) through gaps at the end of each neuron. The nerve cells that transmit the stimuli from ganglia to effectors organs are called motor nerves. The nerve cells that transmit the stimuli from sensory nerves to ganglia organs are called sensory nerves. (Code word to remember: SAME-Sensory=Afferent neuron; Motor=Effect neuron).

The gap between neurons, or between a neuron and a muscle fiber, is called a **synapse**. Incoming signals (the sight of a predator or the odor of food) are transformed by the neuron into an electrical charge, which then travels down the length of the neuron. These charged particles (called ions) move through channels in the membrane of the neurons (axon conduction). There are four main types of channels to allow different ions to move along the neuron: **sodium channels, potassium channels, calcium channels, and chloride channels**. These channels contain gates that open or close in response to a certain stimulus, an important mechanism through which pesticides act. Action potential is said to be 'All-or-none" process that implies for a specific neuron, it is characterized with constant amplitude, variable frequency.

When an electrical charge reaches the neuron end, it stimulates a chemical messenger, called a neurotransmitter, to be released from the neuron ends and this neurotransmitter crosses the synapse and binds to a receptor on the receiving end of the next/succeeding neuron, then neurotransmitter is resorbed back into its originating neuron, and the nerve cell will be in the resting stage till the signal is received.

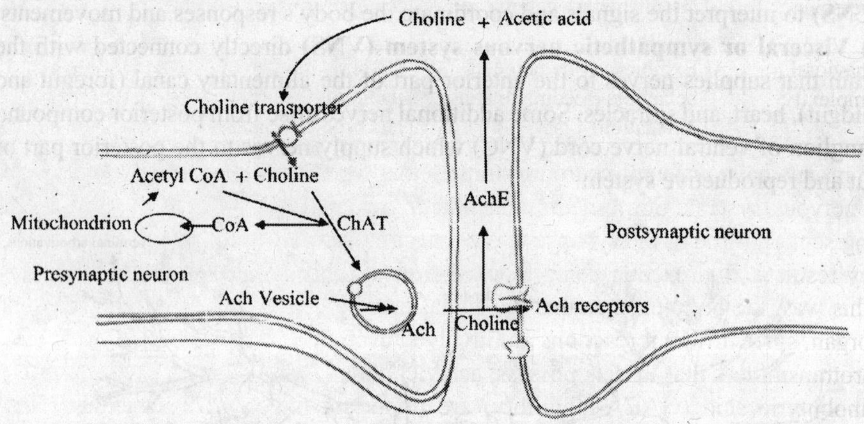

Acetylcholine cycle in nervous system at synaptic cleft

This process repeats over and over till the signal reaches the CNS (a series of ganglia, or nerve bundles, in the insect) to be interpreted. The impulses from the CNS to the peripheral nervous system continue in the same way until the signal reaches the appropriate muscles or organs.

Difference between insect nervous system and mammalian nervous system

Aspect	Insect Nervous System	Mammalian Nervous System
Anatomy and Complexity	Relatively simple, ganglia connected by nerves forming a ventral nerve cord	Highly complex, consisting of brain, spinal cord, and peripheral nerves
Brain Size and Organization	Small brain (ganglion) located dorsally in the head	Large and highly developed brain, divided into distinct regions such as cerebral cortex, cerebellum, and brainstem
Sensory Systems	Specialized sensory organs for light, sound, touch, and chemicals	Wide range of sensory organs including eyes, ears, nose, tongue, and skin
Neural Plasticity	Limited neural plasticity	High degree of neural plasticity, especially in the cerebral cortex
Neurotransmitter Systems	Primarily acetylcholine, GABA, glutamate and various neuropeptides	Acetylcholine, GABA, dopamine, serotonin, glutamate, and other neurotransmitters
Integration and Processing	Relies on simple neural circuits within ganglia and brain	Highly complex neural circuits within the brain and spinal cord
Behavioral Complexity	Exhibits a wide range of behaviors (feeding, mating, navigation)	Displays diverse and complex behaviors including social interactions, tool use, and problem-solving

The insects have different neurotransmitters that work at different sites throughout the nervous system. Some neurotransmitters are **excitatory** (result in the signal being sent on through the synapse to a connecting neuron), and few are **inhibitory** (they result in the reaction being blocked from traveling to a connecting neuron). In this way, the body ensures that the signal has the desired effect in each muscle or organ, since different reactions are involved even in a simple action. Among the neurotransmitters that insects possess, acetylcholine (Ach-excitatory) and gamma-aminobutyric acid (GABA-inhibitory) are important targets of insecticides. The difference between excitatory and inhibitory neurotransmitters are explained below.

Difference between excitatory neurotransmitters and inhibitory neurotransmitters

Aspect	Excitatory Neurotransmitters	Inhibitory Neurotransmitters
Definition	Neurotransmitters that promote the generation of action potentials in postsynaptic neurons	Neurotransmitters that inhibit the generation of action potentials in postsynaptic neurons
Action on Postsynaptic Neurons	Increase the likelihood of an action potential being generated by depolarizing the postsynaptic membrane	Decrease the likelihood of an action potential being generated by hyperpolarizing the postsynaptic membrane

Aspect	Excitatory Neurotransmitters	Inhibitory Neurotransmitters
Examples	glutamate, acetylcholine	Gamma-aminobutyric acid (GABA)
Receptors	Acts on ionotropic receptors (ligand-gated ion channels) and metabotropic receptors (G-protein coupled receptors)	Acts primarily on ionotropic receptors (ligand-gated ion channels)
Function	Enhance neuronal excitability and synaptic transmission	Reduce neuronal excitability and slow down the synaptic transmission
Effects	Can lead to neuronal activation, initiation of action potentials, and transmission of signals	Can lead to neuronal inhibition, prevention of action potentials, and modulation of signal transmission

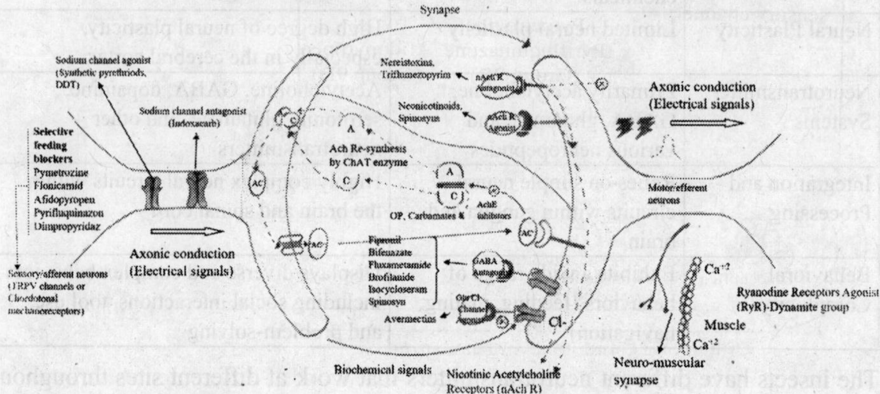

Mode of actions of insecticides in nervous system and muscles

1. Nerve & Muscle

A. Nerve

Mode of actions of insecticides in nervous system

i) Selective feeding blockers: Vanilloid-type transient receptor potential or Transient Receptors Vanilloid Potential (TRPV) channels play an important role in insect stretch receptor signaling (Chordotonal Organ). TRPV channel modulators affect chordotonal organs, stretch receptors in the insect joints that sense relative position of the body parts (proprioception). The modulation of TRPV channels generate continuous chordotonal nerve signals independent of joint movement. This makes insects deaf and uncoordinated, resulting in rapid feeding cessation, leading to starvation and ultimately death. This mode of action effectively works against sucking insect pest complex *viz*., thrips, whitefly, hoppers, aphids etc. The insecticides that are promising selective feeding blockers are Flonicamid, Pymetrozine, Pyrifluquinazon, Afidopyropen, and Dimpropyridaz. Flonicamid metabolite TFNA-AM indirectly activates TRPV channels and the mode of action is different from Pymetrozine, Pyrifluquinazon, and Afidopyropen that act precisely on nicotinamidase (NAAM) as inhibitors; it will stop the conversion process of nicotinamide to nicotinic acid. Pymetrozine, Pyrifluquinazon, and Afidopyropen act as TRPV channel modulators, but Dimpropyridaz is a Chordotonal organ modulator-an undefined target site.

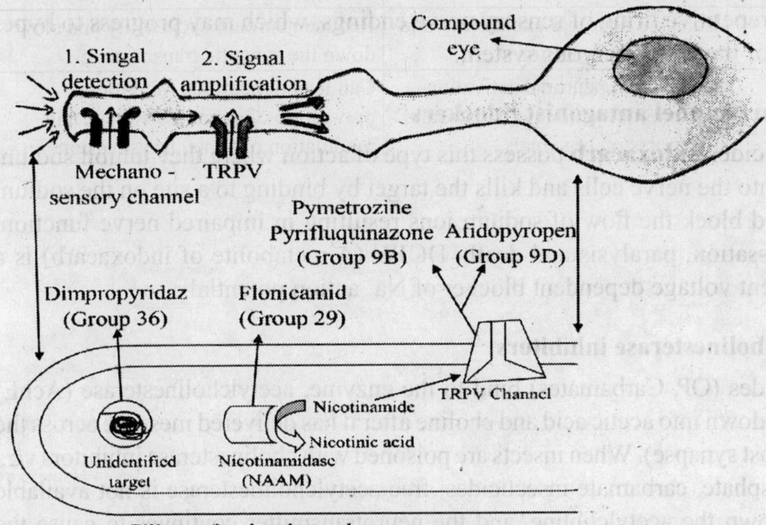

ii) Sodium channel modulators/agonists

Organochlorines and pyrethriods group show this type of action. It acts on the sodium channel to cause "leakage" of sodium ions and eventually make the neurons fire impulses spontaneously, causing the muscles to twitch causing symptoms called as "**DDT jitters**" followed by convulsions and death. DDT shows negative temperature correlation i.e. the lower the surrounding temperature, higher the toxic to insects.

Pyrethroids target sodium channels along the axon and that is the reason they are been called "**sodium channel modulators/agonists**" or **axonic nerve poisons**. In neuronal cells, the generation of an action potential by membrane depolarization involves the opening of cell membrane sodium channels and a rapid increase in sodium influx. The closure of sodium channels begins the process of action potential inactivation and a delayed sodium channel closure thus increases cell membrane excitability.

Pyrethroids modify the gating characteristics of voltage-sensitive sodium channels in insects thus delaying their closure. They are dissolved in the lipid phase of the membrane and bind to a receptor site on the alpha sub-unit of the sodium channel. The interaction of pyrethroids with sodium channels is highly stereospecific, with the 1R and 1S cis isomers binding competitively to one site and the 1R and 1S *trans* isomers binding non-competitively to another. The 1S forms do not modify channel function but do block the effect of the 1R isomers. The prolonged opening of sodium channels by the neurotoxic isomers of pyrethroids produces a protracted sodium influx, which is referred to as a sodium "**tail current**". This lowers the threshold of sensory nerve fibres for the activation of further action potential, leading to repetitive firing of sensory nerve endings, which may progress to hyper excitation of the entire nervous system.

iii) Sodium channel antagonist /blockers

The insecticide, **Indoxacarb** possess this type of action where they inhibit sodium ion entry into the nerve cells and kills the target by binding to a site on the sodium channel and block the flow of sodium ions resulting in impaired nerve function, feeding cessation, paralysis and death. DCJW (A metabolite of indoxacarb) is a highly potent voltage dependent blocker of Na^+ action potential.

iv) Acetylcholinesterase inhibitors

The pesticides (OP, Carbamates) bind to the enzyme, acetylcholinesterase (AchE) that breaks down into acetic acid and choline after it has delivered message across the synapse (post synapse). When insects are poisoned with cholinesterase inhibitors *viz.*, organophosphate, carbamate insecticides, free acetylcholinesterase is not available to break down the acetylcholine, and the neurotransmitter continues to cause the neuron to "fire," or send its electrical charge. This causes overstimulation of the nervous system, and the insect succumbs to death. Further, upon each exposure to an organophosphate or carbamate insecticide, more cholinesterase becomes bound and free AchE is unavailable to do its normal job. Although cholinesterase inhibition by carbamates is somewhat reversible, organophosphate inhibition is not reversible, where it means that the insecticide does not release the bound acetylcholinesterase.

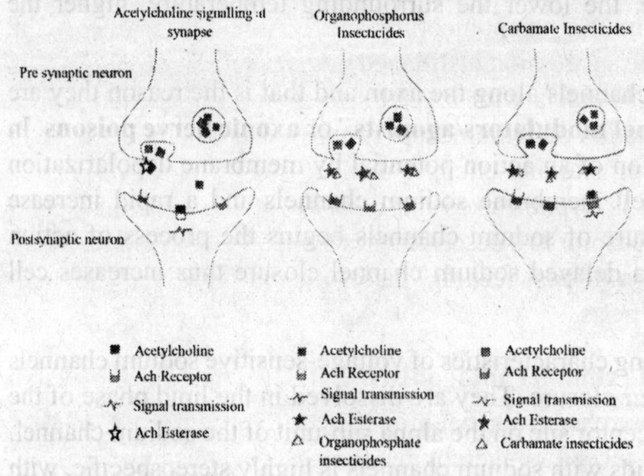

Organophosphates, carbamates mode of action

v) Nicotinic acetylcholine receptor (nAChR) agonists

There are various receptors present on the postsynaptic neuron and one important receptor is nicotinic acetylcholine receptor (**nAChR**). The insecticide group, **neonicotinoids** (imidacloprid, thiamethoxam, acetamiprid, clothianidin, thiacloprid, dinotefuran) act as agonists of the acetylcholine receptor. *i.e.* mimic the action of the neurotransmitter, acetylcholine (Ach). Though acetylcholinesterase is not affected by neonicotinoids, the nerve is continually stimulated and the end result is similar to that caused by acetylcholinesterase inhibitors i.e. overstimulation of the nervous system leading to poisoning and death.

The neonicotinoids show a closer mimic to the insects acetylcholine than for human acetylcholine giving this insecticides group more specificity to insects and less ability to poison humans. The fundamental differences between the nAChRs of insects and mammals confer remarkable selectivity for the neonicotinoids. Nicotinoids, with no ionization have poor binding affinity with the insect nAChR but the presence of an electron donating atom (as seen in neonicotinoids), increases the binding affinity with the insect nAChR. Thus, the neonicotinoids have higher toxicity to insects in contrast to nicotinoids. On the contrary, vertebrate nAChR prefers presence of one unit of positive charge for stronger binding. The ionized nicotine binds at an anionic subsite in the mammalian nAChR, thus resulting in high toxicity for mammals of nicotinoids.

Insect nicotinic acetylcholine receptors (nAChR) are distinct from those observed with neonicotinoids. Sulfoxaflor is a high-efficacy nAChR agonist with low affinity for the imidacloprid binding site. There is no cross-resistance with neonicotinoids.

vi) Nicotinic acetylcholine receptor (nAChR) antagonists

Bensultap, cartap hydrochloride, thiocyclam hydroxide are converted into nereis toxins, and being more toxic to insects, act as nicotinic acetylcholine receptors antagonist. Triflumezopyrim is a potent nicotinic acetylcholine receptor (nAChR) inhibitor that blocks neurotransmission in affected insects. Triflumezopyrim is the insecticide that primarily interacts with nAChRs to inhibit rather than activate receptors. The inhibitory action of Triflumezopyrim is unlike other product acting at the acetylcholine-binding site and elicits a distinct physiological response at the α2-nAChR subunit. It is hypothesized that nAChR inhibition occurs due to rapid modification of the receptor to a desensitized state resulting in reduced nerve stimulation, rapid cessation of pest feeding, increased mortality, and ultimately good control of important pests.

vii) GABA-gated chloride channel antagonists- BHC, Cyclodienes, Fipronil

GABA antagonists are those chemicals that bind to but do not activate GAMMA-AMINOBUTYRIC ACID receptors, thereby blocking the actions of endogenous GABA or GABA antagonists. As GABA cannot bind to the receptor, it does not activate channel (does not open) thereby it interferes/prevents the passage of chlorine ions through the channel, nothing but excitatory state continued the leads to convulsions and ultimately death will occur.

The insecticides namely, **BHC** (*gamma* isomer) is a neurotoxicant whose effects are normally seen within hours as increased activity, tremors, and convulsions leading to prostration. It too, exhibits a negative temperature correlation, but not as pronounced as that of DDT.

Similarly, another insecticides group, **Cyclodienes (Endosulfan)** also acts on the inhibitory mechanism called the GABA (gamma-aminobutyric acid) receptor. This receptor operates by increasing chloride ion permeability of neurons, i.e. prevent chloride ions from entering the neurons, and thereby antagonize the "calming" effect of GABA. Cyclodienes have a positive temperature correlation i.e. their toxicity increases with increase in the surrounding temperature.

Another insecticide group, **Fipronil** also blocks the gamma-aminobutyric acid (GABA) regulated chloride channel in neurons, thus antagonizing the "calming" effects of GABA.

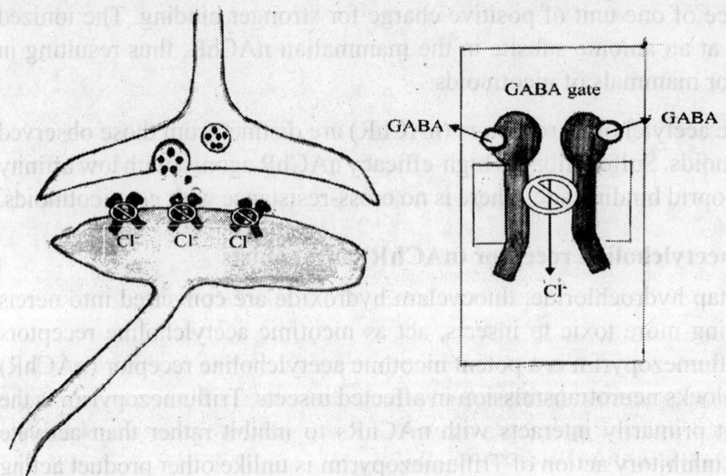

Fipronil prevents GABA from binding to GABA receptor

A new insecticide, Broflanilide also acts as a GABA-gated chloride channel allosteric modulator but its action site was suggested to be different from that of conventional GABA gated chloride channel blockers. Fluxametamide also is an allosteric modulator of GABA gated chloride channel. **Isocycloseram,** a new member of the isoxazoline class of insecticides, acts as a noncompetitive antagonist of the invertebrate γ-aminobutyric acid (GABA) receptor at a site distinct to that of fiproles and cyclodienes and displays an excellent efficacy and selectivity against invertebrate pests.

Bifenazate (acaricide) affects the GABA-gated chloride channel as an agonist, and make ways to open the GABA gates that leads to chloride ion influx due to this, nerve impulses unable to travel (inhibitory effect) eventually casing flaccid paralysis and death. The mode of action is similar to glutamate gated chloride channel activators.

viii) Nicotinic acetylcholine receptor (nAChR) agonists and GABA-gated chloride channel antagonists

Spinosad and spinetoram acts by disrupting binding of acetylcholine in nicotinic acetylcholine receptors at the postsynaptic cell. Spinosad and spinetoram are acetylcholine receptor agonists as well as GABA antagonists. The exact mechanism of spinosad is somewhat different from that of the neonicotinoid class, but the result is same. Insecticides from Spinosyn group will **increase the excitatory state** of insect as **nAChR agonist** and **GABA antagonist** (No inhibitory function), casing nerve failure and death.

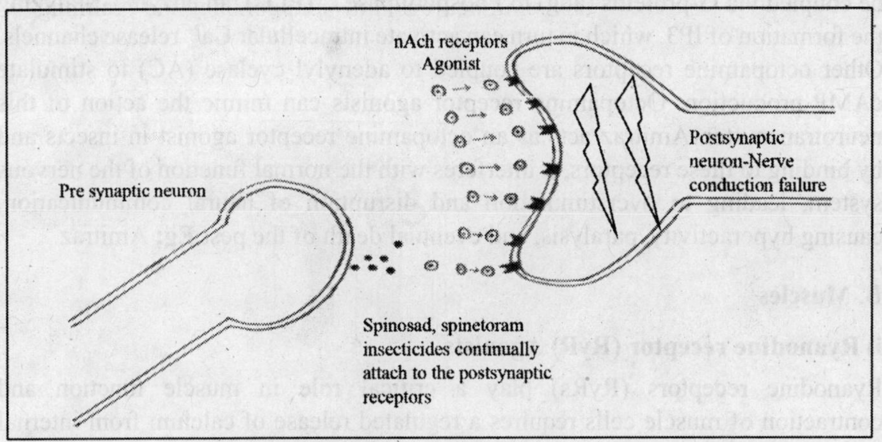

Spinosyn group binding to nicotinic acetylcholine receptors and GABA receptors

ix) Glutamate-gated chloride channel (GluCl) agonists/Chloride channel activators/Glutamate-gated chloride channel (GluCl) allosteric modulators

Avermectins (Abamectin and Emamectin benzoate) act on glutamate gated chloride channels and activate the chloride influx that leads to inhibitory function of the insect or mite. It cannot reach the threshold required for action potential and causes membrane to hyperpolarize, making it less excitatory, and decreasing nerve transmission, finally resulting in flaccid paralysis and death. The avermectin treated insects lose visible activity, such as feeding and egg laying stops shortly after exposure, though mortality may not occur for several days.

x) Monoamine oxidase enzyme inhibitors

Monoamine oxidase (MAO) is an essential enzyme involved in the regulation of neurotransmitter levels and has an important role in the physiology, metabolism, and behavior of many organisms. Monoamine oxidases (MAO) (EC 1.4.3.4) are a family of enzymes that catalyze the oxidation of monoamines, employing oxygen to clip off their amine group and found bound to the outer membrane of mitochondria in most cell types of the body. **Eg:** Chlordimeform are inhibitors of the enzyme monoamine oxidase, which is responsible for degrading the neurotransmitters, norepinephrine and serotonin that results in the accumulation of these compounds, which are known as biogenic amines. The affected insects turn quiescent and die.

xi) Octopamine receptor agonists

Octopamine is a major modulatory neurotransmitter in insects it binds to G-protein-coupled Receptors (GPCRs) on the postsynaptic membrane, which can be coupled via G-proteins (abg) to Phospholipase C (PLC), an enzyme catalyzing the formation of IP3, which in turn can activate intracellular Ca^{2+} release channels. Other octopamine receptors are coupled to adenylyl cyclase (AC) to stimulate cAMP production. Octopamine receptor agonists can mimic the action of this neurotransmitter. Amitraz acts as an octopamine receptor agonist in insects and by binding to these receptors, it interferes with the normal function of the nervous system, leading to overstimulation and disruption of neural communication, causing hyperactivity, paralysis, and eventual death of the pest **Eg;** Amitraz

B. Muscles

i) Ryanodine receptor (RyR) Agonists

Ryanodine receptors (RyRs) play a critical role in muscle function and contraction of muscle cells requires a regulated release of calcium from internal stores (sarcoplasmic reticulum of muscle cells) into the cell cytoplasm. To be more specific, ryanodine receptors act as selective ion channels, modulating the release of calcium. One commercialized insecticide group is **Rynaxypyr (chlorantraniliprole)** that binds to the ryanodine receptors, causing uncontrolled release and depletion of internal calcium, preventing further muscle contraction thereby treated pests exhibit rapid cessation of feeding, lethargy, regurgitation and muscle paralysis, ultimately leading to death. In simple words, Rynaxypyr controls insect pests through a new mode of action i.e. activation of insect ryanodine receptors (RyRs). Similar action by impairing muscle function, was observed with **Cyantraniliprole (Cyazypyr),** but is also called cross spectrum insecticide as it kills pests that belong to Hemipteran and Lepidoptera orders. Tetraniliprole and flubendiamide, cyclaniliprole also falls under this group.

2. Growth and Development

Insect's endocrine system produces the hormones needed for growth and development into an adult form. These hormones are

- Juvenile hormones- Prevent insect from moulting to next instar
- Moulting hormones or ecdysones- Involved in re-absorption of old cuticle and also in deposition, hardening and tanning of new cuticle

The chemicals that interfere with growth, development, disrupt metamorphosis and reproduction in insects are known as Insect growth regulators (IGRs). Insects poisoned with IGRs cannot molt or reproduce, and eventually they die. In a normal insect, juvenile hormone is circulated throughout the insect's body and "tells" the insect to stay in its current stage. After a certain amount of time, the insect stops producing juvenile hormone, and the insect metamorphoses, or changes, into its next life stage. When an insect is poisoned by an IGRs commercially available as JH mimics viz. methoprene, fenoxycarb, pyriproxifen that mimics juvenile hormone, the insect doesn't receive the signal to metamorphose because, even though the insect may have stopped producing juvenile hormone, the IGR is still circulating throughout its body and sending the signal to stay in the current stage i.e. juvenile hormones keep insects at specific instars when present and allow maturation when absent.

Another hormone important in metamorphosis is ecdysone. The ecdysone receptor agonists group namely diacylhydrazines (Methoxyfenozide, Halofenozide, Tebufenozide, Chromafenozide); pyrimidines (Benzpyrimoxan) interfere with the production of ecdysone, causing the insect to be unable to molting. The new insecticide molecule developed, benzpyrimoxan is an ecdysone titer disruptor and it possess a novel MoA with unique ecdysis disruption symptoms and does not show cross-resistance with existing molecules currently used against plant hoppers.

Chitin synthase enzyme inhibitors

Chitin synthase enzyme, a vital and almost indestructible part of the insect exoskeleton. The benzoylureas group of compounds (diflubenzuron, flufenoxuron, lufenuron, novaluron, teflubenzuron, buprofezin) act on the larval stages by inhibiting or blocking the synthesis of chitin to be specific, inhibits chitin synthase enzyme, a vital and almost indestructible part of the insect exoskeleton. The typical effects on developing larvae are the rupture of malformed cuticle or death by starvation.

3. Respiration and Energy

Electron transport system (ETS) and oxidative phosphorylation

The steps in the respiratory process occur in order to release and utilize the energy stored in NADH + H^+ and $FADH_2$. This is accomplished when they are oxidized through the electron transport system and the electrons passed on to O_2 resulting in the formation of H_2O. The metabolic pathway through which an electron passes from one carrier to another is called the electron transport system (ETS) and it takes place in the inner mitochondrial membrane. Electrons from NADH produced in the mitochondrial matrix during the citric acid cycle are oxidized by an NADH dehydrogenase (complex I). Electrons are then transferred to ubiquinone located within the inner membrane. Ubiquinone also receives reducing equivalents via $FADH_2$ (complex II) that is generated during oxidation of succinate in the citric acid cycle. The reduced ubiquinone is then oxidized with the cytochrome c, a small protein attached to the outer surface of the inner membrane and acts as a mobile carrier (bridge) for the transfer of electrons between complex III and complex IV. Complex IV refers to cytochrome c oxidase complex containing cytochromes, α and α_3, and two copper centres. The four main protein complexes (I, II, III, and IV) and two mobile carriers (ubiquinone and cytochrome c) facilitate electron transfer.

When the electrons pass from one carrier to another via complex I to IV in the electron transport chain, they are coupled to ATP synthase (complex V) for the production of ATP from ADP and inorganic phosphate by ATP synthase enzyme. Unlike photophosphorylation where the light energy is utilized for the production of proton gradient required for phosphorylation, in respiration it is the energy of oxidation-reduction utilized for the same process. It is for this reason that the process is called oxidative phosphorylation.

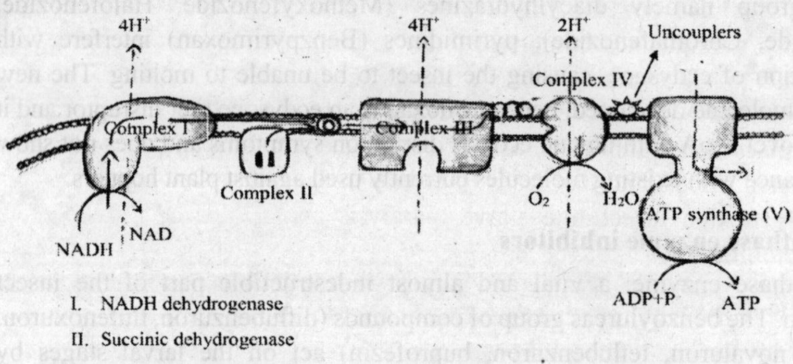

I. NADH dehydrogenase
II. Succinic dehydrogenase
III. Cytochrome C reductase
IV. Cytochrome C oxidase

a) **Mitochondrial complex I electron transport inhibitors:** Electrons from NADH produced in the mitochondrial matrix during the citric acid cycle are oxidized by an NADH dehydrogenase (complex I). These electrons are not transferred to ubiquinone located within the inner membrane because of mitochondrial complex I electron transport inhibitors and in this way, the electron transport chain stops and doesn't enter to complex II, eventually no energy production and insect/mite starve leading to paralysis and death.

The pesticides that fall in this category are fenazaquin, fenpyroximate, pyridaben, tolfenpyrad, pyflubumide, rotenone.

b) **Mitochondrial complex III electron transport inhibitors :** The reduced ubiquinone is oxidized with the cytochrome c, a small protein attached to the outer surface of the inner membrane and acts as a mobile carrier (bridge) for the transfer of electrons between complex III and complex IV. The presence of complex III electron transport inhibitors doesn't allow the electrons to be transferred to complex IV, that ultimately cause insect to starve, paralysis and death. Eg: Bifenazate, Acequinocyl

c) **Mitochondrial complex IV electron transport inhibitors**: Complex IV refers to cytochrome c oxidase complex containing cytochromes, α and α_3, and two copper centres. The mitochondrial complex IV electron transport inhibitor's major role is to stop the next step oxidative phosphorylation in turn ATP (energy) production and finally make insect/mite to starve and cause mortality. Examples: Aluminium Phosphide, Zinc phosphide,

d) **Uncoupler of oxidative phosphorylation:** It is an "uncoupler" or inhibitor of oxidative phosphorylation, preventing the formation of the crucial energy molecule, adenosine triphosphate (ATP). In this situation, energy is not produced that lead to starvation of insect and eventually paralysis and death. It is the connection between electron transport chain and oxidative phosphorylation. Eg : Chlorfenapyr, Sulfuramid

e) **ATP synthase enzyme inhibitors:** The passing of electrons from one carrier to another via complex I to IV in the electron transport chain, are coupled to ATP synthase (complex V) for the production of ATP from ADP and inorganic phosphate by ATP synthase enzyme. This process doesn't take place due to inhibition of ATP synthase enzyme by specific pesticides poisoned insects and that leads to failure in energy production (ATP) and finally treated insect/mite starve, paralysis and death. Eg : Propargite, Diafenthiuron

4. Metabolic Processes

i. Acetyl CoA carboxylase inhibitors (Disturbing fatty acid biosynthesis) : Acetyl-CoA carboxylase (ACC) catalyses the first step in fatty acid biosynthesis or lipid biosynthesis where malonyl-CoA is produced from acetyl-CoA in a two-step reaction. ACCase inhibitors prevent biosynthesis of fats needed for growth and development resulting in incomplete molts and desiccation of the insect.

Eg: Spirodiclofen, Spiromesifen, Spirotetramat, Spiropidion

5. Unknown or Non-Specific (multisite inhibitors)

Unlike many other insecticides and miticides, non-specific or multi-site inhibitors do not act on a distinct target site but likely disrupt important physiological functions and compounds with the unknown designation may act on a distinct target site, but the mechanism of action has not been conclusively determined. The compounds are Borax, Azadirachrin, Dicofol, Pyridalyl.

Mode of action of Insecticides

Insecticides that target nervous system			
Target system/ process	**Target site**	**Mode of action**	**Insecticide group-examples**
Nervous system	Nerve Axon	Sodium Channel modulators/agonist	DDT, BHC, Lindane, synthetic pyrethroids
		Sodium Channel Blockers/ antagonist	Indoxacarb, Metaflumizone
	Nerve synapse	Acetycholine esterase inhibitors	Organophosphates, Carbamates
	Post Nerve synapse	Nicotinic acetylcholine receptor agonists	Neonicotinoids
		Nicotinic acetylcholine receptor antagonists	Neriestoxin analogues- Bensulftap, Triflumezopyrim
		GABA receptor blockage/ antagonist	Fipronil, Fluxametamide, Endofulfan, Bifenazate, Broflanilide, Isocycloseram
		Nicotinic acetylcholine receptor agonists but slight differs from neonicotinoids and GABA antagonist	Spinosad, Spinetoram
		Glutamate receptor stimulation/Chloride Channel activators	Emamectin benzoate, abamectin, milbemectin
	Biogenic amines	Octopaminergic agonists	Amitraz
Insecticides that do not target the nervous system			
Target system/ process	**Target site**	**Mode of action**	**Insecticides**
Muscles	Muscle stimulation	Muscular calcium channel	flubendiamide chlorantraniliprole, Tetraniliprole, cyantraniliprole
Growth and Development	Juvenile Hormones	Mimics JH action -JH degradative enzyme/ receptor	Methoprene, Kinoprene, Hydroprene
	Moulting Hormones	Ecdysone receptor agonists/ moulting disruptor	Methoxyfenozide Tebufenozide, Halofenozide, Benzpyrimoxan
	Chitin synthesis	Block chitin formation	Diflubenzuron, lufenuron, novaluron, buprofezin,

Respiration & Energy production	Inhibit energy production	Inhibitors mitochondrial electron transport chain by binding with complex I	Fenazaquin, Fenpyroximate Pyridaben Tolfenpyrad, Rotenone
		Inhibitors mitochondrial electron transport chain by binding with complex III	Bifenazate, Acequinocyl
		Inhibitors mitochondrial electron transport chain by binding with complex IV	Aluminium Phosphide, Zinc phosphide
		Uncouplers of oxidative phosphorylation via disruption of proton gradient.	Chlorfenapyr
		Inhibition of oxidative phosphorylation or ATP formation i.e. ATP synthase ihibitors	Diafenthiuron, propargite
Metabolic processes	-	Inhibitors of lipid synthesis	Spiromesifen, spirodiclofen
		Insect midgut membrane disruptor	*Bacillus thuringiensis*
Unknown or Non-Specific (multisite inhibitors).	-	-	Borax, Azadirachrin, Dicofol, Pyridalyl,

Main Group and Primary Site of Action	**Sub-group, class or exemplifying Active Ingredient**	**Active Ingredients**
1 Acetylcholinesterase (AChE) inhibitors	1A	Carbamates
	1B	Organophosphates
2 GABA-gated chloride channel blockers	2A	Cyclodiene Organochlorines
	2B	Phenylpyrazoles(Fiproles)
3 Sodium channel modulators	3A	PyrethroidsPyrethrins
	3B	DDT, Methoxychlor
	3B	DDT, Methoxychlor
4 Nicotinic acetylcholine receptor (nAChR) competitive modulators	4A	Neonicotinoids
	4B	Nicotine
	4C	Sulfoxaflor
	4D	Flupyradifurone
	4E	Dicloromezotiaz, Fenmezoditiaz, Triflumezopyrim
	4F	Flupyrimin

Main Group and Primary Site of Action	Sub-group, class or exemplifying Active Ingredient	Active Ingredients
5 Nicotinic acetylcholine receptor (nAChR) allostericmodulators – Site I	Spinosyns	Spinetoram, Spinosad
6 Glutamate-gated chloride channel (GluCl) allosteric modulators	Avermectins, Milbemycins	Abamectin, Emamectin benzoate, Lepimectin, Milbemectin
7 Juvenile hormone receptor modulators	7A	Hydroprene, Kinoprene, Methoprene
	7B	Fenoxycarb
	7C	Pyriproxyfen
8 * Miscellaneous non-specific (multi-site) inhibitors	8A	1,3-Dichloropropene, Methyl bromide and otheralkyl halides
	8B	Chloropicrin
	8C	Cryolite (Sodium aluminum fluoride), Sulfurylfluoride
	8D	Borax, Boric acid, Disodium octaborate, Sodiumborate, Sodium metaborate
	8E	Tartar emetic
	8F	Dazomet, Metam, Methyl isothiocyanate
9 Chordotonal organ TRPV channel modulators	9B	Pymetrozine, Pyrifluquinazon
	9D	Afidopyropen
10 Mite growth inhibitors affecting CHS1	10A	Clofentezine, Diflovidazin, Hexythiazox
	10B	Etoxazole
11 Microbial disruptors of insect midgut membranes	11A	*Bacillus thuringiensis* and the insecticidal proteins they produce
	11B	*Bacillus sphaericus*
12 Inhibitors of mitochondrial ATP synthase	12A	Diafenthiuron
	12B	Azocyclotin, Cyhexatin, Fenbutatin oxide
	12C	Propargite
	12D	Tetradifon
13 * Uncouplers of oxidative phosphorylation via disruption of the protongradient Energy metabolism		Chlorfenpyr, DNOC, Sulfluramid, Nereis toxin analogues

Main Group and Primary Site of Action	Sub-group, class or exemplifying Active Ingredient	Active Ingredients
14 Nicotinic acetylcholine receptor (nAChR) channel blockers Nerve action {Compounds affect the function of this protein, but it is not clearthat this is what leads to biological activity}		
15 Inhibitors of chitin biosynthesis affecting CHS1 Growth regulation {Strong evidence that action atone or more of this class of proteins is responsible for insecticidal effects}	Benzoylureas	Bistrifluron, Chlorfluazuron, Diflubenzuron, Flucycloxuron, Flufenoxuron, Hexaflumuron,Lufenuron, Novaluron, Noviflumuron, Teflubenzuron, Triflumuron
16 Inhibitors of chitin biosynthesis, type 1 Growth regulation {Target protein responsible for biological activity is unknown, oruncharacterized}	Buprofezin	Buprofezin
17 Moulting disruptors, Dipteran Growth regulation {Target protein responsible for biological activity is unknown, oruncharacterized}	Cyromazine	Cyromazine
18 Ecdysone receptor agonists Growth regulation {Strong evidence that action at this protein is responsible for insecticidal effects}	Diacylhydrazines	Chromafenozide, Halofenozide, Methoxyfenozide, Tebufenozide
19 Octopamine receptor agonists Nerve action {Good evidence that action atone or more of this class of protein is responsible for insecticidal effects}	Amitraz	Amitraz
20 Mitochondrial complex III electron transport inhibitors – Qo site Energy metabolism {Good evidence that action atthis protein complex is responsible for insecticidal effects}	20A	Hydramethylnon
	20B	Acequinocyl
	20C	Fluacrypyrim
	20D	Bifenazate

Main Group and Primary Site of Action	Sub-group, class or exemplifying Active Ingredient	Active Ingredients
21 Mitochondrial complex I electron transport inhibitors Energy metabolism {Good evidence that action atthis protein complex is responsible for insecticidal effects}	21A	Fenazaquin, Fenpyroximate, Pyridaben,Pyrimidifen, Tebufenpyrad, Tolfenpyrad
	21B	Rotenone (Derris)
22 Voltage-dependent sodium channel blockers Nerve action {Good evidence that action atthis protein complex is responsible for insecticidal effects}	22A	Indoxacarb
	22B	Metaflumizone
23 Inhibitors of acetyl-CoA carboxylase Lipid synthesis, growth regulation {Good evidence that action at this protein is responsible for insecticidal effects}	Tetronic and Tetramic acid derivatives	Spidoxamat Spirodiclofen, Spiromesifen,Spiropidion, Spirotetramat
24 Mitochondrial complex IV electron transport inhibitors Energy metabolism {Good evidence that action atthis protein complex is responsible for insecticidal effects}	24A	Aluminium phosphide, Calcium phosphide, Phosphine, Zinc phosphide
	24B	Calcium cyanide, Potassium cyanide, Sodiumcyanide
25 Mitochondrial complex II electron transport inhibitors Energy metabolism {Good evidence that action atthis protein complex is responsible for insecticidal effects}	25A	Cyenopyrafen, Cyflumetofen
	25B	Pyflubumide
28 Ryanodine receptormodulators Nerve and muscle action {Strong evidence that action atthis protein complex is responsible for insecticidal effects}	Diamides	Chlorantraniliprole, Cyantraniliprole, Cyclaniliprole Flubendiamide, Tetraniliprole

Main Group and Primary Site of Action	Sub-group, class or exemplifying Active Ingredient	Active Ingredients
29 Chordotonal organ nicotinamidase inhibitors Nerve action {Strong evidence that action at this protein is responsible for insecticidal effects}	Flonicamid	Flonicamid
30 GABA-gated chloride channel allosteric modulators Nerve action {Strong evidence that action atthis protein complex is responsible for insecticidal effects}	Isoxazolines Meta-diamides	Fluxametamide, IsocycloseramBroflanilide
31 Baculoviruses Host-specific occluded pathogenic viruses {Midgut epithelial columnar cellmembrane target site – undefined}	Granuloviruses (GVs) Nucleopolyhedroviruses (NPVs)	*Cydia pomonella* GV *Thaumatotibia leucotreta* GV *Anticarsia gemmatalis* MNPV *Helicoverpa armigera* NPV
32 Nicotinic Acetylcholine Receptor (nAChR) Allosteric Modulators - Site II Nerve action {Strong evidence that action atone or more of this class of protein is responsible for insecticidal effects}	GS-omega/kappa HXTX-Hv1a peptide	GS-omega/kappa HXTX-Hv1a peptide
33 Calcium-activated potassium channel (KCa2) modulators Nerve action {Strong evidence that action at this protein is responsible for insecticidal effects}	Acynonapyr	Acynonapyr
34 Mitochondrial complex III electron transport inhibitors – Qi site Energy metabolism {Modulation of this protein complex has been clearly demonstrated and the specific target site responsible for biological activity is distinct fromGroup 20}	Flometoquin	Flometoquin

Main Group and Primary Site of Action	Sub-group, class or exemplifying Active Ingredient	Active Ingredients
35 RNA Interference mediated target suppressors Activation of the RNAi mechanism which specifically reduces abundance of the target messenger RNA (mRNA) resulting in the reduction of the protein encoded by the mRNA.	Ledprona	Ledprona
36 Chordotonal organ modulators – undefined targetsite Nerve action {Modulation of chordotonal organ function has been clearly demonstrated, but the specific target protein(s) responsible for biological activity are distinct from Group 9 and Group 29 andremain undefined}	Pyridazine pyrazolecarboxamides	Dimpropyridaz
37 Vesicular acetylcholine transporter (VAChT) in hibitor Nerve action Bind to VAChTs, causing cholinergic synaptic transmission block resulting in nervous system shutdown and paralysis. VAChTs are involved in loading acetylcholine into synaptic vesicles	Oxazosulfyl	Oxazosulfyl
UN* Compounds of unknown or uncertain MoA	Azadirachtin	Azadirachtin
	Benzoximate	Benzoximate
	Benzpyrimoxan	Benzpyrimoxan
	Bromopropylate	Bromopropylate
	Chinomethionat	Chinomethionat
	Dicofol	Dicofol
	Lime sulfur	Lime sulfur
	Mancozeb	Mancozeb

The following notes provide additional information about particular sub-groups.

Sub-groups	Notes
3A & 3B	Because DDT is no longer used in agriculture, this is only applicable for the control of insect vectors of human disease such as mosquitoes.
4A, 4B, 4C, 4D, 4E & 4F	Although these compounds are believed to have the same target site, current evidence indicates that the risk of metabolic cross-resistance between subgroups is low.
10A	Hexythiazox is grouped with clofentezine because they exhibit cross-resistance, even though they are 01structurally distinct. Diflovidazin has been added to thisgroup because it is a close analogue of clofentezine and is expected to have the same mode of action.
11A	Different *Bacillus thuringiensis* products that target different insect orders may be used together without compromising their resistance management. Rotation between certain specific *Bacillus thuringiensis* microbial products may provide resistance management benefits for some pests. Consult product-specific recommendations. *B.t.* Crop Proteins: Where there are differences among the specific receptors within the midguts of target insects, transgenic crops containing certain combinations of the listed proteins provide resistance management benefits.
22A & 22B	Although these compounds are believed to have the same target site, current evidence indicates that the risk of metabolic cross-resistance between subgroups is low.
25A & 25B	Although these compounds are believed to have the same target site, current evidence indicates that the risk of metabolic cross-resistance between subgroups is low.

Penetration of insecticides in to plant

After spraying, the water droplet evaporate and active ingredient reach into plant system.

Routes of Penetration

1. Foliar Penetration

Cuticular Penetration: Insecticides applied to leaves must penetrate the cuticle, the waxy outer layer that protects the plant. The cuticle is composed of waxes and cutin, which can affect the absorption of insecticides.

Stomatal Penetration: Stomata are small openings on the leaf surface that allow gas exchange. Insecticides can enter through these openings, especially if they are in a liquid form.

2. Root Uptake

Soil Application: Insecticides applied to the soil are absorbed by plant roots. From the roots, they can be transported throughout the plant via the xylem, the vascular tissue responsible for water and nutrient transport.

Seed Treatment: **Systemic Insecticides**: Insecticides applied to seeds can be taken up by the growing plant and transported throughout its tissues. This provides protection against pests that attack young seedlings.

Factors Affecting Penetration of insecticides in to the plant system

1. Physicochemical Properties of the Insecticide

Solubility: Water-soluble insecticides are more easily absorbed and transported within the plant. Lipophilic insecticides may penetrate the cuticle more effectively but may face challenges in moving through aqueous plant tissues.

Molecular Size: Smaller molecules especially nano pesticides generally penetrate plant tissues more easily than larger ones.

2. Plant Characteristics

Cuticle Thickness: A thicker cuticle can act as a stronger barrier to penetration.

Leaf Surface: The presence of trichomes (hair-like structures) and the waxy composition of the leaf surface can influence the absorption of insecticides.

Root Structure: The root system's health and structure can affect the uptake of soil-applied insecticides.

Penetration of insecticides in to insects

The penetration of insecticides on insects refers to the process by which these chemical substances enter the insect's body. This penetration is crucial for the effectiveness of insecticides, as it determines how well the substance can reach its target site within the insect and exert its toxic effects. There are several routes through which insecticides can penetrate an insect's body *viz.*, cuticle, mouth or ingestion, spiracle etc.

Factors affecting the penetration of insecticides

1. Chemical nature of insecticides
2. Components of cuticle
3. Nature of the carriers and solvents

1. **Chemical nature of insecticides:** Apolar or non polar insecticides penetrate faster than polar.
2. **Components of cuticle:** The insect cuticle possess of two phases lipophilic-hydrophilic system. The outermost phase is waxy and hydrophobic (lipophilic) in nature and most pesticides being nonpolar, the first barrier improves contact action and as a result, an insecticide's contact toxicity is identical to its oral toxicity. In contrast, mammalian skin is generally resistant to pesticide entry, acute oral toxicity is substantially higher in animals than contact toxicity.

3. **Nature of the carriers and solvents:** The carriers *viz.* detergents easily penetrate the outer most cement layer of the epicuticle. Oil as solvents in formulation dissolves the epicuticular wax that allow the insecticides to enter and disrupt the internal protein of the cuticle.

9

Synergism and Potentiation

Synergists are chemicals that significantly increase the toxicity of an insecticide while being nearly non-toxic on their own. Thus, non-toxic chemicals that contribute to the increased toxicity of insecticides are referred to as synergists. Many chemical substances have the potential to synergize with poisons. Some enzyme inhibitors work well synergistically by inhibiting enzymes that metabolize pesticides, particularly in insecticide-resistant strains. When mixed with other insecticides, some insecticides increase the toxicity of the latter at non-toxic dosages. When a mixture is more harmful than expected based on the sum of its separate efficacies when administered alone, the interaction between substances is known as synergy. Antagonistic interactions among components reduce a mixture's efficacy.

Synergism can also develop between two or more poisons. Several statistical methods have been proposed to quantify the synergy of both non-toxic and hazardous substances. Toxin interactions are assessed using bioassays, which involve mixing toxicants (two or more) in a specific ratio. The combination is then used to prepare serial dilutions. Alternatively, the poisons can be supplied sequentially in the same amounts as in the mixture. Bioassays are also used to determine the dose-mortality response of each toxin on the same insect strain. The data are subjected to probit analysis, and the toxin interactions can be evaluated using the statistical methods listed below.

The following is a short list of enzyme inhibitors that are frequently used to synergize the toxicity of specific insecticides in resistant insect strains, and hence provide useful indicators of metabolic resistance mechanisms.

Table 1: List of synergists

Synergists	Enzyme systems	Reference
Piperonyl butoxide (PBO)	Mixed function oxidases	Chadwick, 1963
Sesamex	Mixed function oxidases	Beroza and Barthel, 1957
Propargloxy pthalimide (PGP)	Mixed function oxidases	Casida, 1970
1,2,4-trichloro-3-propynyloxybenzene (TCPB)	Mixed function oxidases	Casida, 1970
Triphenyl phosphate (TPP)	Hydrolase	Casida, 1970
s,s,s-tributylphosphoro-trithioate (DEF)	Hydrolase	Casida, 1970
Diethyl maleate (DEM)	Glutathione S-transferase	Motoyama et al., 1990
4-4'-dichloro-α-methylbenzhydrol (DMC)	DDTase	Casida, 1970

Combination index (CI)

Chou and Talalay, 1984 described methods to derive a combination index (CI) that is useful to determine the additive, synergistic or antagonistic relationship of components in a mixture.

$$\mathrm{CI_X} = \frac{LD_x^{a(m)}}{LD_x^a} + \frac{LD_x^{b(m)}}{LD_x^b} + \frac{LD_x^{a(m)}}{LD_x^a} + \frac{LD_x^{b(m)}}{LD_x^b}$$

In the given mixture, $LD_x^{a(m)}$ and $LD_x^{b(m)}$ represent the dosages of toxicants 'a' and 'b' that caused mortality x. LD_x^a and LD_x^b represent the lethal concentrations of toxicants 'a' and 'b' that, when used individually, result in mortality 'x'. **If the two insecticides exhibit an additive effect (potentiation), the value of the combination index is 1; otherwise, it is ≤1 synergistic; and ≥1, antagonistic.**

For instance, if the LD_{50} (lethal dose for 50% of the population) of a pyrethroid was 0.35 and 0.42 for an OP (organophosphate) compound, the mixture can be classified as synergistic (CI = 0.77) when the pyrethroids and OP were used at a dose of 0.1 + 0.15 in the mixture. It would be considered additive (CI = 1.0) if their dose was 0.14 + 0.18, and antagonistic (CI = 1.69) if it was 0.2 + 0.3.

Where LDx py (m) and LDx op(m) are lethal doses of parathyroid and OP used in the mixture giving the mortality the mortality x; LDx py & LDx OP are lethal doses of parathyroid and organophosphates required to produce the same mortality x when used alone. When two insecticides have an additive effect, the combination index value CI = 1, when two insecticides are synergistic the lethal dose of the mixture is lower than expected and CI = 1, when two insecticides are synergistic the lethal dose of the mixture is lower than expected and CI <1, when two insecticides are antagonistic toxicity of the mixture is higher than expected and CI > 1. To apply this method, the Calcusyn® software developed by BIOSOFT of USA (Chou and Hay ball, 1996) was used. The combination index value (CI) integrating the toxicity of each product in a mixture and their interaction were calculated at mortality levels of 50 and 90 %. The degree of synergism based on CI value are given in table 2.

Table 2: Degree of synergism based on combination index value as given by Chou and Talalay (194)

Range of CI	Symbol	Description
< 0.1	+++++	Very strong synergism
0.1-0.3	++++	Strong synergism
0.3-0.7	+++	Synergism
0.7-0.85	++	Moderate synergism
0.85-0.9	+	Slight synergism
0.9-1.10	±	Nearly additive
1.10-1.20	-	Slight antagonism
1.20-1.45	--	Moderate antagonism

Range of CI	Symbol	Description
1.45-3.3	- - -	Antagonism
3.3-10	- - - -	Strong antagonism
>10	- - - - -	Very strong antagonism

Dose Reduction Index

A measure of how much the dose of each pesticide in a synergistic combination may be reduced at a given effect level compared with the doses of each pesticide alone

Intrinsic toxicity and synergistic mortalities

Toxicity of a mixture originates from the sum of the insecticide intrinsic toxicities (% M_{py} for parathyroid and %M_{op} for OP and any synergistic effect. Intrinsic toxicity was calculated using probit mortality relationship obtained with the insecticide used alone. The toxicity because of the synergism of mixture was estimated as % $M_{(py+op)}$ = 100-(%M_{py}+%M_{op}).

Tabashnik, 1992, proposed the following simple similar-action model to evaluate synergism among toxins with similar modes of action.

Expected $LD_{50\,(m)} = [\, r_a / LD_{50(a)} + r_b / LD_{50(b)} + r_c / LD_{50(c)} \,]^{-1}$

The variables a, b, and c represent the components of the mixture. LD_{50} (m) refers to the median lethal dose of the mixture. The variables r_a, r_b, and r_c represent the relative proportions of components a, b, and c utilized in the mixture. $LD_{50(a)}$, $LD_{50(b)}$, and $LD_{50(c)}$ represent the median lethal doses of components a, b, and c, respectively. The calculated LD_{50} value is subsequently compared to the measured LD_{50} value to determine if it falls within the 95% confidence intervals. If the anticipated LD_{50} exceeds the upper threshold, the toxins exhibit synergistic effects, and if it falls below the lower threshold, the toxins display antagonistic effects.

For instance, the lethal dose 50 (LD_{50}) of a pyrethroid is 0.35, while the LD_{50} of an organophosphate (OP) compound is 0.42. These two poisons were combined in a mixture with a ratio of 0.4 for the pyrethroid and 0.6 for the OP compound. As a consequence, the LD_{50} (95% FL) of the mixture was determined to be 0.25, with a range of 0.14 to 0.42. The LD_{50} value of 0.388 falls within the range of fiducial limits, indicating that the mixture is not synergistic but potentially additive. It should be noted that the equations proposed by Chou and Talalay indicate a state of synergism based on the same set of data.

Synergists play a crucial role in clarifying resistance processes, particularly when they are targeted inhibitors of a specific mechanism that causes resistance, such as a detoxifying enzyme. The chemical is administered at a non-toxic dosage with incremental dilutions of the insecticide to evaluate the alterations in toxicity between susceptible and resistant strains. The LD_{50} values of the insecticide, both alone and in conjunction with the synergist, are determined using the bioassay data for both the susceptible and resistant strains. The presence of improved metabolic detoxification is evident when the synergist has a minimal impact on the toxicity of the insecticide in susceptible strains, but significantly increases toxicity in the resistant strain to a greater degree. The resistance mechanism is subsequently inferred by analyzing the enzyme-inhibition features of the synergist.

For instance, if the combination of sesamex and cypermethrin on a strain that is resistant to pyrethroids produces LD_{50} values that are similar to those of the susceptible strain, it can be concluded that resistance is caused by monooxygenases. This inference is made because sesamex specifically inhibits the cytochrome P450 group of monooxygenases. Below, we give a selection of representative data.

The effectiveness of a synergist in overcoming resistance mechanisms is represented by

$$\text{Synergist ratio "SR"} = \frac{LD50r}{LD50r^{+x}}$$

Where LD_{50r+x} is the LD_{50} of the insecticide + synergist 'x' of resistant strain and LD_{50r} is the LD_{50} of the insecticide on the resistant strain.

Synergism is calculated by using a fixed non-toxic dose of a synergist in bioassays with an insecticide to generate LD_{50} values.

$$\%\ \text{Synergism} = \frac{LD_{50r}}{LD_{50r}^{+x}} \times 100$$

Where LD_{50r+x} is the LD_{50} of the insecticide + synergist 'x' of resistant strain and LD_{50s} is the LD_{50} of the insecticide on the susceptible strain.

The interpretation of results obtained from a diagnostic dosage is more straightforward. The diagnostic dose is employed in conjunction with a predetermined non-harmful dose of the synergist. The mortality increase caused by the synergist is calculated and represents synergism. For instance, if the mortality rate is 24% when using cypermethrin at a concentration of 0.1 µg, and 88% when using cypermethrin at a concentration of 0.1 µg in combination with PBO at a concentration of 20.0 µg, then the synergistic effect can be calculated as 88% - 24% = 64%. Nevertheless, the aforementioned approaches tend to exaggerate the level of synergism caused by the inhibition of enzymes linked with resistance, as they fail to consider the degree of synergism in the strain that is sensitive to the treatment. From a practical standpoint, the significance may be minimal as it only demonstrates the degree to which the synergist enhances the toxicity of insecticide in a resistant strain, compared to the toxicity of the insecticide on a susceptible strain. The current resistance synergy is determined using the following formula:

$$\%\ \text{Actual resistance related synergism 'Sr'} = \frac{LD_{50r}}{LD_{50r}^{+x}} \times 100$$

Wherein LD_{50r+x} is the LD_{50} of the insecticide + synergist 'x' of resistant strain and LD_{50s}^{+x} is the LD_{50} of the insecticide + synergist 'x' of the susceptible strain.

Potentiation-Potentiation in combination products of pesticides refers to the phenomenon where the combined effect of two or more pesticides results in a greater impact than the sum of their individual effects.

Combination Insecticide Products

The combination insecticide products were introduced with a view to delay or prolong the insect resistance compared to individual insecticides used in pest management. Most of the combination products developed/registered in India were a mixture of organophosphate and synthetic pyrethroids. Later on the combination insecticides (pre mix or ready mix insecticides) developed and marketed has compounds from different groups [Eg., organophosphate (acephate) + neonicotinoid (imidacloprid)] as one of their component

Few Insecticide combinations showing dual mode of action

Insecticide combination	Dual mode of action
Chlorpyrifos 50%+ Cypermethrin 5%EC (Nurelle D 505®)	Acetyl choline esterase enzyme inhibitor (Nerve Synapse) + Sodium channel agonist (Nerve axon)
Acephate 50% + Imidacloprid 1.8% SP (Lancer gold®)	Acetyl choline esterase enzyme inhibitor (Nerve Synapse) + nAch receptor agonist (Nerve post synaptic)
Acephate 35% + Buprofezin 15 % WP (Tapuz®)	Acetyl choline esterase enzyme inhibitor (Nerve Synapse) + Chitin synthesis inhibitor
Buprofezin 5.65% + Deltamethrin 0.72% EC (Dacedi®)	Chitin synthesis inhibitor+ Sodium channel agonist (Nerve axon)
Betacyfluthrin 8.49% + Imidacloprid 19.81% OD (Solomon®)	Sodium channel agonist (Nerve axon) + nAch receptor agonist (Nerve post synaptic)
Thiamethoxam 12.6% + Lambda cyhalothrin 9.5 ZC (Alika®)	nAch receptor agonist (Nerve post synaptic)+ Sodium channel agonist (Nerve axon)
Ethiprole 40%+ Imidacloprid 40%WG (Glamore®)	GABA antagonist+ Sodium channel agonist (Nerve axon)
Chlorantaniliprole 9.3 %+Lamdacyhalothrin 4.6 % (Ampligo 150ZC®)	Ryanodine receptor agonist (Muscle)+ Sodium channel agonist (Nerve axon)
Teflubenzuron 75 G/L + Alphacypermethrin 75 G/L SC (Imunit®)	Inhibiting the formation and growth of the cuticle (exoskeleton)+Sodium channel agonist (Nerve axon)
Methoxyfenozide 20 % + Chlorantraniliprole 5 % SC (Engage®)	Ecdysone receptor agonist (Moulting hormone) + Ryanodine receptor agonist (Muscle)
Novaluron 5.25% + Indoxacarb 4.5% SC (Pleothora®)	Inhibiting the formation and growth of the cuticle (exoskeleton)+Sodium channel antagonist (Nerve axon)

Co-toxicity Coefficient (CTC)

The Co-toxicity Coefficient (CTC) is a parameter used in toxicology and pest management to evaluate the combined effects of two or more substances (such as pesticides) when they are used together. It helps determine whether the combination has synergistic, antagonistic, or additive effects on the target organism (insects).

The Co-toxicity Coefficient is calculated using the following formula,

CTC = (Observed toxicity ÷ Expected toxicity) × 100

- Observed toxicity: Actual effect (e.g., mortality percentage) observed when two substances are used together.
- Expected toxicity: Effect predicted based on the individual toxicities of the substances when used separately.

CTC Scale

CTC Value	Effect	Interpretation
CTC > 120	Synergistic Effect	The combined toxicity is greater than the sum of the individual toxicities.
80 ≤ CTC ≤ 120	Additive Effect	The combined toxicity is approximately equal to the sum of the individual toxicities.
CTC < 80	Antagonistic Effect	The combined toxicity is less than the sum of the individual toxicities.

Example

- Pesticide A causes 40% mortality, and Pesticide B causes 30% mortality when applied alone.
- When combined, they cause 70% mortality.
- **Expected toxicity** = (40 + 30) = 70%.
- **Observed toxicity** = 70%.
- **CTC** = (70 ÷ 70) × 100 = 100 (Additive effect).

Pseudo synergism: A phenomenon in toxicology and pharmacology where the perceived enhancement of a combined effect (synergism) between two or more agents is not due to an actual interaction between the agents themselves but is instead caused by external factors. These external factors create the illusion of synergism, even though the agents are not inherently synergistic.

How It Differs from True Synergism

- **True Synergism**: Two or more agents directly interact, enhancing each other's effects.
- **Pseudosynergism**: The enhanced effect is due to external factors rather than a true interaction between the agents.

Pseudo synergism must be carefully distinguished from true synergism to ensure accurate interpretation of data in toxicology

10

Insects Resistance to Insecticides

Insecticide resistance is a growing concern for those who need insecticides for medical, veterinary, and agricultural control of insect pests. In response to the introduction of DDT resistance in 1947, the incidence of resistance has increased annually by an alarming amount. Resistance is "the inherited ability of a strain of some organism to survive doses of a toxicant that would kill the majority of individuals in a normal population of the same species" (WHO, 1957).

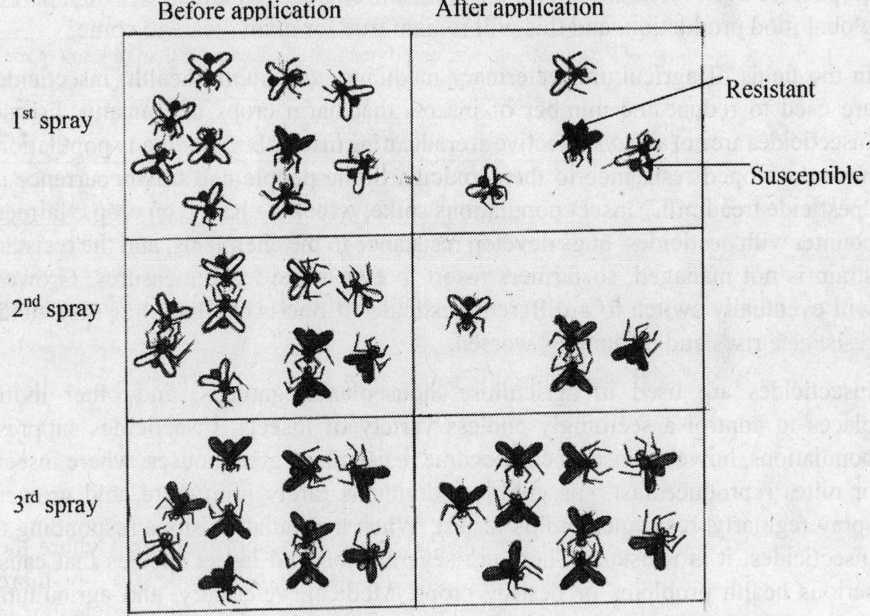

Fig. 1: Resistance development in insects

As a result of continued insecticide application, the proportion of resistant insects increases compared to that of susceptible and the population becomes increasingly difficult to control

Above photograph theoretical example illustrating the increase of insecticide resistance levels in a pest population. Some individuals (black) with genetic traits allowing them to survive insecticide applications can reproduce; if the selection pressure is frequent, they easily become the preponderant part of the population (Fig. 1).

An adaptation to a pesticide leads to reduced susceptibility to it among the pest population targeted by that pesticide. In nature, pests develop resistance to chemicals through natural selection. In the most resistant organisms, their genetic traits are passed on to their descendants. Since 1945, at least 500 species of pests have developed resistance to pesticides (Anonymous, 2007), but other sources estimate as many as 1000 species (Miller, 2004).

Almost all pesticides used against malaria-transmitting mosquitoes have now become ineffective. In addition to causing malaria, the organisms are also resistant to drugs used to treat it in humans. Several pesticides cannot control the corn earworm, which attacks many crop plants worldwide, including cotton, tomatoes, tobacco, and peanuts. While alternative methods to control pests and diseases in crops have been researched for many years, pesticides remain an essential part of global food production, and this will remain true for many years to come.

In the fields of agriculture, veterinary medicine, and public health, insecticides are used to reduce the number of insects that harm crops or transmit disease. Insecticides are not always effective at eradicating insects because many populations have developed resistance to their toxicity. Some people call this occurrence as "pesticide treadmill." Insect populations spike, wreaking havoc on crops; farmers counter with pesticides; bugs develop resistance to the chemicals, and the resistant strain is not managed, so farmers resort to ever more toxic measures. Growers will eventually switch to a different pesticide (if one is available) as insecticide resistance rises and difficulties worsen.

Insecticides are used in agriculture, houseplants, gardens, and other living places to control a seemingly endless variety of insects. Insecticides suppress populations, however insects can become resistant. In greenhouses, where insects or mites reproduce fast, susceptible individuals rarely immigrate, and growers spray regularly, resistance builds fastest. When a population stops responding to insecticides, it is resistant. There are several thousand insect species that cause serious health problems or destroy crops. Medical, veterinary, and agricultural insect pest control is increasingly threatened by pesticide resistance.

1. History of insecticide resistance

A. L. Melander initially documented pesticide resistance in 1914. A lime sulphur, an inorganic insecticide used against the San Jose scale (*Quadraspidiotus perniciousus*), an orchard pest in Clarkston valley, Washington destroyed all scales in one week in conventional orchards but 90% survived after two weeks in a resistant orchard.

Similarly, codling moth in 1928 to lead arsenate; two species of ticks to sodium arsenite; and three scales of black scale, *Saisseria oleae* (Oliver) to HCN in 1912; California red scale, *Aonidiella aurantii* (Maskell) in 1913; citricola scale, *Coccus pseudomugnoliurum* (Kuwana) in 1925). Before 1944, pesticide resistance was rare, but introduction of DDT and other synthetic organic insecticides increased the frequency enormously and then insects start showing resistant to inorganics, DDT, cyclodienes, organophosphates, carbamates, pyrethroids, juvenile hormone analogues, chitin synthesis inhibitors, avermectins, neonicotinoids, and microbials.

After the introduction of synthetic organic insecticides i.e. DDT, new resistant populations appeared at a rapid rate. Every 2-5 years before 1946, a new resistant species arose. Many insects had problems with all major pesticide classes. Each year since 1947, DDT resistance has developed alarmingly. Between 1946 (1 year after DDT was introduced) and 1954, the rate rise to 1 or 2 species per year; between 1954 and 1960, it was 17 per year, and it remained constant until 1980, where at least 428 species from 14 orders and 83 families of insects and acarines showed resistance. In 1980, more species were resistant to cyclodienes (including lindane) than to DDT, organophosphorus compounds, carbamates, and pyrethroids. However, carbamate, organophosphorus, and pyrethroid compounds have seen the greatest percentage increases in resistant species in recent years due to the introduction of new materials and the discontinuation of chlorinated hydrocarbons in many regions due to resistance and environmental concerns. At least 447 pesticide-resistant arthropod species exist worldwide (Callaghan, 1991). Many insects have acquired pesticide resistance to new insecticides with distinct modes of action from the major four classes including neonicotinoids.

Resistance occurred in thirteen insect orders, but more than 90% of arthropod species with resistant populations are Diptera (35%), Lepidoptera (15%), Coleoptera (14%) Hemiptera (in the broad sense, 14%) and mites. The high number of resistant insects belong to order Diptera is due to intensive insecticide use against disease-carrying mosquitoes. Medical and veterinary pests make up 41% of dangerous resistant species, whereas agricultural pests make up 59%. Many insecticide-resistant populations exist in many species. Statistical investigations show that crop pests with four to ten generations per year that eat or sucked plant cell contents and gain resistance faster.

More than twenty times as many resistant pest species exist as resistant predators and parasitoids. This trend is likely a result of the biological distinctions between beneficials and pests and the lack of focus on resistance in beneficials. The current body of research refutes the claim that natural enemies evolve resistance more slowly than pests do because their inherent levels of detoxifying enzymes are lower than those of the pests. More evidence points to the opposite theory that natural enemies take longer to develop resistance after being exposed to insecticides because they experience food shortages as a result of the widespread use of these chemicals.

Georghiou (1986) found pesticide resistance in 100 plant diseases, 55 weeds, 5 rodents, and 2 worms. Many economically significant pests exhibit group, cross, and multiple resistance. Toxicological and phenotypical approaches have been developed to detect pesticide resistance, and based on 100 years of experience, insecticide resistance prevention systems have been developed. Resistance monitoring and plant protection techniques reduce harmful impact on agroecosystems in these systems.

All over the world, wherever insecticides are used, has led to an increase in vector-borne diseases, an increase in pesticide hazards in the environment, crop losses, a decrease in product quality, an increase in production costs, higher number of secondary pests, and a variety of socioeconomic impacts. In the course of developing chemical control, much has been learned about mechanisms of resistance and genes, biology, and operational parameters that enable pest populations to thwart man's best efforts. As a result, we need to develop management strategies based on this knowledge to prevent the emergence of multi resistance, which threatens the successful management of numerous arthropod pests at present, and to avoid similar problems with other pest species in the future. The following are a list of the resistant arthropods that live in agriculture and urban areas (Table 1.)

Table 1: Top 20 Resistant arthropods in agricultural and urban ecosystems

Rank	Common name	Scientific Name	Number*	Ecosystem
1	Two-spotted spider mite	*Tetranychus urticae*	94	Agricultural
2	Diamondback moth	*Plutella xylostella*	92	Agricultural
3	Green peach aphid	*Myzus persicae*	76	Agricultural
4	House fly	*Musca domestica*	62	Urban
5	Colorado potato beetle	*Leptinotarsa decemlineata*	55	Agricultural
5	Sweetpotato whitefly	*Bemisia tabaci*	55	Agricultural
7	Southern cattle tick	*Rhipicephalus microplus*	50	Agricultural
8	Cotton aphid	*Aphis gossypii*	49	Agricultural
9	Corn bollworm	*Helicoverpa armigera*	48	Agricultural
9	European red mite	*Panonychus ulmi*	48	Agricultural
11	German cockroach	*Blattella germanica*	42	Urban
12	Southern house mosquito	*Culex quinquefasciatus*	40	Urban
13	Beet armyworm	*Spodoptera exigua*	38	Agricultural
13	Oriental leafworm moth	*Spodoptera litura*	38	Agricultural
15	House mosquito	*Culex pipiens pipiens*	36	Urban
16	Yellow fever mosquito	*Aedes aegypti*	35	Urban
16	Tobacco budworm	*Heliothis virescens*	35	Agricultural
18	Hop aphid	*Phorodon humuli*	34	Agricultural
19	Red flour beetle	*Tribolium castaneum*	33	Urban
20	African cotton leafworm	*Spodopotera littoralis*	30	Agricultural

* Number of active ingredients to which the pest has exhibited documented resistance

(Source: Mark Whalon and David Mota-Sanchez)

2. Biochemistry of resistance in different group of insecticides

2.1 Resistance to DDT

The physiological mechanism that distinguishes DDT-resistant insects from susceptible ones has been extensively studied. It appears to be unrelated to DDT cuticle penetration or cell components viz. lipids. Enzymatic detoxication appears to be the most major DDT-resistance mechanism. Dehydrochlorination of DDT to DDE has been found in various insect species, and this mechanism is thought to make mosquitoes and houseflies resistant to DDT. DDT dehydrochlorination may be one of several insect detoxication methods. Thus, the human body louse appears to break down DDT in vitro without producing DDE. *Drosophila* that resist DDT can oxidize it to kelthane (1,1-bis(p-chlorophenyl)2,2,2 trichloroethanol).

2.2 Resistance to Hexachlorocyclohexane (BHC)

Resistant houseflies metabolize BHC alpha, beta, gamma, and delta-isomers faster than susceptible ones. The alpha- and gamma-isomers of BHC rapidly convert to water-soluble metabolites, removing 4 or more chlorine atoms per molecule. First discovered detoxication product was pentachlorocyclohexene. C-S bonds were inferred from fly metabolism of BHC to produce water-soluble metabolites. Since reduced glutathione is needed to convert alpha or gamma-BHC into water-soluble metabolites in vitro, it appears to provide the C-S bond sulphur. The metabolite may be a pentachlorocyclohexyl-cysteine compound that produces dichlorothiophenols when alkali is applied. BHC appears to poison insects' central nervous systems, and resistant insects' ganglions can tolerate direct gamma-BHC crystal application.

2.3 Resistance to Pyrethrins

Pyrethrins cause reversible paralysis, suggesting insects have insecticide detoxication systems. Their toxic effects are reversible without synergists, and their paralytic action temperature coefficient is positive, but low temperatures increase fatality.

When pyrethrins are given to a small part of an insect's body, the insect loses water quickly all over its surface, even if the insect's muscles aren't overactive and it isn't breathing faster. Most people, and probably rightly so, think that pyrethrins are quickly broken down in the insect body into non-toxic materials. They think this is why symptoms go away quickly after an effective but not overpowering amount. Some work has been made in finding the things that break down. There are also both indirect and direct signs that some synergists work by stopping pyrethrins and allethrins from breaking down in the body. Since the many effective synergists have very different chemical structures, it is hard to imagine that they could all work through the same biochemical process.

2.4 Resistance to organophosphorus insecticides

Until recently, resistance to chemicals of this class was slow to develop and small in the field and lab, except for greenhouse mites. Newer phosphate pesticides have replaced chlorinated hydrocarbons in fly control, and reports of increasing resistance. Metcalf (1959) detailed the metabolic pathways that oxidize organophosphorus insecticides to active antiesterases. Parathion, Malathion, Phorate and Diazinon oxidize P=S to P=O groups. Dipterex (O,O-dimethyl-2,2,2-trichloro-l-hydroxyethyl phosphonate) is an exception: dehydrochlorination converts it to dichlorvos, a strong cholinesterase inhibitor. However, phosphatases and carboxyesterases can convert organophosphorus pesticides to inactive metabolites. Organisms are vulnerable or resistant to organophosphorus insecticides depending on enzyme activation and deactivation. The main way organophosphorus insecticides detoxicate insects is phosphoryl hydrolysis.

Malathion, S-(1,2-bis(ethoxycarbonyl)ethyl) O,O-dimethyl phosphorodithioate, is detoxified through hydrolysis of the P-S and S-C links, as well as hydrolysis of the carboxylic ester linkages before or after oxidation. Phosphoryl hydrolysis appears to predominate in the housefly, and the rate of excretion of harmless malathion metabolites is the same in susceptible and resistant houseflies. Resistant houseflies collect parathion (O,O-diethyl O-p-nitrophenyl phosphorothioate) to a lesser level than susceptible strains. Furthermore, the resistant strain degraded paraoxon more quickly than the susceptible strain. Because parathion resistance is linked to paraoxon resistance, it is likely that a quicker rate of paraoxon excretion is also implicated in parathion resistance. Without attempting to establish a hypothesis that encompasses all organophosphorus insecticides, it appears highly likely that resistance is linked to the development of effective hydrolytic detoxication systems.

2.5 Resistance to carbamates

Carbamate insecticides are phenol esters containing N-methyl- or N, N-dimethylcarbamic acid. They have received a lot of attention, especially since Sevin (1-naphthyl N-methyl carbamate) was introduced as an agricultural pesticide. Carbamates are structural counterparts of acetylcholine, and as such, many of them are cholinesterase inhibitors. Although resistant flies absorbed this carbamate faster, four times as much unmodified insecticide accumulated internally in vulnerable flies as in resistant flies within two hours of administration. Furthermore, resistant flies digested 85% of the ingested substance while susceptible flies metabolized only 23%. Piperonyl butoxide inhibited the detoxication mechanisms, resulting in a pattern of pesticide absorption and internal buildup identical to that reported in susceptible flies. Piperonyl butoxide's extraordinary action is similar to its inhibition of pyrethrin detoxication.

3. Factors affecting evolution of resistance

The development of pesticide resistance, which is preadaptive in nature, is dependent on the presence of genes required for resistance to the chemical administered within the population. The evolution of resistance in a population occurs in stages. There is a latent phase during which the resistance genes become segregated and connected with other genes that govern factors favorable to the pest's survival; this may be accompanied by a notable increase in tolerance to the chemical.

As a result, the resistance of the population rapidly increases, contributing to multiple factors. When the initial pesticide fails, these associated resistance features are inherited, enabling rapid development of resistance to unrelated types of chemicals. Many factors influence resistance development, including the species' nature and the chemical control techniques used. Genetic, biological, and operational factors influence the kinetics of resistance, according to Georghiou and Taylor. Genetic parameters include population frequency and dominance of resistant alleles; biological parameters include generation time and offspring per generation; and behavioural parameters include population isolation, mobility, and migration.

A number of soil-inhabiting species have been exposed repeatedly to soil applications of aldrin/dieldrin with varying levels of resistance. A corn rootworm [*Diabrotica longicornis* (Say)], which produces only one generation per year, becomes resistant after eight to ten years of exposure (*Hylemya* spp.).

Genetic and biological factors differ from pest to pest and are independent of our actions. On the other hand, we can directly control operational factors that affect resistance development, such as the chemical's type and persistence in the environment or its application, such as treatment frequency and crop coverage. These factors determine the selection pressure, which in turn affects the evolution of resistance. In extreme situations, resistance may evolve most rapidly under the following conditions:

a. A prolonged persistence of the compound in the environment
b. Slow-release formulations
c. Low density population thresholds
d. High percentage of the population is reached and selected for treatment
e. Selection against larvae, or worse, against both larvae and adults
f. Coverage of a large geographical area
g. Populations are selected against every generation

In the field, when control failure is recognized (by which time resistance is well along in its evolutionary path), reapplication of the treatment is usually followed by increasing dosage and shortening application intervals. After the limits of practicability of these remedial actions have been reached, the pesticide is usually replaced with another more effective material, providing that one is available. After a while, the process is usually started all over again. This scenario has played out throughout history, and the toll, which cannot be measured in money alone, has been devastating. Pesticide resistance has caused tremendous losses and even complete crop failures, as well as deteriorating product quality in many cases. In some cases, production costs have increased because of the requirement to use higher dosages of chemicals and to apply them more frequently, as well as because soil fertilization and irrigation practices have increased in an effort to save crops whose pest control has been unsatisfactory. The resurgence of pest populations can occur when the ecological disturbances of pest-beneficial species density relationships are disrupted while the pest is still susceptible, but in some cases, it has been brought on by increasing dosages and application numbers to combat resistance. Increased pesticide usage has also resulted in many minor or secondary pests reaching major economic status.

Pesticide resistance can lead to (i) disease recurrence and its effects on population health and welfare when insect vector control deteriorates or fails, as has happened for malaria in several countries; (ii) increased pesticide burdens to humans and the environment; and (iii) socioeconomic disruptions in agricultural communities.

The formation of resistance to a single compound may involve the selection of mechanisms that give cross-resistance not only to some or all related pesticides, but also to chemicals of very different types. Because resistance genes may persist long time, it is often not possible to reuse materials after groups have lost their resistance. If you keep doing this with a series of different chemicals, you may lose the ability to use chemicals to control the pest in the long run.

IPM programmes might be destroyed by resistance. Such solutions normally demand a lot of time and resources and may employ one pesticide that is non-disruptive to the system when used at the right amount. Increasing pesticide dosage, frequency, or using a pesticide with inadequate performance data could make the programme unusable.

Despite these issues, chemical control is still established in most cases, although time may be running out. The current drop in pesticide introduction shows that resistance may exceed the discovery of effective new materials. Since DDT, more than one million chemicals have been tested for pesticide activity, and despite recent relaxation of federal registration procedures, it will be difficult and expensive to design, develop, and introduce new pesticides to stop the rise of resistant pests to uncontrollable levels, which appears imminent for several serious pests. Understanding the factors that influence resistance evolution and the mechanisms behind pesticide resistance is crucial for addressing current issues and preventing future ones.

Biological, genetic, and operational factors in resistance development

Factor	Potential for resistance development	
	Lower	Upper
Biological factors		
Population size	Small	Large
Reproductive potential	Low	High
Generation turnover	One or less generations per year	Many generations per year
Type of reproduction	Sexual	Asexual
Dispersal	Little	Much
Pesticide metabolism	Difficult	Easy
Number of target sites of the pesticide	Multiple sites	Single, specific
Pest host range	Narrow	Wide
Genetic factors		
Occurrence of resistance genes	Absent	Present
Number of resistance mechanisms	One	Several
Gene frequency	Low	High
Dominance of resistance genes	Recessive	Dominant
Fitness of "R" individuals	Poor	Good
Protection provided by "R" gene	Poor	Good
Cross resistance	Negative or none	Positive
Past selection	None	Significant
Modifying genes	Absent	Present
Operational factors		
Activity spectrum of the pesticide	Narrow spectrum	Broad spectrum
Pesticide application rate	Label rate; heterozygote's killed (If R gene is incompletely dominant)	More than label rate; only some homozygous resistant individuals survive and reproduce (especially if there is little immigration)
Application coverage	Good	Poor
Treatment frequency	Low	High
Presence of secondary pests	Absent (only the target pest is treated)	Present (non targeted pests are also treated)
Life stages treated with related pesticides	Single	Multiple
Persistence	Short	Long
Number of crops treated	One	Many

Factor	Potential for resistance development	
	Lower	Upper
Pest control tactics	Multiple control tactics (chemical, biological, cultural)	Continuous use of single method or compound
Non target effects	Selective activity, no effect on natural enemies	Non selective, natural enemies also killed

3.1 Biological factors

3.1.1 Population size

Population size increases resistance. Greater insect populations increase resistance. Even if the fraction of resistant individuals is low, a huge group may survive following a pesticide application. If repeated pesticide treatments kill most susceptible individuals, some resistant survivors may find mates and pass on the resistance genes. If the pest population is limited, the few resistant survivors may have poor mating opportunities and delayed resistance development.

The extent of pest management problem depends on number of resistant individuals. Even with considerable resistance, a low insect invasion is unlikely to cause a control problem. However, a heavy pest infestation will make field control difficult even with moderate resistance.

3.1.2 Reproductive potential

The amount of offspring, seeds, or spores that each "parent" can make has a big effect on how quickly pest populations become resistant. All pests that reproduce sexually and are killed by pesticides will have more resistant individuals of insects.

In insects, raising a high number of progeny increases the likelihood of more individuals carrying the resistance gene, and hence, if pesticide use continues, the likelihood of choosing individuals containing one or two resistant alleles. The bigger the number of survivors containing resistance genes, the more likely it is that heterozygote or homozygote individuals will mate. As a result, the frequency of resistance genes in the population may grow.

3.1.3 Generation turnover

The rate of resistance development is influenced by generation turnover. Because the pest population is picked just once a year, resistance development for insects, weeds, and plant pathogens will be relatively slower if there is only one generation each year rather than numerous.

3.1.4 Type of reproduction

Resistance can occur through both sexual and asexual reproduction. Sexual reproduction drives genomic rearrangement. However, once resistance has been selected, it is more likely to spread fast through asexual reproduction. For example, in aphids, most reproduction is asexual throughout the year; if a portion of an aphid population has a resistance gene, this portion will survive while the susceptible portion of the population is eliminated by intensive pesticide application, and the surviving resistant population may rapidly increase and leading to a rapid development of resistance unless a resistance management programme is implemented.

3.1.5 Dispersal

Short and long-range insect movement can alter a population's vulnerability in a field or area. Insects can fly, and heterozygous or susceptible individuals can breed with treatment survivors, diluting resistance in the population. Refugia preserve sensitive pest populations for this reason. Migrants from resistant areas can potentially introduce resistance genes into a population. A greenhouse-selected insect strain may spread to nearby farms and introduce the resistance gene.

When resistance is functionally recessive, only a few homozygous resistant (RR) individuals can survive through an insecticide treatment and as homozygous susceptible (SS) individuals arrive into the area and mate with the survivors, many of their offspring will be heterozygotes (RS) or (SS) susceptible individuals.

3.1.6 Pesticide metabolism

Some insects and mites withstand pesticides by increasing metabolic breakdown. Resistance development is more likely to reduce the efficacy of pesticides that are easily metabolised by biotransformation mechanisms.

3.1.7 Number of target sites of the pesticide:

Single-target pesticides build resistance faster compared to pesticides with multiple target sites require which the insect should develop resistance at all sites of action. A single gene mutation can cause pesticide resistance if it has just one target site.

3.1.8 Pest host range

Pests having a broad host range that infest more crops may be more likely to acquire resistance than pests that are highly crop-specific. This is especially important for insecticides because a single insect species can infest multiple crops. In many cases, tactics have been developed for a specific crop without concern for nearby or rotating host crops or insect movement, resulting in an underestimation of the number of treatments received by the insects. A certain insect pest species, for example, may receive three to four treatments on cotton and four to five applications on a nearby or subsequent vegetable crop. Cotton specialists consider three to four selections, while vegetable specialists notice four to five selections but these insects in total possess seven to nine selections. That is why, particularly for insecticides, it is critical to develop control techniques that are area-focused rather than crop-focused.

3.2 Genetic factors

3.2.1 Occurrence of resistance genes

To select resistance in a pest population, pests must contain a resistance gene. The pest-protecting gene(s) determine the population's resistance and speed the resistance development. Faster resistance selection is due to better gene(s) protection, lower fitness cost, and higher resistance gene frequency.

3.2.2 Number of resistance mechanisms

There are various ways that allow agricultural pests to survive toxicant exposure, and resistance develops more quickly when an organism possesses more than one of these mechanisms. In many cases, pests, particularly insects, will use more than one pathway to build resistance, even if one mechanism is more evident than the others.

Two mechanisms can considerably increase resistance to pesticides. If an insect is 10-fold resistant to a pesticide through enzymatic detoxification and 2-fold resistant via reduced penetration, its overall resistance could be 20-fold. For fungicides, such difficulties are irrelevant. If the same organism has multiple resistance mechanisms, it may develop resistance to more than one class of pesticide.

3.2.3 Gene frequency

Gene frequency (or allele frequency) is the number of gene copies for a given gene variant. Resistance is strongly influenced by resistance allele frequency. The frequency of homozygote individuals resistant to a new pesticide is usually lower, but heterozygous ones may be greater. The higher resistance gene frequency leads to faster resistance development.

3.2.4 Dominance of resistance gene(s)

Genes for resistance can be dominant, semi-dominant, or heterozygous. If a trait is dominant or semi-dominant, only one parent insect needs to have a resistance gene, but their offspring should have it fully or partly. If it is a recessive gene, it means that both parents must have it. If the resistance is strong genetically, it can spread quickly through a population and can be hard to control. Most resistance mechanisms, *viz. kdr* are controlled by recessive or semi-dominant genes that makes easier to keep resistant groups under control.

When insects are exposed to lower pesticide rates, incompletely recessive or dominant genes can become functionally dominant. This decreased dose can occur from deliberately using a low rate, inadequate plant or area coverage, or pesticide residues degrading on the treated surface. After mating with susceptible or heterozygote individuals, heterozygotes survive and pass on the resistance gene.

3.2.5 Fitness of "R" individuals

Resistance genes may cause lower vigour and/or a timing difference in the life cycle, making mating harder. Resistance genes can accumulate quickly in a population if their fitness cost is modest. Only in the presence of the pesticide, resistant individuals show a significant advantage over susceptible ones if the fitness cost is high and in absence of pesticides, resistant forms may be non-competitive and quickly lost. The pesticide rotation is an effective resistance management strategy.

3.2.6 Protection provided by the "R" gene

Individuals harbouring the resistance gene have a very high chance of surviving a pesticide application and transmitting the resistance gene on to the next generation if the resistance gene shows a high level of protection against the pesticide. Individuals with resistant genes will be protected against lesser doses of the pesticide but not from high doses if the resistance gene only gives a moderate level of protection. This is one another reason to apply full label rates of a pesticide to attain the best coverage feasible. Lower doses and insufficient coverage allow resistance genes to accumulate in the population.

3.2.7 Cross-resistance

Resistance to one pesticide gives resistance to another, even if the pest has never been exposed to it. Therefore, its presence increases resistance risk. Cross-resistance arises when two or more chemicals act on the same target location or resistance mechanism. Compounds with the same mode of action and chemically linked from the same chemical group generally develop cross-resistance. It can be complete or partial (if many mechanisms cause resistance).

Some resistance mechanisms impact chemicals in different chemical classes, but mostly insecticides. DDT and pyrethroids are affected by the kdr gene, which disrupts nerve cell sodium channels. A population that was resistant to DDT may develop pyrethroid resistance from intensive use. Negative cross-resistance arises when a resistance mechanism makes an organism resistant to one pesticide but susceptible to another.

3.2.8 Past selection

Past selection of resistance genes may help generate resistance to novel substances because previous use has likely enhanced resistance gene frequency. That does not mean the new compound will be ineffective or develop resistance soon. It just indicates that resistance development is higher than if no related chemicals were employed. However, if cross-resistance is significant and there was a serious resistance problem in the past, the new compound may develop resistance quickly.

3.2.9 Modifying genes

Modifying genes to affect the development of resistance in insects offers promising opportunities for more effective and sustainable insect pest control. By targeting the genes responsible for resistance mechanisms, genetic modification can enhance the effectiveness and longevity of insecticides and other control methods. However, it is essential to address the ethical concerns, regulatory hurdles, and potential for resistance associated with GM insect technology to ensure its safe and responsible use. Ongoing research and collaboration between scientists, regulators, and stakeholders are crucial for realizing the full potential of genetic modification in insect pest management.

3.3 Operational factors

The resistant insects may overcome their reduced fitness with time and selection. The fitness cost of the resistance gene may be nearly overcome, allowing the insect population to retain the resistant gene and slowly or never revert to the susceptible gene. In other circumstances, the fitness cost cannot be overcome, and reversion happens quickly without selection.

3.3.1 Activity spectrum of the pesticide

Broad spectrum insecticides are more likely to generate resistance than narrow-spectrum pesticides because they kill more insect species and are applied more often in a particular area. The narrow spectrum product will be applied less often and have reduced selection pressure in most cropping conditions with other target pests. The broad spectrum insecticides to be used with caution since they may select resistance in non-target pest species in the treated area at sub-treatment threshold levels. If the secondary pest turns to be a primary problem later and requires a higher rate of control than the primary insect targeted by the treatment, resistance may develop faster.

3.3.2 Pesticide application rate

The pesticide application rate is not based on resistance, but it is important to adhere to the recommended rate and not underdose. As a result, all susceptible individuals and essentially all heterozygous resistant individuals would be eliminated from the pest population while the pest population would be reduced below the economic threshold. The susceptible individuals will be removed if the dose is too low, but the partially resistant heterozygotes will survive. A dose that is too low will also make the resistance gene functionally dominant, and resistance may develop quickly. Attempting to eliminate heterozygote individuals, on the other hand, is most effective if the population is not extremely large, consists mostly of susceptible individuals, and is subject to immigration by susceptible individuals; then highly resistant homozygous resistant individuals should be rare and will likely suffer from reduced fitness due to the resistance genes.

It is also not recommended to use higher than specified application rates. This is because any survivors from a high rate are most likely to be homozygous resistant. High dose rates, in particular when there is no immigration of susceptible individuals, are particularly likely to accelerate the establishment of resistance. Higher doses will also kill more natural enemies, perhaps leading to an increase in insect population.

3.3.3 Application coverage

Coverage of the target area (e.g. crop, commodity) being treated is very important. Pests are more likely to die if the body of insect is well-covered with enough pesticide. Poor coverage, with some parts receiving more pesticide and others less or none, will have the same effect as below-label rates. Selection of homozygous individuals will promote resistance.

3.3.4 Treatment frequency

To reduce the selection pressure on pest populations, pesticide treatment frequency should be limited to the number of treatments necessary to protect the crop or control the pest. Frequent treatments at inadequate rates can soon lead to resistance. Only susceptible individuals will be eliminated, while functionally resistant heterozygotes will be selected with homozygote resistant ones. The use of frequent applications can provide spectacular pest control (temporarily), but will usually be followed by a very serious resistance issue. Rotation of pesticide treatments with unrelated compounds is recommended in situations where an area is continually being invaded by untreated individuals, thereby reducing the selection pressure on the pest population.

3.3.5 Presence of secondary pests

The presence of non-target and susceptible pests in crops at sub-economic levels while another pest has reached a treatment threshold, as discussed in the insect host spectrum, also has to be considered. The insecticide treatment targets both the first and second pest species. It is for this reason that detailed pesticide application records should be kept and reviewed. When designing a pest control program, crop specialists often fail to take into account the treatments applied when the pest was at sub-economic levels.

3.3.6 Life stages treated with related insecticides

Resistance is less likely to happen if insects are treated when they are susceptible to pesticides (for example, in Lepidoptera, neonates, first-instar larvae, and adult males are much less able to metabolise insecticides) or if different life stages can be treated with different insecticides. As a result, if some individuals prove resistant to a pesticide at one stage, they are likely to be eliminated at the next stage. This approach is generally difficult to achieve, however, unless the generations are very synchronous or the larvae and adults live in different environments. In a field situation, there is usually a mixture of life stages.

3.3.7 Persistence

Resistance is less likely to develop in the presence of less persistent pesticides because the selection pressure is reduced. In many instances, however, long residual products are desired because they require fewer applications only cause resistance due to strong selection pressure. Persistence and frequency of application play a greater role in development of resistance. Short residual insecticides tend to exert less selection pressure, resulting in slower resistance development and needed more frequency of applications than longer residual products.

3.3.8 Number of crops treated

When multiple crops are treated with the same pesticide, the likelihood of resistance development increases, especially for pests with a broad host range. Pesticides may select insects on subsequently grown crops, and refuges with susceptible individuals will be smaller.

3.3.9 Pest control tactics

The danger of pesticide resistance increases with the continuing use of a single pesticide or an exclusive dependence on chemical control. This is why plans for reducing the likelihood of resistance typically call for a combination of chemical, biological, and cultural methods of suppression.

Selective insecticides and other forms of integrated pest management are two strategies of insect pest control that have been shown to reduce the rate at which resistance to insecticides spreads. Reason being, if resistance is not already widespread, the frequency of resistant genes in the population will be reduced due to natural enemies killing both resistant and susceptible insects.

3.3.10 Crop sequence

If crops cultivated in the same area are separated in time (e.g., with fallow periods between successive cropping cycles) or if they are grown in different geographic regions, the risk of resistance development will be reduced. Conversely, if continuous cultivation is practised, the number of selection events caused by a pesticide may be high, and resistance will develop more rapidly.

3.3.11 Systemicity

The use of systemic insecticides rather than contact pesticides can both speed and slow resistance development. In general, systemic insecticides have a significantly smaller impact on beneficial insects that are connected with the pest. Thus, predators are still present after a pesticide treatment and may remove many of the remaining pests, preventing further spread of their resistance genes to the pest population. However, systemic chemicals have several disadvantages.

Systemicity affects compounds differently. Systemic insecticides can reach pests under leaves that would have been shielded from foliar contact pesticide application and deliver a more equal dose. Although this is beneficial for pest control, it can enhance resistance by preventing sensitive pests from escaping treatment and contributing their genes to the pest population. Systemicity permits insecticides to infiltrate plants as leaves expand, sparing untreated tissue. Systemic chemicals stay in the plant longer than contact insecticides, creating higher selection pressure for resistance, especially if the pest is introduced continuously.

There is a strong desire to utilise systemic insecticides as curative treatments to control insect infestations (leaf miner) that have established themselves in plant tissue. This is not acceptable practise in general, and most resistance management guidelines specifically warn against using curative applications because of the enhanced selection pressure they present.

4. Mechanisms of insecticide resistance in insects

Resistance is defined as a decrease in a population's sensitivity, as evidenced by a product's repeated failure to achieve the expected level of control when used according to the label recommendations for that pest species, and where problems with product storage, application, and unusual climatic or environmental conditions can be eliminated. Insects can develop resistance to crop protection chemicals in a variety of ways, and pests frequently demonstrate more than one of these mechanisms at the same time.

1. Behavioral resistance
2. Penetration resistance
3. Metabolic resistance
4. Altered target-site resistance

4.1 Behavioral resistance: Insects that are resistant to toxins may recognize and avoid their dangers. Several insecticide classes, including organochlorines, organophosphates, carbamates, and pyrethroids, have shown this resistance mechanism. When insects encounter certain insecticides, they may stop feeding, or move away from the area sprayed (for example, they may move to underside of sprayed leaves, or move deeper into the crop canopy, or fly away). One such example is cockroaches that are resistant to gel-baits where they avoid to eat bait and difficult to be controlled with insecticide.

4.2 Penetration resistance: A susceptible insect may absorb the toxin faster than a resistant insect. A penetration resistance occurs when an insect's outer cuticle develops barriers that prevent pesticides from entering its body. Insects are protected from various group of insecticides by adopting this approach. Often, penetration resistance occurs along with other forms of resistance, and reduced penetration enhances their effects.

4.3 Metabolic resistance: Insects that are resistant to the poison may be able to eliminate it from their bodies more quickly through metabolic activities and this is known as metabolic resistance. Insects have enzyme systems that help them metabolise pesticides. Some resistant bacteria in the gut of insect may produce enzymes can metabolise a wide variety of pesticides.

4.3.1 Process of xenobiotic metabolism occur in insects

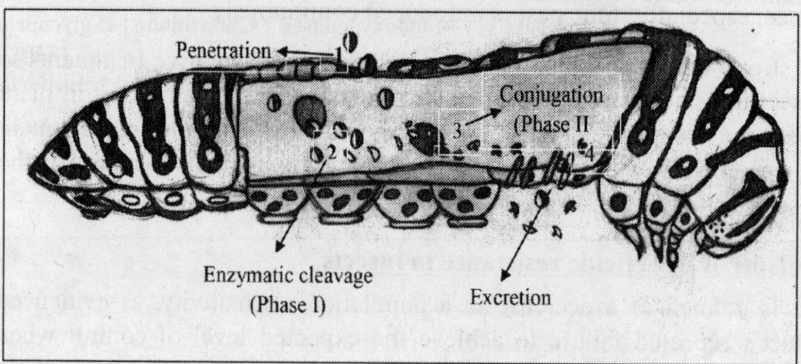

Fig. 2: Process of xenobiotic metabolism in insects

Once xenobiotic enter the insect cuticle, esterase enzymatically cleaves them into alcohol and acid upon adding water (Phase I). Some insecticides will conjugate with glutathione for easy excretion (Phase II) (Fig 2).

4.3.2 Phases of metabolic pathways

Phase I (primary) processes	Phase II (secondary) processes
• These reactions introduce a reactive polar group into lipophilic xenobiotics • It is called **functionalization reaction** • Phase I reactions are mainly carried out by two major groups of enzymes, **1. Oxidoreductases-** The oxidoreductases comprise of the cytochrome P450 dependent superfamily of monooxygenases, which introduce oxygen into or remove electrons from their substrates. **2. Hydrolases**	• Reactions involving the conjugation with endogenous substrates of phase I products and other xenobiotics that contain functional groups such as hydroxyl, amino, carboxyl, epoxide, or halogen • It is called **conjugation reaction** • Phase II reactions are mainly carried out by the **transferases**. • In these reactions, the functional group introduced in phase I is combined with highly water-soluble endogenous metabolite such as glutathione to give rise to the conjugation products which are readily excreted from the insect body.

4.3.3 The families of enzymes involved in metabolic resistance

1. Esterase (ESTs)-Phase I metabolic enzymes
2. Monooxygenases (P450s) / Mixed function oxidases (MFOs) - Phase I metabolic enzymes

3. Glutathione S-transferases (GSTs)- Phase II metabolic enzymes
4. ATP Binding Cassette (ABC) transporters comes under Phase III, also called P-glycoprotein (Pgp) (Panini *et al.*, 2016)

Relationship between metabolic resistance mechanism and important insecticide groups

Metabolic Resistance Mechanism				
	Esterase's	**Monooxygenases (Cytochrome P450/Mixed function oxidases)**	**Glutathione S transferases**	**P-glycoprotein (ABC transporters)**
Pyrethroids	✓	✓	✓	✓
DDT (Organochlorines)	-	✓	✓	✓
Carbamates	✓	✓	✓	✓
Organophosphate	✓	✓	✓	✓
Neonicotinoids	✓	✓	✓	✓

- Symbol indicates the presence of metabolic resistance mechanisms in important insecticide groups.

4.3.3.1 Esterases are a large group of phases I metabolic enzymes and normally located in the fat body. Since many insecticides, especially organophosphates and carbamates, contain ester bonds by hydrolysis into their corresponding acid and alcohol compounds (Plate 6). This will increase the polarity of the metabolites which can be easily excreted from the insect body. It is not surprising that the mechanism of resistance in many cases is caused by elevated levels of esterases and these esterase levels can be elevated by either gene amplification or alteration in the structural gene or altered gene expression.

The relationship among expression, amplification, and methylation of FE4 esterase genes in Italian populations of *Myzus persicae* (Sulzer) (Homoptera: Aphididae) was studied by Bizzaro *et al.,* (2005).

Esterases resistance mechanisms

1. Gene amplification in esterase mechanism
2. Structural change in the gene

4.3.3.2 Monooxygenases (P450s)- Organochlorine insecticides

Mixed function oxidases (MFOs), or microsomal oxidases, are a large family of phase 1 enzymes involved not only in the detoxification of xenobiotics but also in the metabolism of endogenous substances such as hormones, pheromones or fatty acids. They can convert lipophilic compounds into polar metabolites that can be easily eliminated from the body; and for this reason they are mainly located in the digestive system (Feyereisen, 2015).

Cytochrome P450 monooxygenases (P450s) are microsomal oxidases that belong to the group of heme thiolate proteins, so named because they show a characteristic absorbance peak at 450 nm in their reduced form when complexed with carbon monoxide (Plate 12). They are abundant in fat bodies, malpighian tubules and the insects midgut.

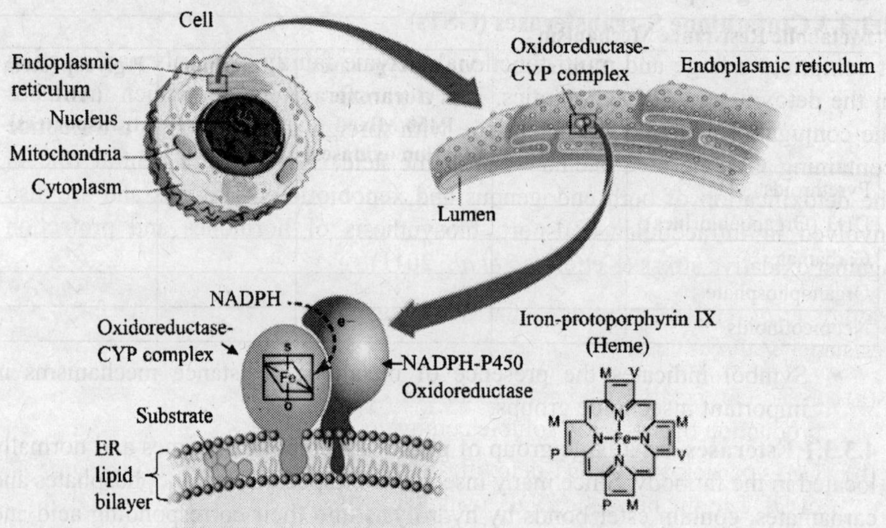

Location in the cell: Endoplasmic reticulum of eukaryotic cells

The above picture shows an oxidation system that requires NADP (Nicotinamide Adenine Dinucleotide phosphate). It catalyses the transfer of one atom of molecular oxygen to a substrate and the reduction of the second atom of oxygen to form water; the process requires the transfer of two electrons provided by NADPH cytochrome P450 reductase (Feyereisen, 2005).

$$\text{(Substrate) RH} + \text{NADPH} + H^{+} + O_2 \rightarrow \text{(Substrate) ROH} + \text{NADP}^{+} + H_2O$$

Many cases of resistance correlated to

1. Overexpression of P450s
2. Point mutation

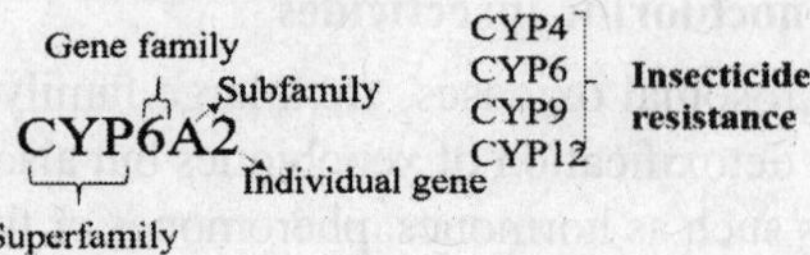

The CYPs nomenclature were introduced by Nebert and Nelson (1991) and presently it is universally accepted that describe all the gene members of the CYP superfamily with a prefix CYP, which followed by a number for the family, a letter used to denote the subfamily, and again number for the individual gene, for example, "*CYP6A2*". If it related to a gene or cDNA sequence, it is denoted in italics (*CYP6A2*) (Plate 4), however, the gene product such as mRNA and enzyme are shown in capitals (Feyereisen, 2011). For instance, CYP6A2 refers to a cytochrome p450 gene, family 6, subfamily A and gene number 2.

4.3.3.3 Glutathione S-transferases (GSTs)

It comprises a large and multi-functional enzyme family primarily participating in the detoxification of xenobiotics. It is a transferase enzyme which facilitates the conjugation reaction of glutathione with foreign materials. It is a tripeptide containing Glycine + Cysteine + Glutamic acid. They play a central role in the detoxification of both endogenous and xenobiotic compounds and are also involved in intracellular transport, biosynthesis of hormones and protection against oxidative stress (Ketterman *et al.*, 2011).

Pavlidi *et al.* (2018) mentioned the mechanisms of GST mediated insecticide resistance is three types.

(a) GSTs can catalyze the conjugation of GSH to insecticides leading to the production of less toxic soluble conjugate.

(b) GSTs participate in DDT detoxification by catalyzing the direct metabolism of DDT to the nontoxic DDE, using GSH as a co-factor.

(c) GSTs displaying peroxidase activity reduce the toxic peroxides produced by oxidative stress caused by insecticide intake.

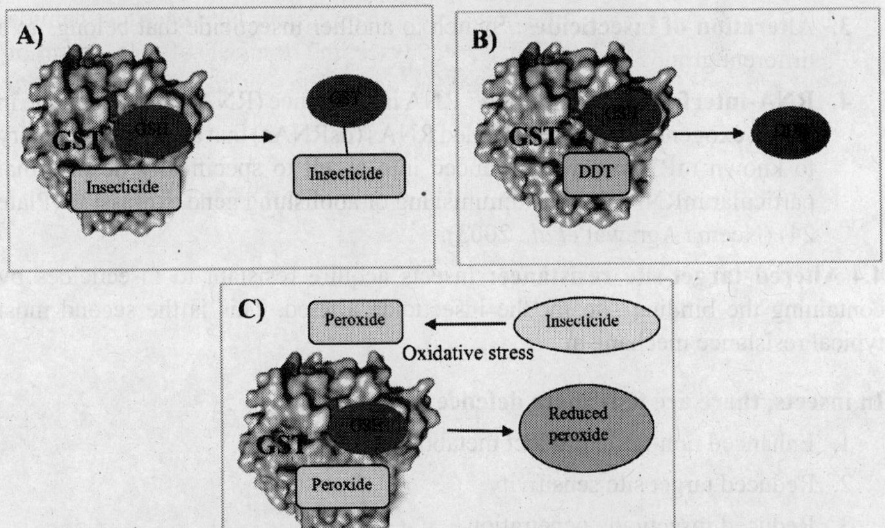

Three types of mechanisms of GST mediated insecticide resistance

4.3.3.4 P-glycoprotein (Pgp) transporters or (ATP binding cassette) superfamily

In phase III, the polar compounds or conjugates can be transported out of the cell by cellular transporters i.e. ABC transporters that play an important role. ABC transporters are two types namely Importers and Exporters

The basic structure of ABC transporters contains mainly two parts

A. Transmembrane domains (TMD)

B. Nucleotide-binding domains (NBD)

Location of Pgp transporters: Pgp transporters were found in the malpighian tubules and midgut & hindgut

Metabolic resistance management

1. Use of synergist piperonyl butoxide: Synergists (Piperonyl butoxide) are non-toxic compounds that enhance the efficacy of insecticides and reduced metabolic resistance.
2. **Temporal synergism concept:** If an insect is treated with a synergist a few hours before exposure to an insecticide, it allows time for the synergist to cross the insect cuticle and inhibit those enzymes involved in metabolic resistance. It creates sensitivity or hypersensitivity in insects. It can be achieved by the split application of synergist and insecticide. Use of formulation technologies, which allow a differential time release of synergist and insecticide (Bingham *et al.*, 2008).
3. **Alteration of insecticides:** Switch to another insecticide that belongs to a different group
4. **RNA-interference technology:** RNA interference (RNAi), is a technique in which exogenous, double-stranded RNAs (dsRNAs) that are complementary to known mRNAs, are introduced into a cell to specifically destroy that particular mRNA, thereby diminishing or abolishing gene expression (Plate 24) (Neema Agrawal *et al.*, 2003).

4.4 Altered target-site resistance: Insects acquire resistant to insecticides by containing the binding site for the insecticide altered. This is the second most typical resistance mechanism.

In insects, there are four main defence systems

1. Enhanced non-toxic product metabolism
2. Reduced target site sensitivity
3. Reduced insecticide penetration
4. Excretion of insecticides increases

The first two groups are the most crucial

The place of action has been changed to reduce toxic attack sensitivity. Alterations of amino acids that bind insecticides at their site of action reduce or eliminate their efficacy. Organophosphorus (OPs) and carbamate (propoxur, carbaryl) insecticides target nerve synapses' acetylcholinesterase, while organochlorines (DDT) and synthetic pyrethroids target nerve sheath sodium channels. DDT-pyrethroid cross resistance can be caused by single amino acid alterations in the axonal sodium channel insecticide-binding site. This cross-resistance shifts the sodium current activation curve and reduces pyrethroid sensitivity. Similar to cyclodiene (dieldrin) resistance, single nucleotide mutations within a GABA receptor gene codon shows resistance. At least five point mutations in the acetylcholinesterase insecticide-binding region diminish OP and carbamate insecticide sensitivity.

Insects have four target site resistance types

- Knock-down resistance disrupts the sodium channels of the nerve cell. Pyrethroids, one of the broadest-spectrum insecticides, control almost all arthropods of agricultural and medical importance. *Chrysanthemum* spp. extracts yield a vast class of structurally varied synthetic pyrethrins. Pyrethroid induces knockdown resistance (kdr) due to mutations in the voltage-gated sodium channel, the target location for DDT and pyrethroids. This is a frequent DDT and pyrethroid resistance mechanism in *Anopheles gambiae* and *Blattella germanica.* Many mutations cause kdr and super kdr. Among the several pesticide resistance mechanisms proposed, the kdr and metabolic resistance due to insecticide detoxification are the most important.
- The structure of acetylcholinesterase is altered by MACE (modified acetylcholinesterase), resulting in its resistance to the insecticide's efficacy. The mechanism underlying pirimicarb resistance in *Phorodon humuli* and its association with resistance in *Tetranychus urticae.*
- The Rdl mutation, characterised by a point mutation, is associated with a diminished affinity of dieldrin towards the GABA receptor. Eg: Dieldrin resistance in *Anopheles quadrimaculatus* mosquitoes.
- The AChE-R gene variant results in an altered form of the acetylcholinesterase (AChE) enzyme, leading to varying levels of resistance. Different alleles of this gene are associated with distinct levels of resistance. Three point mutations have been found in the acetylcholinesterase gene (ace) of *Bactrocera dorsalis*, commonly known as the oriental fruit fly.
- The regulatory element known as the "Barbie Box" facilitates the activation of genes responsible for insecticide-detoxifying oxidase and esterase resistance. Multiple mutations resulting in amino acid substitutions have been identified in the CYP6A2 gene of a resistant strain in *Drosophila melanogaster*, namely in the P450 mono-oxygenases gene.

- *Bt* resistance is caused by cadherin loss, which helps cells adhere to tissues. This is seen in *Bt*-resistant diamondback moths (*Plutella xylostella*).

Mechanisms of resistance to major insecticide groups

Mechanism (s)	Insecticide group to which resistance evolved
Detoxication by	
a) Carboxylesterases	Carbamates, organophosphates (OPs), pyrethroids
b) Cytochrome P450/mixed-function oxidases	Carbamates, OPs, pyrethroids, neonicotinoids
c) Glutathione S-transferases	Organochlorines, OPs, pyrethroids
d) DDT dehydrochlorinases	DDT
Disruption of GABA-gated chloride channels	Avermectin, cyclodiene, phenylpyrazoles, spinosyn
Disruption of sodium ion channels	Organochlorines, pyrethroids
Inhibition of adenosine *triphosphatase*	Thiourea insecticide/acaricide
Inhibition of chitin synthase	Phenylurea-insect growth regulators
Insensitive *acetylcholinesterases*	Carbamates, OPs
Insensitive nicotinic acetylcholine receptors	Neonicotinoids
Uncoupling oxidative phosphorylation	Pyrroles, fluorine-based insecticides

5. Insecticide resistance detection/verification methods

Target site sensitivity reduction and metabolism or sequestration detoxification are the main resistance mechanisms. Target site alterations that reduce insecticide binding or damage lower target site sensitivity. Enzymes quickly bind and convert insecticides to less effective compounds during metabolism and slow or no processing enzymes or other compounds sequester quickly. Resistance strategies include cuticle penetration reduction and behavioural modifications that decrease insecticide exposure. Individual insects species can have multiple resistance development processes that cooperate to give great resistance in insects. Standard bioassay, biochemical, immunological, and molecular approaches can determine resistance (covered in later chapters).

5.1 Conventional Detection Methods

The conventional way of detection is to raise insects from the field to the following generation. Testing larvae or adults for pesticide resistance involves measuring their mortality following exposure to different doses. LD_{50} or LC_{50} values were calculated using probit analysis for susceptible and field populations. The results are compared with those from standard susceptible populations. A another classic way of determining pesticide resistance is to expose individual insects to a diagnostic single dosage in a chamber or filter paper soaked in the insecticide for a specified time. However, these tests basically indicate resistance presence and frequency and provide little information about the resistance mechanism. Most resistance evolution is driven by a few key genes. Because resistance genes tend to decrease fitness in the absence of the pesticide, they are uncommon in a susceptible population prior to exposure. The low fitness costs associated with resistance when the population is not exposed to insecticide are offset by strong selection for resistance when an insecticide is used frequently.

5.2 Discriminating dose assay

The most common approach to field monitoring of resistance, especially for insecticides, has been the use of a discriminating or diagnostic-dose assay. The discriminating-dose assay's end goal is to find out if the susceptibility status of the population has changed, but it's usually not possible to find resistant people until the frequency of the resistance gene is higher than 1%.

The three most critical factors to consider while developing a single, discriminating-dose monitoring programme are:

1. Determining the "**diagnostic dose**" to distinguish between susceptible and resistant individuals
2. Calculating the appropriate sample size for each location.
3. Determining the appropriate response for a discriminating dose survivor.

Bioassays of survivors in the treated area can generate data, presuming the area was not rapidly treated with another pesticide.

These bioassay tests should be developed prior to, or shortly after, the commercialization of a new pesticide compound on the target pest(s) or the sowing of a new transgenic crop. Typically, pesticide manufacturers collaborate with national or regional research institutions to accomplish this. The tests will be used to establish a baseline that can be used to identify the natural variability of susceptibility in the population of pests and corroborate future resistance situations. The tests should be rigorous, quick, and relatively simple to execute. The procedure should be precise and yield results that are realistic, quantitative, reproducible, and easily comprehensible.

5.3 Dose-response test

The dose–response bioassay is the most precise method for assessing the susceptibility of a population to a compound or trait. On a number of population samples, initial dose-response data with a series of doses that generate mortality ranging from 5 to 95% in the case of insecticides should be developed. Only the target population and a known susceptible population need to be tested for insecticides. These data can be utilised to assess the susceptibility range within the population prior to implementing large-scale applications. This information may prove valuable in situations where unexpected lack of control arises in the future.

5.4 Baseline data

It is important to gather baseline data on the susceptibility of the pest organism to the pesticide prior to introducing the product in a specific area. Regardless of the method used for resistance verification, the test results are always compared to the baseline.

Laboratory strains are commonly employed to determine baseline susceptibility data for insecticides. The data is useful as it can provide information on the maximum susceptibility that can be observed. However, laboratory populations often exhibit higher susceptibility compared to field populations due to the detrimental effects of the rearing process. A wide range of baseline values suggests significant genetic diversity within the population of the target organism, potentially leading to a faster development of resistance compared to a population with a narrow range of baseline values.

The insect resistance ratio is defined as the ratio of the median lethal concentration (LC_{50}) or the median effective dose (EC_{50}) of an insecticide for a resistant population to that of a susceptible population. Mathematically, it can be represented as:

$$\text{Resistance ratio }(RR) = \frac{LD_{50}\text{ of resistant population}}{LD_{50}\text{ of susceptible population}}$$

Relationship between bioassay results and field performance

As quickly as possible, a reasonable relation needs to be made between bioassay findings and actual performance in the field and with some compounds, even a moderate difference in susceptibility, as determined by bioassays, can have a significant impact on product field performance, while with others, large differences in susceptibility are required before effects on field performance are observed.

5.5 Biochemical detection of insecticide resistance

The mechanism of resistance may be determined by the use of biochemical assays or other approaches. Some biochemical assays can be used to quantify shifts in resistance gene frequencies in field populations subjected to varying levels of selection when a population has been adequately described.

5.6 Immunological Detection Methods

This method is only available in collaboration with laboratories that have access to the antiserum for specific elevated esterases. For this purpose, no monoclonal antibodies are currently available. Antiserum against E4 carboxylesterase in the aphid *Myzus persicae* has been prepared. This antiserum's affinity-purified 1gG fraction was used in a simple immunoassay to distinguish between the three prevalent resistant variants of *M. persicae* found in field populations.

5.7 Detection of monooxygenase (cytochrome P450) based insecticide resistance

Individual pests have relatively low levels of oxidase activity, and no reliable microtiter plate or dot-blot assay has been developed to measure p450 activity in single insects. The p450s are also a complex family of enzymes, and it appears that distinct cytochome p450s confer insecticide resistance.

6. Management of insecticide resistance

The resistance monitoring plan should no longer rely on measuring the reaction to one insecticide with the goal of switching to another chemical when resistance levels above the threshold that affects insect control. Early discovery of the problem and rapid assimilation of information on the resistant insect population are required for effective resistance management so that sensible pesticide decisions can be made.

How can you control a pest species that has developed resistance to a specific pesticide? One approach is to apply a different pesticide, preferably one from a different chemical class or family of pesticides, with a different mode of action against the insect. Of course, the capacity to employ additional pesticides to avoid or postpone the development of resistance in insect populations is dependent on the availability of a sufficient supply of pesticides with different modes of action.

This strategy may not be the ideal answer, but it permits a problem to be controlled until other pest management tactics can be devised and implemented. Pesticides are frequently utilised in these tactics, but they are used less frequently and at lower application rates.

Resistance management delays pest resistance evolution. The best strategy is to reduce insecticide use. Thus, resistance management is part of integrated pest management, which uses chemical and non-chemical measures to control pest populations safely, economically, and sustainably. Biological control by predators, parasitoids, and pathogens, cultural controls (crop rotation, manipulation of planting dates to limit pest exposure, and use of cultivars that tolerate pest damage), and mechanical controls also work as insecticide alternatives.

Because large-scale resistance studies are expensive, time-consuming, and may increase resistance problems, modelling has been important in developing resistance management strategies. Although models have revealed several techniques to postpone resistance, real resistance management has focused on lowering pesticide treatments and diversifying insecticide kinds. In Australia, Israel, and the US, numerous programmes limit the number and duration of insecticide use against cotton pests.

Early resistance detection methods are needed for resistance management. Reducing pesticide selection pressure on pest populations can manage pesticide resistance.

Thus, avoid chemicals that kill all pests except the most resistant. Avoiding unnecessary pesticide applications, adopting non-chemical management methods, and allowing untreated refuges where susceptible pests can survive. Integrated pest management (IPM) helps manage resistance. Pesticide rotation manages resistance when pesticides are the only or main control technique. Alternating pesticide classes with distinct mechanisms of action delays or mitigates insect resistance. Rotate product classes and modes of action, evaluate pesticide effects on beneficial insects, and use products at indicated rates and spray intervals.

Different pesticide classes affect pests differently. Pesticide manufacturers may mandate a certain number of consecutive pesticide class applications before switching classes on product labels. Tank mixing pesticides improves application results and delays or mitigates insect resistance by combining two or more pesticides with different mechanisms of action. Some insecticides in tank mixtures require that the water be adjusted to a pH between 6 and 8. Marking the swath widths with permanent markers is recommended for aerial application. The pressure that sprayers can withstand should be appropriate for effective coverage, and nozzles should be checked for obstruction and wear. Calibration and maintenance of spray equipment is essential. In addition, intensive and well-executed pruning will improve both canopy penetration and tree coverage in tree-borne fruits. Follow the guidelines provided by the manufacturer and local experts regarding application rates and methods.

To minimise harm to beneficial arthropod populations, it is advisable to choose insecticides that cause little to no harm. Additionally, applying insecticides in a band over the row, rather than broadcasting or using them in-furrow, can help preserve natural enemies.

In certain programmes, unsprayed sections within treated fields, nearby "refuge" fields, or habitat attractions that enable immigration are used to preserve sensitive insects in the target population. Susceptible individuals may outcompete and reproduce with resistant ones, diminishing resistance. Destroying crop leftovers can starve insects and destroy overwintering sites. This cultural practise kills pesticide-resistant and susceptible insects and prevents their offspring from becoming resistant. Farmers should check soil conservation regulations before removing residue.

Insecticide Resistance Management (IRM)

1. Management by moderation

Leave the untreated area in the field to preserve the susceptible population

Use of shorter persistence insecticide, lower dose of insecticides, less frequency application and use of adulticide rather than larvicide.

However, not ideal for high value crop and newly introduced pest.

2. Management by saturation

Use of high doses of insecticides and use of synergists to kill resistant population.

3. Management by multiple attack

The method is more practical. Control by many separately operating stressors, including pesticides, each exerting selection pressure below resistance levels.

The strategies

1. Rotating or changing insecticides for treatment.
2. Combination insecticides are used.
3. Suppressing the insect detoxication pathway by the use of synergists
4. Use of multiple components of IPM in combination with insecticides.

11

Insect Resurgence against Insecticides

Farmers still rely heavily on insecticides because of their efficiency, portability, and speedy returns on investment. Although synthetic organic insecticides are highly efficient, their widespread usage has led to toxicity to natural enemies, toxic residues in plants and the environment, insect resistance, and a resurgence in insect populations. Ripper (1956) first time recognized the problem of resurgence in plant protection.

Metcalf (1986) distinguished between two types of resurgence: primary insect resurgence, in which populations of target insects that had been suppressed by insecticide application quickly recover to excessive levels (Fig. 1.A), and secondary insect outbreak, in which non-target species develop into serious insect s after insecticide application (Fig. 1.B).

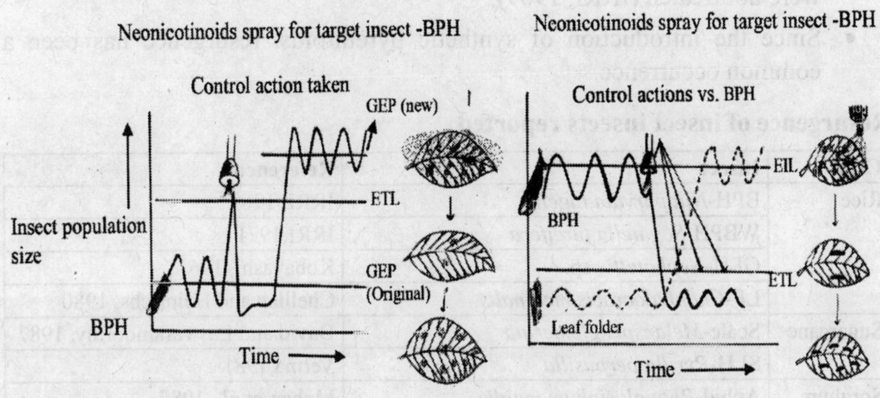

Fig. 1: A) Primary insect resurgence B) Secondary insect resurgence

Primary insect resurgence

Primary insect resurgence occurs when the target insect population increases to at least as high (Hajek, 2004) or higher (Hardin et al., 1995) than in an untreated control or higher than before the treatment (Pedigo and Rice, 2006).

Secondary insect resurgence/ secondary insect outbreak (Type II resurgence)

The phenomenon known as replacement of a primary insect with a secondary insect, or secondary insect outbreak, occurs when the population of a non-target insect, which is harmful to the crop, experiences a surge following the application of a insecticide aimed at controlling the population of the primary insect. A second insect outbreak is also called resurgence. A secondary insect outbreak is the rise of a non-target species after an insecticide has been used.

Historical Background

- In 1951, Bartlett and Ewart reported instances of insecticide-induced outbreaks of insect insects, specifically the walnut brown soft scale *Coccus hesperidium,* which were attributed to the use of parathion.
- Ripper wrote in 1956 that the use of insecticides like calcium arsenate, cryolite, DDT, HCH, aldrin, toxaphene, and parathion had led to a rise in the number of more than 50 species of plant-eating arthropods.
- IRRI was the location where the brown planthopper (BPH) was found to be the first incidence of resurgence in rice.
- In the plots that were treated with gamma HCH, the population of BPH was found to be three times larger than in the plots that served as the control and were not treated (IRRI, 1969).
- Since the introduction of synthetic pyrethroids, resurgence has been a common occurrence.

Resurgence of insect insects reported

Crop	Insect	Reference
Rice	BPH-*Nilaparvata lugens*	IRRI, 1969
	WBPH-*Sogatella furcifera*	IRRI,1971
	GLH-*Nephotettix* sp.	Kobayashi, 1961
	LF-*Cnaphalocrocis medinalis*	Chelliah and Heinrichs, 1980
Sugarcane	Scale-*Melanspis glomerata*	David and Easwaramoorthy, 1987
	SLH-*Pyrilla perpusilla*	Verma,1981
Sorghum	Aphid-*Rhopalosiphum maidis*	Mohan *et al.*, 1987
Brinjal	Aphid-*Myzus persicae*	Subbarami reddy *et al.*, 1987
Chilli	Yellow mite- *Polyphagotarsonemus latus*	David and Easwaramoorthy,1987
Cotton	Whitefly-*Bemisia tabaci*	Jayaswal and singh, 1987
	Aphid-*Aphis gossypii*	Natarajan *et al.*,1987
	Spider mite-*Tetranychus cinnabarinus*	Patil,1987
	LH-*Amrasca biguttula biguttula*	Singh *et al.*,1987

Resurgence is a complex mechanism and major reasons are host-related, insect related and insecticide-related factors. Host related factors include resistance level and altered biochemistry of the host plant. Insect related factors include suppression of natural enemies and sublethal effects. Insecticide related factors include insecticide classes, insecticide rate, timing, frequency and method of application (Dhaliwal and Arora, 2006).

Formula normally used for efficacy of insecticides (Henderson and Tilton, 1955)

$$\text{Corrected \%} = \left\{1 - \frac{\text{Number in control before treatment} \times \text{Number in Treatmentaafter treatment}}{\text{Number in control after treatment} \times \text{Number in Treatmentaafter before treatment}}\right\}$$

Percent resurgence will be calculated using **Henderson and Tilton (1955)** formula with modification by **Jayaraj and Regupathy (1985).**

$$\text{Percent resurgence} = \left\{\frac{Ts \times C_F}{Cs \times 1_F} - 1\right\} \times 100$$

Where, Infestation in the treated plot during first count = T_F; Infestation in the treated plot during subsequent count = T_S; Infestation in the untreated check plot during first count = C_F; Infestation in the untreated check plot during subsequent count = C_S.

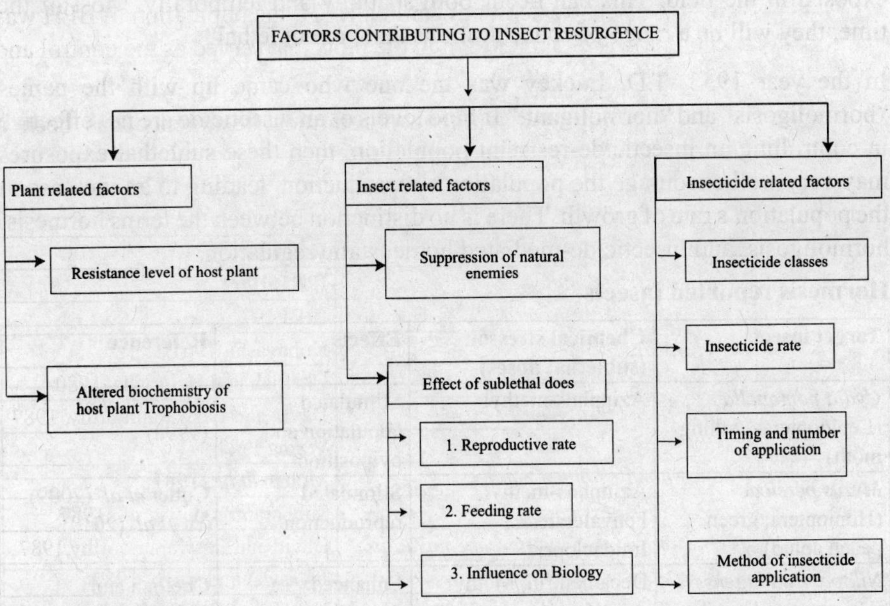

I) Plant-related factors:

Resistance level of host plant:

Resistance level of host plant is one of the factors for resurgence of insects. Susceptible host plant will lead to more resurgence percent, when compared to resistance host plants.

Altered biochemistry of host plant-Trophobiosis

The use of insecticide causes changes in the biochemistry of the host plant, which makes the plant more susceptible to assault by insects. As a result, the insecticide acts as a stimulant, not only for the host plant but also for the insect, resulting in the eventual appearance of a insect population that has been revived.

II) Insect-related factors

Suppression of natural enemies

The elimination of insects' natural predators is one of the key contributing factors to the proliferation of insect population. The recurrence of BPH was caused by a decrease in the population of the mirid predator *Cyrtorhinus lividipennis*, which occurred as a result of repeated spraying with methyl parathion 50 EC.

Sub-lethal doses of insecticides-Hormoligosis

Growers make an effort to apply insecticides in an equitable manner at concentrations that are designed to kill target insect s; however, various biotic and abiotic factors will vary the dose of insecticide to which an insect is actually exposed in the field. This can occur both spatially and temporally. Most of the time, they will be a range of concentrations that are not lethal.

In the year 1955, T.D. Luckey was the one who came up with the names "hormoligosis" and "hormoligant." If field levels of an insecticide are not effective in controlling an insecticide-resistant population, then these sublethal exposures may very well encourage the population's reproduction, leading to an increase in the population's rate of growth. There is no distinction between the terms hormesis, hormoligosis, and insecticide-mediated homeostatic regulation.

Hormesis reported insects

Target insect	Chemical stressor (sublethal doses)	Effects	Reference
Cydia pomonella (Lepidoptera; codling moth)	Azinphos-methyl	Stimulated copulation and oviposition	Abivardi *et al.* (1998)
Myzus persicae (Homoptera; green peach aphid)	Azinphos-methyl, Fenvalerate, Imidacloprid	Stimulated reproduction	Cutler *et al.* (2009); Yu *et al.* (2010)
Nilaparvata lugens (Homoptera; brown planthopper)	Decamethrin, Methyl parathion	Enhanced fecundity	Chelliah and Heinrichs (1980)
Plutella xylostella (Lepidoptera; diamondback moth)	Fenvalerate, Methomyl	Increased fecundity	Fujiwara *et al.* (2002)
Scirtothrips citri (Thysanoptera; citrus thrips)	Dicofol, Fluvalinate, Formetanate, malathion	Stimulated reproduction	Morse and Zareh (1991)

The research on hormesis is expanding rapidly. Insect toxicologists have paid little attention to the phenomena despite the possible consequences of hormesis-rooted stimulations in insect management and the opportunity insect-insecticide models can give in basic investigations.

Insect resurgences following insecticide applications have been observed from the beginning of the insecticide discoveries, and it was hypothesized even then that a insecticide's favourable effect on arthropods could explain some of these resurgences (Ripper 1956). However, entomologists appear to be unaware of the hormetic dose-response phenomena, and no toxicological explanation for such occurrences was presented.

Why is it vital to research on concept of insect hormesis? There are at least two primary motivations for this: (1) the quest of basic hormesis knowledge and (2) practical usefulness in insect management.

General experimental design for hormesis study

When studying the stimulatory effects of insecticides on insects, insect toxicologists typically choose 3-4 test concentrations below the LC_{50}, such as LC_{5}, LC_{10}, or LC_{25}, with little or no reason provided. Sublethal insecticide effects should include a large number of duplicates and doses.

Insecticide-plant-insect interactions

When doing dose response or hormesis experiments with insects and plants, it is important to pay attention to the route of exposure and think about whether the stimulatory effects seen are caused by the insecticide's direct interaction with the insect or by the insecticide's indirect effect on the plant tissue that the insect is eating.

The nature of the stressor

Hormesis might not always happen with chemicals that cause a toxicological reaction. This may be especially true for substances that cause hormesis by directly activating a receptor-mediated mechanism. Even if the molecular structure of two chemicals is very similar, they may not have the same power to cause hormesis.

Hormesis in insecticide resistant populations

Insecticide resistance is still a big problem when it comes to getting rid of bug insects that hurt crops. If field amounts of an insecticide aren't enough to control an insecticide-resistant population, sublethal exposures could make the population grow thereby enable them more likely to reproduce and increase the number of resistant alleles in the population, indicating that the problem of resistance in the due course to be worse.

The problem could be especially important in populations of insects that are resistant to insecticides, where a normal field rate could expose people to the "hormetic-zone" of the dose-response curve. This would help resistant populations reproduce more and make resistant alleles more common.

Insect behaviour

Insect resurgence may also be changes in an insect behaviour. According to research on the behavioural hormoligosis of *Bemisia tabaci,* oviposition preference on cotton, the higher concentrations of total sugars and free amino acids in cotton resulted in more eggs being laid per leaf.

Insect biology & Demographic parameters

Insect biology & demographic parameters are important in insect resurgence. In case of biology of insect, decrease in longevity of instars and increase in the longevity of adult pre oviposition, oviposition, post oviposition periods and fecundity play important role. In demographic studies, intrinsic rate of increase, finite rate of increase, net reproductive rate (offspring/individual), mean generation time and gross reproductive rate also influence the insect resurgence and one should consider these points during insect resurgence studies

Reproductive rate

Recent studies demonstrate that an increase in the expression of the protein vitellin is responsible for the physiological revival caused by activation of the RNA transcript level in fat bodies. *Nilaparvata lugens* females exposed to permethrin developed ovaries one day after reaching adulthood.

III) Insecticide-related factors

Insecticide class

In case of rice brown planthopper, *Nilaparvata lugens* higher resurgence occurred in synthetic pyrethroids when compared to organophosphates and carbamates and that indicates different insecticide class also can play a major role in resurgence.

Insecticide rate

Deltamethrin application at the rate 30 g a.i./ha, produced a considerably higher BPH population in rice compared to lower dose of 20 and 10 g a.i./ha. In the same way, the BPH population was significantly higher in methyl parathion application @ 750 g a.i./ha compared to 500 and 250 g a.i./ha where it was proved that the insecticide dosages have an effect on the insect population rebound.

Timing and number of applications

Methyl parathion 0.075 kg of a.i./ha

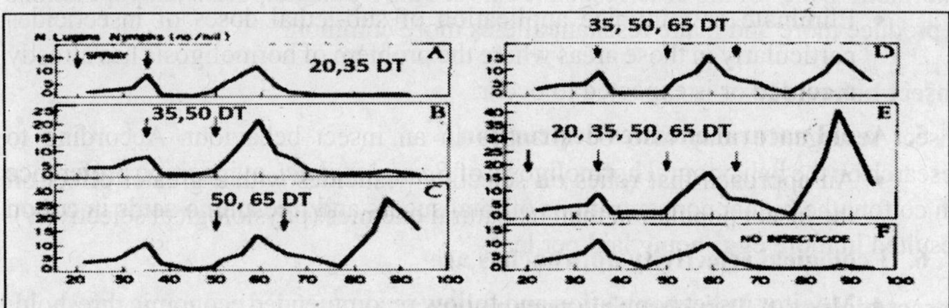

Timing and number of applications of methyl parathion 0.075 kg of a.i./ha.

S.No.	Days after transplanting (DT)	Peak status in graph	BPH population (No./hill)
1.	20, 35	No peak	-
2	35, 50	No peak	-
3	50, 65	Peak @ 80 DT	30
4	35, 50, 65	Peak @ 80 DT	15
5	20, 35, 50, 65	Peak @ 80 DT	40

If methyl parathion@ 0.075 kg a.i./ha applied at 20 and 35 days after transplanting reveals no peak, means no higher population of BPH and same was observed at 35 and 50 days after transplanting, but not the case at 80 DAT where highest population/peak occured if application was carried out at 50 and 65 DAT (2 applications), 35,50, and 65 DAT (3 applications) and 20,35,50,65 DAT (4 applications).

Method of insecticide application

Heinrichs *et al.* (1982) found that foliar spraying induced higher BPH population i.e. 149 per hill and 97.0% hopper burn than root zone placement and broadcasting indicating that the method of application also plays a significant role.

III. Management of Insect Resurgence

1. Resurgence can be easily avoided by not spraying same group of insecticides continuously
2. **Conservation of natural enemies:** Insecticides are the main cause of resurgence due to the mortality of natural enemies.
3. **Inoculative release of natural enemies:**
   - As a result of the application of insecticide, natural enemies are being released inoculative to repopulate the area.

- Choosing insecticides that won't harm natural enemies

4. **Avoid Hormoligosis**
   - Eliminate or minimize application of sub-lethal doses of insecticides particularly in those areas where the problem of hormoligosis has already recorded or is expected to occur.
5. **Avoid natural enemies destruction**
   - An approach that relies on selective pesticides with a greater effect on the insect population than its natural enemies (physiological selectivity).
6. **Ecological selectivity approaches are**
   - Monitor insect population and follow recommended economic threshold.
   - Avoid treating broad areas. Untreated areas can serve as refugia for natural enemies.
7. **Agronomic practices**
   - Insecticide induced resurgence can be reduced by judicious use of fertilizer and irrigation water, as well as date of sowing.
   - The early sowing of a crop helps to avert insect damage and minimizes the need for insecticides.
   - Due to increased fertilizer and irrigation use, insect s multiply more easily in the crop.
   - In the cotton crop, the whitefly resurgence was more pronounced with repeated use of synthetic pyrethroids.
8. **Use of combination of insecticides**
   - Private companies have developed the insecticides combination in the Indian pesticides market.
   - Using these combination insecticides will decrease the resurgence percentage because of different mode of action.

Points to be considered while taking research on insect resurgence

1. Resistance level of host plant plays a major role, that's way selection of susceptible variety is first most important.
2. To conduct biology studies, a easily mass multiplication of insect under laboratory conditions is required
3. A minimum of at least one or two sublethal dose to be used as they give higher resurgence percentage
4. Host biochemical constituents also to be considered for correlation
5. A minimum number of insecticides to be selected and need to add one or two sublethal doses in addition to recommended dose.

Important basic questions related to insect pest resurgence

1. Why resurgence happen more in sucking pests than others insect pests?

Mostly sucking pests are subjected to resurgence. Normally all sucking pests are phloem feeders and they got direct food from phloem as sugars and free amino acids and proteins. Phloem sap is nutrient rich compared with many other plant products and generally lacking in toxins and feeding deterrents.

Two important mechanisms occur in phloem feeders for resurgence, first contain symbiotic micro-organisms which provide them with essential amino acids. Second, the insect tolerance of the very high sugar content and osmotic pressure of phloem sap is promoted by their possession in the gut of sucrase-trans glucosidase activity, which transforms excess ingested sugar into long-chain oligosaccharides voided via honeydew. After treating continues spray of same chemical (Pesticides) which increase the total sugars and essential amino acids, and proteins in plants will directly enters in to phloem feeders and symbiotic microorganisms will helpful for resurgence mechanism (Douglas, 2006).

2. Why synthetic pyrethroid group creating more resurgence in insects?

As general rule, pyrethroids are break down most quickly in direct sunlight, usually just a few days after application which act as sub-lethal dose. Synthetic pyrethroids are not systemic and do not have translaminar action, so they are not effective against sucking pests like aphids and mealy bugs resulting in the resurgence of these pests. Type II pyrethroid cause more resurgence problem insects because they have cyano group at the alpha carbon of phenoxy benzyl alcohol position will increase the total sugars and free amino acids in plants (Rehman *et al.*, 2014).

Calculations: Dose calculation for pesticides in order to study the lethal and sub lethal dose effects on insects

$$\text{Active ingredient per ha} = \frac{\text{Formulation}}{100} \times \frac{\text{Dosage}}{\text{ha}} (\text{Formulati on (gm/ml)})$$

Note: Formulation commercially available %; Dosage/ha information is available in at www.cibrc.nic.in

Eg: Find the g active ingredient per ha of Diafenthiuron 50 % WP

$$\text{g active ingredient pr ha} \frac{50}{100} \times 600 = 300$$

Diafenthiuron 50 % WP

Crop	Common Name of the pest	Dosage/ha			Waiting period (days)
		a.i (gm)	Formulation (gm/ml)	Dilution in Water (Liter)	
Cotton	whiteflies, aphids, thrips, jassids	300	600	500 – 1000	21

If you want per mL dose, then use the formula.

$$\text{To find per mL dose of diafenthou ron 50 WP} = \frac{\text{Formulation (gm/ml)}}{\text{Dilution in water}}$$

$$\text{To find per mL dose of diafenthou ron 50 WP} = \frac{600}{500} = 1.2\text{g/L}$$

Fixation of lethal dose

From recommended dose of 1.2 g/L reduce to 10 times then it is called lethal dose = 0.12 g/L

Fixation of sub-lethal dose

From the lethal dose 0.12 g/L reduce to 10 times then it is called sub-lethal dose = 0.012 g/L

12

Toxicology Studies for Pesticide Data Generation and Safety

The role of toxicology in the evaluation of new agrochemicals

The widespread use of agrochemicals for crop protection may expose humans, animals, and the environment to toxic chemicals. Humans may be exposed during manufacture or application, and to a lesser extent as consumers of food products containing trace amounts of residues.

Strategies of toxicity testing

No-Observable-Adverse- Effect-Level (NOAEL)

A level in which the total diet that causes no adverse effect in treated animals when compared to untreated animals maintained under identical conditions. This NOAEL is expressed on a mg/kg of body weight/day basis. Chronic and sub chronic studies are need to be carried out in different species to determine NOAEL values.

It depends on how much of the substance an organism is exposed to, but all chemicals are toxic to some degree. Toxicology studies try to figure out the exact types and amounts of harmful effects, as well as the lowest doses that won't have any negative effects (called the "No Observed Adverse Effect Level" or NOAEL). Most data come from studies with lab animals and cells that have been grown in a lab. Using these numbers and extrapolation or uncertainty factors, it is possible to figure out an Acceptable Daily Intake (ADI). This is the maximum amount of a substance that a person can take in every day without putting themselves at a very high risk of getting sick. These numbers will be compared to how people might be exposed to a compound in the environment, including through food. This will show if the use of an agrochemical is potentially allowed.

Animal studies

The goal of toxicity studies, which are typically conducted on laboratory animal species, is to determine the toxicological profile of a test chemical. The selection of the proper animal test species is critical, and ideally species that respond to a hazardous stimuli in a manner comparable to that predicted of humans. Typically, rodents (rats, mice, guinea pigs), canines, and nonhuman primates are employed.

The choice of test animal species is also influenced by practical issues such as availability, ease of handling and housing, and the availability of background data.

Good Laboratory Practice (GLP) rules say how tests on animals should be done. This set of guidelines is used all over the world. It explains what test facilities are needed, how to plan a study protocol, how to run experiments, how to record and report data, who is responsible for running a study, and how to make sure the quality of the data. These rules are always being changed to reflect the newest discoveries in chemistry and toxicity.

Prior to animal research, test substances must be evaluated for chemical identification, purity, and stability. A variety of toxicological studies have been designed to characterize a compound:

Pharmacokinetic and metabolism studies- A compound's absorption, distribution, and excretion in the animal body are determined by these experiments. Foreign compounds are often transformed into metabolites that are more easily excreted from the body, therefore their nature must be determined. Determine whether a test agent or metabolite accumulates in the body at certain places and disrupts metabolic pathways. The companion paper 'Metabolism of Pesticides in Plants and Livestock' describes pesticide metabolism studies in plants and livestock.

(Sub) acute toxicity studies-(Sub) acute studies are conducted to identify potential harmful effects caused by an unexpectedly high exposure to a chemical. Animals given a single or several doses of the test substance via oral, dermal, or inhalation are checked for signs of acute toxicity, skin or ocular irritation, or sensitization. These findings are especially important for persons involved in the production or use of agrochemicals.

Long term toxicity studies- These studies are intended to explore the effects of a chemical on animals exposed for a significant portion of their typical life span. Blood and urine are analyzed, food and water intake, changes in body and organ weights, organ functioning are assessed, and tissues and organs are histopathologically examined. This research may provide insights into the processes underlying the observed toxicity. When correctly conducted, these studies should provide information on toxic responses generated by escalating dose levels of the test substance and may reveal a dose level that produces no apparent adverse effects (NOAEL).

Reproduction studies- Reproduction studies are conducted to ensure that agrochemicals do not impact the reproductive capacity of male and female animals, nor do they affect newborns or young animals. To this goal, one or more generations of studies may be conducted, with a focus on potential deleterious effects on male and female fertility, resorption and abortion rates, litter size, sex ratio, birth weight, and newborn growth.

Carcinogenicity studies- Specific animal investigations are done to determine how well a test chemical grows tumors. Typically, these investigations are conducted on two types of mice who receive varying amounts of the test chemical orally over the course of their lives. In addition to physiological, clinical, and anatomical characteristics, tumors are recognized based on the type of tissue from which they originate. Studies on carcinogenicity are conducted when significant levels of pesticide residues are expected in foods.

1. The structural similarity between the test chemical or its metabolites and other recognized carcinogens has been observed.
2. Experimental evidence indicates that the test chemical may cause early symptoms of carcinogenicity.
3. The test chemical was shown to interact with cellular DNA or chromosomes (mutagenic effects).
4. The test chemical has new structural features.

Table 1: Gaitonde committee guidelines for toxicology study details

Study title	Duration	No. of animals	Observations	Remarks
Acute oral rat	14 days	40 (20m+ 20f) 1. Control-5+5 2. Low Dose-5+5 3. Inter Dose-5+5 4. High Dose-5+5	Toxicity signs, mortality, gross pathology, LD_{50} determination	Limit test can be performed with 5000mg/kg b.w. alone if the product is non toxic
Acute oral mice	14 days	40 (20m+ 20f) 1. Control-5+5 2. Low Dose-5+5 3. Inter Dose-5+5 4. High Dose-5+5	Toxicity signs, mortality, gross pathology, LD_{50} determination	Limit test can be performed with 5000mg/kg b.w. alone if the product is non toxic
Acute inhalation rat	14 days	40 (20m+ 20f) 1. Control-5+5 2. Low Dose-5+5 3. Inter Dose-5+5 4. High Dose-5+5	Toxicity signs, mortality, gross pathology, LC_{50} determination	Limit test can be performed with 5mg/l alone if the product is non toxic
PSI-Rat	3-14 days generally, 72h	6-Rabbits-3m-3f	Erythema, odema, and eschar formation as per Drize classification	-
IMM-Rabbit	3- 21days generally, 72h	6-Rabbits- 3m-3f	Eye irritation as per Drize classification	-
Acute dermal rabbit	14 days (Limit test)/21 days full test	24 (12m+ 12f) 1. Control-6 2. Low Dose-6 3. Inter Dose-6 4. High Dose-6	Toxicity signs, mortality, gross pathology, LD_{50} determination	Limit test can be performed with 2000mg/kg b.w. alone if the product is non toxic

Study title	Duration	No. of animals	Observations	Remarks
Synergism and potentiation in rat	14 days	40 (20m+ 20f) 1. Control-5+5 2. Low Dose-5+5 3. Inter Dose-5+5 4. High Dose-5+5	Toxicity signs, mortality, gross pathology, LD_{50} determination Separate LD_{50} determinations of two products and mixture should be determined. Mixture should be 1:1 ratio. Predicted and observed LD_{50} should be determined to calculate the synergistic/potentiation effect	-

Study title	duration	No.of animals	Observations	Remarks
Acute oral chicken	14 days (Limit test)/21 days full test	24 (12m+ 12f) 1. Control-6 2. Low Dose-6 3. Inter Dose-6 4. High Dose-6	Toxicity signs, mortality, gross pathology, LD_{50} determination	Limit test can be performed with 5000mg/kg b.w. alone if the product is non toxic
Acute oral pigeon	14 days (Limit test)/21 days full test	24 (12m+ 12f) 1.Control-6 2. Low Dose-6 3. Inter Dose-6 4. High Dose-6	Toxicity signs, mortality, gross pathology, LD_{50} determination	Limit test can be performed with 5000mg/kg b.w. alone if the product is non toxic
Acute fish toxicity	21 days	40 1. Control-10 2. Low Dose-10 3. Inter Dose-10 4. High Dose-10	Toxicity signs, mortality, gross pathology, LC_{50} determination	Limit test can be performed with 100mg/kg b.w. alone if the product is non toxic
Acute contact in *Apis mellifera* (A dult worker bees)	24/48h	No. of bees/concentration: 30 No. of replications/ concentration: 30	Toxicity signs, mortality, LD_{50} determination	

Study title	duration	No.of animals	Observations	Remarks
Acute oral in *Apis mellifera* (Adult worker bees)	24/48h	No. of bees/concentration: 30 No. of replications/ concentration: 30	Toxicity signs, mortality, LD_{50} determination	
Subacute oral rat	118 days 90 day exposure 28 days non-exposure	120 (60m+ 60f) 1. Control-20+20 2. Low Dose-20+20 3. Inter Dose-20+20 4. High Dose-20+20	Toxicity signs, mortality, gross pathology, Hematology & biochemistry on day 0, 45, 90 and 119. Gross and HP on day 91 and 119	NOEL should be arrived. Highest dose should produce severe toxicity, inter dose should produce moderate toxicity and low dose should produce any toxicity
Subacute inhalation	28 days 14 days exposure 14 days non-exposure	120 (60m+ 60f) 1. Control-20+20 2. Low Dose-20+20 3. Inter Dose-20+20 4. High Dose-20+20	Toxicity signs, mortality, gross pathology, Hematology & biochemistry on day 0, 14, and 29. Gross and HP on day 91 and 119	NOEL should be arrived. Highest dose should produce severe toxicity, inter dose should produce moderate toxicity and low dose should produce any toxicity

Mutagenicity studies-Specific tests have been developed to screen for the genotoxic potency of compounds, which is a substance's ability to react with the genetic material of live cells (DNA, chromosomes), potentially leading to tumor formation or heritable abnormalities. To detect these types of detrimental effects, a combination (battery) of several tests are typically employed with microbes, animals, and cells of both animal and human origin. The findings of these studies are utilized as supplemental information for determining a compound's carcinogenic risk.

Specialized toxicity studies- There have been special studies set up to find out how a chemical might hurt the immune and nervous systems, as well as to see if chemicals can cause spine defects and other changes in fetuses while they are still developing (teratogenic effects). When the structural properties of a substance or data from toxicity studies point to possible bad effects, or when compounds with completely new structures are found, these studies are done.

Ecotoxicological assessment

Eco-toxicology: is the branch of science that deals with the nature, effects and interactions of substances that are harmful to the environment.

The *Eisenia fetida* is an invertebrate that is currently used for ecotoxicological assessment of substances in soil. It is the recommended earthworm test species by the Organization for Economic Cooperation and Development (OECD) and the International Standardization Organization (ISO).

Study title	duration	No. of animals	Observations	Remarks
Subacute dermal	35 days 21 days exposure 14 days non-exposure	24 (12m+ 12f) 1. Control-3+3 2. Low Dose-3+3 3. Inter Dose-3+3 4. High Dose-3+3	Toxicity signs, mortality, gross pathology, Hematology & biochemistry on day 0, 21, and 36 Gross and HP on day 91 and 119	NOEL should be arrived. Highest dose should produce severe toxicity, inter dose should produce moderate toxicity and low dose should produce any toxicity

Registration Categories

Category	9(3)	9(4)
Manufacture	FIM	FIM
	TIM	TIM
	TIM vs TI	
	TIM vs FI	
	FIM vs FIT	
	LE	
Import	TI	TI
	FI	FI
	TI (New source)	
	FI-WRT	
	TI vs TIM	
	Import for export	
Export	9(3) Export only (Fast Track)	
	9(3) Export only (Start Export)	
FIM – Formulation Indigenous Manufacture		
TIM- Technical Indigenous Manufacture		
TI-Technical Import		
FI-Formulation Import		
LE-Label Expansion		

Table 2: Toxicology data generation for registration of pesticides in India

Sl.No.	Parameter	9(3B)			9(3)								9(4)
	Toxicity	**TI**	**TIM**	**FIM**	**TI**	**TIM**	**FI**	**FIM**	**TI Vs TIM**	**TIM Vs TI**	**TI (New Source)**	**NF** (IM)**	**TIM* (AR) Vs TI**
1	2	3	4	5	6	7	8	9	10	11	12	13	14
1.	Acute oral in rat & mice	R	R	R	R	R	R	R	R	R	R	R	R
2.	Acute dermal	R	R	R	R	R	R	R	R	R	R	R	R
3.	Acute inhalation	R	R	R	R	R	R	R	R	R	R	R	NR
4.	Primary skin irritation	R	R	R	R	R	R	R	R	R	R	R	R
5.	Irritation to mucous membrane	R	R	R	R	R	R	R	R	R	R	R	R
6.	Sub-acute oral rat	R	R	NR/R	R	R	NR/R	NR/R	NR	NR	R	NR/R	NR
7.	Sub-acute oral dog	R*	R*	NR/R*	R*	R*	NR/ R*	NR/ R*	NR	NR	R*	NR/ R*	NR
8.	Sub-acute dermal	R	R	NR/R	R	R	NR/R	NR/R	NR	NR	R	NR/R	NR
9.	Sub-acute inhalation	R	R	NR/R	R	R	NR/R	NR/R	NR	NR	R	NR/R	NR
10.	Neuro-toxicity	NR	NR	NR	R	R	NR/R	NR/R	NR	NR	NR	NR/R	NR
11.	Synergism & potentiation	NR	NR	NR	R	R	NR/R	NR/R	NR	NR	NR	NR	NR
12.	Teratogenicity	NR	NR	NR	R	R	NR	NR	NR	NR	NR	NR	NR
13.	Effect on reproduction	NR	NR	NR	R	R	NR	NR	NR	NR	NR	NR	NR
14.	Carcinogenicity	NR	NR	NR	R	R	NR	NR	NR	NR	NR	NR	NR
15.	Metabolism	NR	NR	NR	R	R	NR	NR	NR	NR	NR	NR	NR
16.	Mutagenicity	NR	NR	NR	R	R	NR	NR	NR/R	NR/R	R	NR	NR
17.	Toxicity to birds (two)	R	R	R	R	R	R	R	NR	NR	R	NR	NR
18.	Toxicity to fish (Fresh water)	R	R	R	R	R	R	R	NR	NR	R	NR	NR
19.	Toxicity to honeybees	R	R	R	R	R	R	R	NR	NR	R	NR	NR
20.	Toxicity to live stock	R	R	NR	R	R	NR	NR	NR	NR	R	NR	NR
21.	Medical data	R	R	R	R	R	R	R	R	R	R	NR	R

Sl.No.	Parameter	9(3B)			9(3)								9(4)
	Toxicity	**TI**	**TIM**	**FIM**	**TI**	**TIM**	**FI**	**FIM**	**TI Vs TIM**	**TIM Vs TI**	**TI (New Source)**	**NF** (IM)**	**TIM* (AR) Vs TI**
22.	Human toxicity information from foreign countries	R	R	R	R	NR	NR	NR	NR	NR	R	NR	NR
23.	Observation in man (Health records of spray operators)	NR	NR	NR	NR	NR	NR^^	NR^^	NR	NR	NR	NR^^	NR
24.	Health records of Industrial workers.	NR	NR	NR	R	NR	R	NR	R	NR	R	NR	NR
25.	Toxicity to live stock (Field trial & observation)	NR	NR	NR	NR	NR	NR^^	NR^^	NR	NR	NR	NR	NR
26.	International report on carcinogenicity & genotoxicity status	NR	NR	NR	R/ NR	R/ NR	NR	NR	NR	NR	NR	NR	NR

TI-Technical import, TIM-Technical import manufacture, FIM-Formulation import manufacture, FI-Formulation import

R: Required

NR: Not required

NR^^: Not required as per decision of 318th RC held on 27-04-2011

NF:** In case of wettable powder, if toxicological data is generated for EC formulation applicable as per guidelines, then there is no need to generate data on wettable powder containing the same a.i.

TIM*(AR) Vs TI U/s 9(4): If chemical equivalence full regulatory level studies in the sample submitted by the applicant /in-process samples, required to submit data on toxicity as per the guidelines for TIM vs TI U/s 9(3)

R*: Any peer reviewed published data/information shall be acceptable (approved in 343rd and 344th RC Meeting)

13

Technologies Developed for Better Pest Protection

i) Silicon Technology

Silicon technology in agriculture refers to the use of silicon-based compounds, primarily silicon dioxide (SiO_2), in pesticide formulations and as part of integrated pest management (IPM) strategies. Silicon technology is not a pesticide itself but is used to enhance the effectiveness of pesticides and contribute to sustainable pest management. Here are some ways silicon technology is applied in pesticide use:

1. Silicon-Based Adjuvants: Silicon-based adjuvants are added to pesticide formulations to improve their efficacy. These adjuvants can enhance the spreading, wetting, and sticking properties of pesticides, ensuring better coverage of plant surfaces and improving the retention of the active ingredient. This leads to more effective pest control and reduces the need for higher pesticide concentrations.
2. Induced Resistance: Silicon has been shown to induce plant resistance against pests. When plants take up silicon from the soil and deposit it in their cell walls, it can create physical barriers that make it more difficult for pests to penetrate plant tissues. Additionally, silicon can trigger the production of defensive compounds in plants, making them less susceptible to pest damage.
3. pH Buffering: Silicon compounds can act as pH buffers in pesticide formulations, helping to maintain the stability and efficacy of the active ingredients over a broader pH range. This can be particularly useful when mixing pesticides with other chemicals or using them in different environmental conditions.
4. Soil Amendment: Applying silicon-based amendments to soil can improve its structure and nutrient availability, which can indirectly contribute to healthier plants that are better able to withstand pest pressure.
5. Reduced Pesticide Use: By improving the effectiveness of pesticides and enhancing plant resistance, silicon technology can contribute to reduced pesticide use and minimize the environmental impact associated with pesticide applications.

It's important to note that the effectiveness of silicon technology in pesticide use can vary depending on factors like crop type, pest species, and local environmental conditions. Therefore, it is essential to consider the specific circumstances and conduct proper testing and research to determine the optimal use of silicon-based compounds in pest management strategies. Additionally, regulatory approvals and guidelines may vary by region, so compliance with local regulations is crucial when using such products in agriculture.

Eg: Silicon based non-ionic tank mix adjuvant- PI company- SUPER SPREADER™

ii) Zeon® Technology-Syngenta

Zeon® Technology encapsulation, a Syngenta proprietary formulation technology, protects the active ingredient for a prolonged period and allows it to adhere to the plant surface better than its generic counterpart. When paired with the second active ingredient, thiamethoxam, Endigo ZC provides exceptional protection against a broad spectrum of insect pests.

Enhances the insecticides performance

1. Improved rain fastness
2. Resistance UV degradation
3. Excellent contact activity on surface
4. Long lasting insecticide effects
5. Fast absorption through leaf

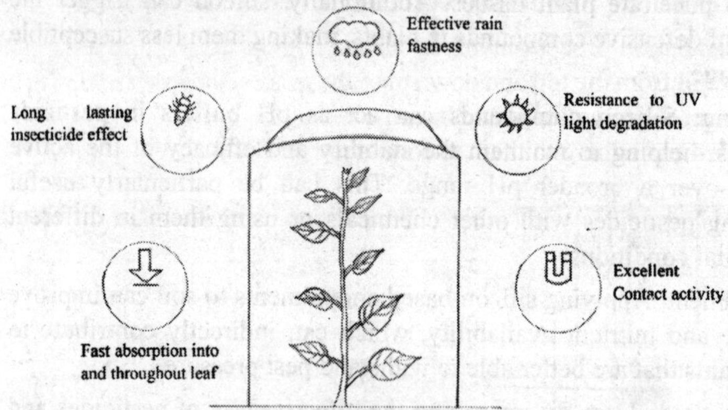

When comparing Zeon® technology to EC (Emulsifiable Concentrate) formulations, Zeon® offers several advantages due to its unique combination of Capsule Suspension (CS) and Suspension Concentrate (SC) formulations. One of the key benefits of Zeon technology is its slow and controlled release mechanism, which is achieved through encapsulation technology. This feature is not present in traditional EC formulations.

Eg: Chlorantraniliprole 9.30 % + Lambda-cyhalothrin 4.60 % ZC

Thiamethoxam 12.60 % + Lambda-cyhalothrin 9.50 % ZC

Lambda-cyhalothrin 4.90 % CS

iii) Prevathon® Technology- Corteva Agriscience:

Prevathon® is a notable insecticide technology developed by DuPont and now part of Corteva Agriscience. The active ingredient in Prevathon® insect control powered by Rynaxypyr® is chlorantraniliprole, which belongs to the diamide class of insecticides. This technology is specifically designed to target lepidopteran pests, such as caterpillars, which are significant pests in various crops including corn, cotton, soybeans, and vegetables.

Prevathon® works by disrupting the insect pests muscle function, leading to paralysis and ultimately death. It exhibits both translaminar and systemic activity, meaning it can penetrate plant tissues and provide protection from pests residing on both sides of the leaves. One of the key advantages of Prevathon® is its long-lasting efficacy. It offers extended residual control of targeted pests, reducing the need for frequent applications and providing growers with reliable pest management solutions. Additionally, Prevathon® has demonstrated low toxicity to beneficial insects, such as pollinators and predatory species, making it an environmentally friendly choice for integrated pest management programs.

Eg : Chlorantraniliprole 18.5% SC

iv) Fluid Bed Technology-Fipronil 80 WG (Bayer Crop Science)

The fluid bed granulation process involves suspending particles in an air stream and spraying a liquid from the top and down onto the fluidized bed. Particles in the path of the spray get slightly wetted and become tacky. The tacky particles collide with other particles and adhere to them to form a granule.

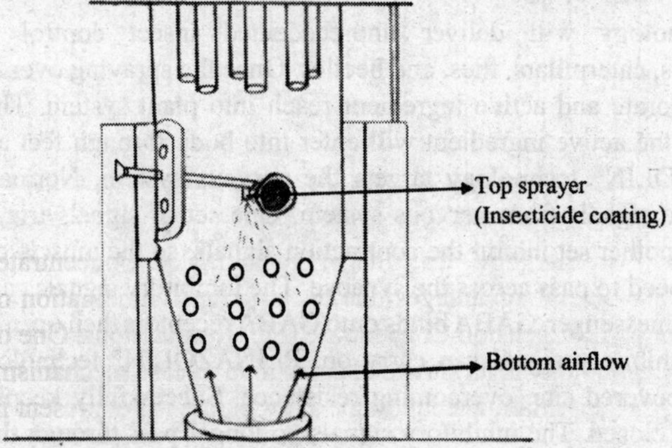

The innovative Fluid Bed Technology offers several advantages over traditional formulations such as suspension concentrate. It provides ease in handling, precise measuring, and accurate dosing, while being free from dust particles. This technology also ensures excellent suspension in water, leading to improved crop coverage. Compared to existing fipronil 5 % suspension concentrate formulations, which have been associated with reports of resistance and resurgence, the fipronil 80% WG formulation using Fluid Bed Technology offers a more effective and reliable solution.

Eg: Fipronil 80 % WG (Jump™)

v) BPX® Technology-Nichino india Pvt. Ltd.

Orchestra is powered by BPX® (Benzpyrimoxan), an innovative technology. BPX is developed by Japanese pioneer & innovator in agrochemicals, our parent company Nihon Nohyaku Corporation. BPX® has a very novel mode of action against the rice BPH - **"Ecdysone Titer Disruptor".** Effective and long duration of control (14-21 days), highly safe to natural enemies and the honeybees. Best partner to beat plant hopper for beautiful paddy fields. It is highly effective on BPH and WBPH including population which have developed resistance against existing products.

Tebufenozide and Methoxyfenozide are ecdysone receptor agonists that disrupt the moulting process in insects by continuously releasing ecdysone hormone, leading to premature moulting. In contrast, BPX® interferes with ecdysone metabolism, causing ecdysone titers to remain elevated during the moulting process. This disrupts ecdysis and leads to the death of the target insect by inhibiting the decline of ecdysone titer at the critical peak time of molting, ultimately resulting in moulting failure.

Eg: Benzpyrimoxan 10 % SC (Orchestra™)

vi) Plinazolin® Technology-Syngenta

PLINAZOLIN® technology will deliver unprecedented insect control on stinkbugs, mites, thrips, caterpillars, flies, and beetles. Once the spraying over the water molecules evaporate and active ingredient reach into plant system. Then in case of caterpillar, the active ingredient will enter into body through feet and mouthparts. PLINAZOLIN® technology targets the nervous system. Normally insect muscles receive signals from nervous system. One set of signals trigger muscle contraction. Another set inhibit the contraction signals, so the muscle can relax. All the signals need to pass across the synapse. The inhibitory signals carry by GABA, a chemical messenger. GABA binds onto GABA receptors then opening the channel so the inhibitory signal can carry on. PLINAZOLIN® technology binds to a newly discovered site, overcoming resistance. Successfully keeping the inhibiting channel closed. The inhibitory signals no longer pass through then muscles only contract cannot relax. This produces muscular creamps. The insect is paralysed and dies.

The PLINAZOLIN® technology targets GABA receptors as antagonists, disrupting signals from the insect's central nervous system (GABA gated chloride channel allosteric modulators. This leads to nervous system collapse, severe muscular cramps, paralysis, and ultimately, the insect's death. While existing insecticides like Fipronil also act on GABA-gated chloride channels, PLINAZOLINE® technology operates differently as GABA-gated chloride channel allosteric modulators. This unique mode of action makes it more potent and effective against Fipronil-resistant insects. PLINAZOLINE® offers a more effective solution for resistance management strategies and outperforms older, less effective chemistries. This technology is a novel properties viz., rain resistant, ultraviolet stability which contribute to its extended residual activity on a range of insect and mite pests.

Eg: Simodis (Isocycloseram 9.2 % DC), Incipio (Isocycloseram 18.1 % SC)

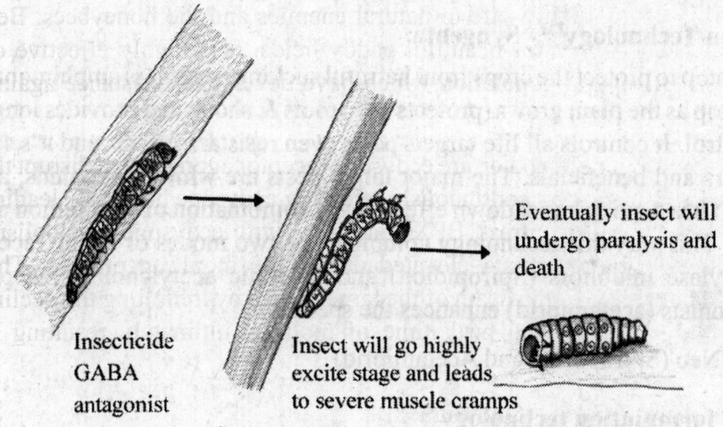

vii) Tymirium® Technology-Syngenta

Tymirium technology provides long-lasting protection against the highly destructive plant-parasitic nematodes and soil-borne diseases-particularly the *Fusarium* species. Key benefits include its ability to safeguard plant roots from attack, its ability to translocate to above-ground parts of the plant and protect against early-season diseases, as well as its simplicity of application and compatibility. The mode of action is unique as inhibitors of mitochondrial complex II electron transport.

Eg : Cyclobutrifluram-Not registered in India

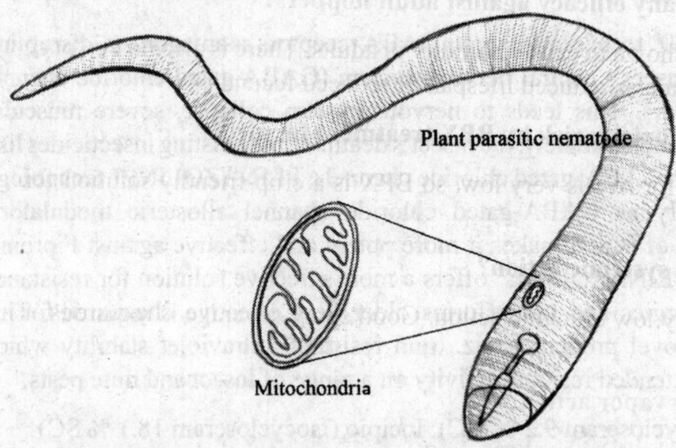

viii) Tinivion Technology™ - Syngenta

It works nonstop to protect the crops from harmful sucking pests. It is amphimobile, works non-stop as the plant grows (protects both roots & shots) and provides longer duration control. It controls all life targets pests even resistant insects and it's safe for pollinators and beneficials. The major target pests are whiteflies, aphids, and mealybugs. It has a quick knockdown effect. It is a combination of spiropidion and acetamiprid. This way this technology combines the two modes of action (acetyl CoA carboxylase inhibitors (Spiropidion) and nicotinic acetylcholine receptor (nAChR) agonists (acetamiprid) enhances the spectrum.

Eg: Elestal® Neo (Spiropidion and Acetamiprid)

ix) O-TEQ® formulation technology

O-TEQ® formulation technology means a vegetable oil in combination with various adjuvants. Bayer Crop Science has developed a new formulation concept to overcome these drawbacks, which is best described as an in-can oil dispersion (OD) that contains adjuvants in combination with vegetable oil to improve retention, spreading and penetration. Furthermore, the formulation assures good rain fastness, limited run-off and, ultimately, a higher efficacy. The use of vegetable oil is one of the most important features of this new technology because it assures a "persistent" liquid layer in close contact with the leaf surface. Due to the low volatility of the oils used, the distribution of the active ingredient between the spray deposit and the leaf tissue can be optimized. The OD formulations recognized by the above features will be marketed under the O-TEQ family brand.

Eg: Cyantraniliprole 10.26 % OD; Spirotetramat 15.31 % OD

Questions and answers

1 Q: Does BPX have any efficacy against adult hoppers?

A: BPX has not been shown to cause mortality in adults. There is also no evidence of sub-lethal effects such as reduced lifespan or reduced fecundity.

2 Q: Is there a phytotoxicity risk by BPX treatment for rice?

A: Phytotoxicity risk for rice is very low, so BPX is a crop-friendly solution for hopper control.

3 Q: Does BPX show systemic action?

A: BPX has no or very low systemic action. Good spray coverage is required for best efficacy.

4 Q: Does BPX show vapor action?

A: No vapor action from BPX has been detected.

5 Q: How many times can BPX be sprayed in one crop season?

A: One application per season is recommended for resistance management purposes.

6 Q. Does PLINAZOLIN® technology target the insect nervous system?

A: Yes, Plinazolin technology targets GABA receptors as antagonists

7 Q. How are Zeon technology-based formulations different from EC formulations?

A: When comparing Zeon® technology to Emulsifiable concentrate formulations, Zeon® offers several advantages due to its unique combination of Capsule Suspension (CS) and Suspension Concentrate (SC) formulations.

8 Q: How is silicon technology effective in pesticide formulation?

A: To improve their efficacy, Silicon-based adjuvants are added to pesticide formulations. These adjuvants can enhance pesticides' spreading, wetting, and sticking properties.

9Q. What is the benefit of O-TEQ® formulation technology

A: The oils used are low-volatility, so the distribution of the active ingredient between the spray deposit and the leaf tissue can be optimized.

10Q.Which insecticides are combined in TINIVION Technology

A: Spiropidion and Acetamiprid

14

Status of Insecticides Banned in India (on 31.03.2024)

Pesticide statistics

Pesticide registered	339
Pesticide banned for manufacture, import and use	49
Pesticides/pesticide formulations banned for use but allowed for export	5
Pesticides refused for registration	18
Pesticides restricted for use in India	16

List of insecticides which are banned, refused registration and restricted in use as on 31.03.2024

Sl.No.	Name of the insecticide	Year	Remarks (Gazette notification)
1	Toxaphene (Camphechlor)	1989	(S.O. 569 (E) dated 25th July 1989
2	Dibromochloropropane (DBCP)** Nematicide	1989	
3	Aldicarb	2001	(S.O. 682 (E) dated 17th July 2001)
4	Chlorbenzilate	2001	
5	Dieldrin	2001	
6	Ethylene Dibromide (EDB)	2001	
7	Aldrin	2001	-
8	Chlordane	2001	-
9	Endrin	2001	-
10	Heptachlor	2001	-
11	Lindane (Gamma-HCH)	2011/2013	S.O. 637(E) Dated 25/03/2011)-Banned for manufacture, import or formulate w.e.f. 25th March, 2011 and banned for use w.e.f. 25thMarch, 2013.
12	Endosulfan	2017	(ad-Interim order of the Supreme Court of India in the Writ Petition (Civil) No. 213 of 2011 dated 13th May, 2011 and finally disposed of dated 10th January, 2017)

Sl.No.	Name of the insecticide	Year	Remarks (Gazette notification)
13	Carbaryl	2018	(S.O 3951(E) dated 8th August, 2018
14	Diazinon	2018	
15	Dichlorvos	2018	
16	Fenthion	2018	
17	Methyl Parathion	2018	
18	Thiometon	2018	
19	Phorate	2018	
20	Phosphamidon	2018	
21	Triazophos	2018	
22	Trichlorfon	2018	
23	Sodium Cyanide*	2018	
24	Dicofol	2023	(S.O. 4294(E) dated 3rd October, 2023)
25	Methomyl	2023	
26	Benzene Hexachloride	-	-
27	Calcium Cyanide	-	-
28	Chlorofenvinphos	-	-
29	Copper Acetoarsenite	-	-
30	Ethyl Parathion	-	-
31	Menazon	-	-
32	Pentachlorophenol	-	-
33	Tetradifon	-	-

* Banned for Insecticidal purpose only)** Nematicide

Insecticides/ Insecticides formulations banned for use but continued to manufacture for export

Sl. No.	Insecticide	Year	Remarks (Gazette orders)
1	Nicotine Sulfate	1992	(S.O. 325 (E) dated 11th May 1992)
2	Dichlorvos	2020	(S.O. 1196 (E) dated 20th March 2020)
3	Phorate	2020	
4	Triazophos	2020	

Insecticides formulations banned for import, manufacture and use

Sl. No.	Insecticide	Year	Remarks (Gazette orders)
1	Carbofuran 50% SP	2001	(vide S.O. 678 (E) dated 17th July 2001)
2	Methomyl 12.5% L	-	-
3	Methomyl 24% formulation	-	
4	Phosphamidon 85% SL	-	

15

Pesticide Residue

Crops are grown with the use of plant protection chemicals to boost production, enhance quality, and lengthen shelf life. Wherein the enormous use of pesticides which are having the longer half life period led to leave the residues in the different bodies of the ecosystem. Pesticide residues refer to the residual presence of pesticide active ingredients, their metabolites, or breakdown products in various environmental components subsequent to their application, accidental release, or disposal.

Analysis of residues gives insight on the type and extent of chemical contamination in an area, as well as how long it has been there. Before governments will approve the use of pesticides, they require comprehensive testing of their efficacy, environmental impact, and toxicology. A possible risk to human health arises from the presence of chemical residues and/or breakdown products in agricultural goods. Selective sampling plans can be used to learn more about pesticide traces in the environment, how they move about, and how quickly they degrade.

Pesticide: The act defines "pesticide" as any substance or other substance or product that is used to stop, kill, repel, or control insects, rodents, fungi, weeds, and other plants and animals that humans don't need.

(or)

A pesticide is defined as any substance or combination of substances that is designed to prevent, eliminate, deter, or reduce the impact of pests. These pests can include insects, rodents, nematodes, fungi, weeds, and other forms of plant or animal life found on land or in water. Additionally, pesticides can also target viruses, bacteria, or other microorganisms that are present in or on living humans or animals. The administrator has the authority to classify any substance or combination of substances as a pest, and pesticides can also be used as plant regulators, defoliants, or desiccants.

(or)

According to the Insecticides Act of 1968, any substance listed in the schedule or any other substances that the central government, in consultation with the Board, may include in the schedule through notification in the official gazette, are subject to regulation. A pesticide is defined as any preparation that contains one or more of the aforementioned substances.

Pesticide residue: Pesticide residue refers to the measurable amount of pesticide and its associated derivatives or metabolites that can be found on or within agricultural produce, animal feed, food items consumed by humans, and the surrounding environment. The residual concentration is typically quantified using units such as parts per million (ppm) (mg/kg), parts per billion (ppb) (μg/kg), or parts per trillion (ppt) (ng/kg), which represent the weight by weight ratio.

Definition by WHO: The term "pesticide residue" refers to any substance or combination of substances found in food intended for human or animal consumption that arises from the application of pesticides. This includes various specified derivatives, such as degradation and conversion products, metabolites, reaction products, and impurities, which are deemed to have toxicological importance.

Definition by FAO (1986): Residues refer to the substances that persist in or on a feed or food commodity, soil, air, or water subsequent to the application of a pesticide. In the context of regulation, the term "regulatory purposes" encompasses the primary compound as well as any designated derivatives, which may include degradation and conversion products, metabolites, and impurities that are deemed to have toxicological importance.

Pattern of pesticide loss in the environment

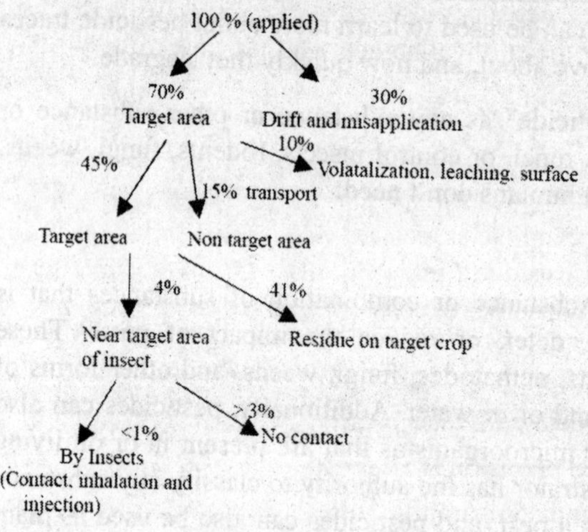

Above flowchart shows the pesticide loss in the environment. Only less than 1 per cent of pesticide reach the target insects and 41 per cent pesticide transports on to target crop as residue.

Pesticide Residue Analysis (PRA)

The process of **qualitative** and **quantitative** analysis of pesticide residues and their toxic metabolites in representative samples drawn from agricultural fields, markets and environment using analytical instruments followed in the quality control laboratories.

Qualitative Analysis: The process of separation and detection of residues based on their chemical properties using an analytical column and detector respectively. Qualitative analysis is a measure of standard peak retention time in minutes (R_t).

Quantitative Analysis: The process of finding concentration of a specific pesticide residue present in the sample, on the basis of peak area or peak height (peak area is usually considered) by following the residue formula as follows,

Standard conc. (ppm): Wt. of CRM (mg) / solvent volume (mL) * % purity of CRM * 1000

$$\text{Sample conc. (ppb)} = \frac{\text{Sample area}}{\text{Standard area}} \times \frac{\text{Standard Conc. (ppb)}}{\text{Sample weight (g)}} \times \text{Final volume}$$

Importance of pesticide residue

The importance of insecticide residue analysis in the agricultural food products lies due to the mutagenic and carcinogenic effects caused by the consumption of pesticide contaminated food or agricultural products. Agricultural food commodities like fruits, vegetables, cereals, pulses as well as drinking water are the very essential requirements for the survival human and animals.

Analytical Quality Control (AQC)

The inclusion of the Analytical Quality Control (AQC) is imperative throughout all stages of insecticide residue analysis. Its purpose is to evaluate the reliability and accuracy of analytical findings and the associated data, which are utilized for verifying adherence to Maximum Residue Limits (MRLs), enforcing regulatory measures, and assessing potential consumer exposure to pesticides.

Objectives of AQC

- To prevent the reporting of data with false positives or false negatives.
- To guarantee that the desired precision is attained.
- To guarantee the accuracy and comparability of analytical findings.
- To establish the consistency of unknown sample result batches.
- To facilitate conformity with ISO/IEC 17025 standards.

The contribution of analytical instrumentation in the analysis of residues as follows

1. Screening of various agricultural food commodities for the presence of pesticide residual contaminations.
2. To find out whether the residue concentration is below or above the MRL.

3. To find out whether it is the residue concentration is below or above the ADI.
4. To study the dissipation patterns and to establish MRL (Supervised Field Trial).

Objectives

- To investigate the persistence in/on soil, plants, and water.
- To establish MRLs by field trials using good agricultural practices (GAP), in addition to the information collected from toxicological investigations.
- To define pre-harvest intervals (PHI) or safe waiting intervals based on multi-location trials.
- To inspect farm products taken from farmers' fields to assess the pesticide usage trends.
- To carry out a market analysis of agricultural products in order to forecast dietary pesticide intake and determine the danger to the general public by comparing it to the permissible daily intake (ADI).
- To test alternative techniques for the efficient removal of pesticide residues.
- To monitor agricultural produce and food commodities for pesticide residues under the Food Safety and Standard Act, 2006.

Types of Residues

1. Surface or Effective Residues: The portion of the pesticides left from the initial deposit.
2. Dislodgeable Residues: Easily detachable and a potential risk indicator for farm workers.
3. Penetrated Residue: The surface residue becomes penetrated residue by migrating into the sub-strata.
4. Terminal Residue: Breakdown products which are stable and create as many problems as the original compound.
5. Conjugated Residue: The products of secondary metabolism due to the reaction of pesticides or its metabolites with endogenous substrates.
6. Bound Residue: Chemical species in soil, plant or animal tissues originating from pesticide usage that cannot be extracted by the methods commonly used in the analysis.

Methods of Residue Analysis

1) **Single Residue Method (SRM):** Generally followed in supervised field trial or any other research purpose.
2) **Multi Residue Method (MRM):** Followed in case of monitoring of pesticides in field produces, market samples and environmental samples.

Challenges of Pesticide Residue Analysis (PRA)

1. Sample variability (matrix)
2. Different compound characteristics
3. Number of samples
4. Number of analytes monitored
5. Low levels detection (<10 ng/g)
6. Fast response required

Regulation of Pesticide Residue

Codex Alimentarius Commission (CAC)

In 1963, the Food and Agriculture Organization and the World Health Organization created the Codex Alimentarius Commission to establish uniform global food standards, guidelines, and codes of practice for the safety of consumers and the integrity of the food trade. The Codex Alimentarius is an international agreement on food safety and consumer protection that has been adopted as a standard by the World Trade Organization.

- Codex Alimentarius technically applies to all types of foods, more emphasis has been placed on those that are sold directly to consumers.
- It also includes overarching guidelines for food safety, storage, packaging, and labeling.

Joint Meeting on Pesticide Residues (JMPR)

The Commission, along with its specialized Committee on Pesticide Residues, depends on the impartial scientific guidance offered by the Joint FAO/WHO Meetings on Pesticide Residues. The determination of maximum residue limits (MRLs) and the subsequent risk assessment process through the Joint Meeting on Pesticide Residues.

CAC (Codex Alimentarius Commission)

The Codex Alimentarius Commission, established by FAO and WHO in 1963 to develop harmonized international food standards, guidelines and codes of practice to protect the health of the consumers and ensure fair trade practices in the food trade.

Codex in India

The central coordinating body for matters related to the Codex Alimentarius Commission within the Indian government is the Food Safety and Standards Authority of India (FSSAI), which operates under the Ministry of Health and Family Welfare (MoH & FW)

Food Safety and Standards Authority of India (FSSAI)

The establishment of the Food Safety and Standards Authority of India (FSSAI) can be attributed to the enactment of the Food Safety and Standards Act in 2006. It is a statutory body, assumes the responsibility of formulating standards for food products based on scientific principles. Additionally, it undertakes the regulation of various aspects of food production, processing, distribution, sale, and importation, all with the overarching goal of safeguarding the provision of safe and nutritious food for human consumption.

Food Safety and Standards Act, 2006 (FSSA, 2006)

1. The primary objective of the Food Safety and Standards Authority of India (FSSAI) is to streamline the management of food safety and standards by shifting from a complex, multi-level, and multi-departmental control system to a centralized chain of command. This transition aims to establish a singular point of reference for addressing all matters related to food safety and standards.
2. The enforcement of the provisions outlined in the Act is the responsibility of both the Food Safety and Standards Authority of India (FSSAI) and the respective State Food Safety Authorities.
3. All the issues related to food safety vis-à-vis pesticide residues, and maximum residue limits (MRLs) etc. are dealt only under FSSA, 2006 and Food Safety and Standards (contaminants, toxins and residues) Regulation, 2011.

National Codex Contact Point (NCCP)

- To maintain communication with the CAC and manage Codex activities in India, the FSSAI established the National Codex Contact Point (NCCP)
- In partnership with the National Codex Committee, it coordinates and promotes Codex activities in India and supports India's participation in Codex work through a formal consultation process

Functions of NCCP-INDIA

- The primary function is to acquire and distribute all conclusive documents (includes standards, codes of practice, guidelines, and other advisory texts) as well as working papers generated during Codex Sessions.
- Encourage Codex Activities in India
- Develop the country's capacity to effectively engage in Codex activity.
- Monitor international food standards work.
- Encourage food makers to increase quality and hygiene management to comply with requirements set forth by international food standards

- The objective is to disseminate information regarding food standards and legal requirements to pertinent government entities, primary producers, manufacturers, exporters, consumers, and relevant organizations.

APEDA

Under the Agricultural and Processed Food Products Export Development Authority Act, the Indian government set up the Agricultural and Processed Food Products Export Development Authority (APEDA) on February 13, 1986.

Functions

- The establishment of standards and specifications for the designated products with the intention of facilitating their exportation
- The task involves conducting inspections on fruits and fruit products, milk and poultry products, as well as meat and meat products. These inspections are carried out in various locations such as slaughterhouses, processing plants, storage facilities, conveyances, or any other premises where these products are stored or handled. The primary objective of these inspections is to ensure the quality of the aforementioned products.

Steps in Insecticide residue analysis

- Sampling
- Sample Preparation
- Extraction of residues
- Clean-up
- Analysis by Mass spectrometer.

1. Sampling

Sample: It is defined as few individuals from whole population or a small part of a whole lot.

Sampling: Sampling is the act, procedure, or method of selecting a representative sample from a population to determine its parameters or characteristics. Always ensure that the sample size and representation are adequate.

Sample size: The size of the sample is usually determined by the need of collecting a representative sample rather than getting enough material for analysis. Thus, sample size depends on the size of plot or commodity from which sample is drawn. The sample collection containers must be thoroughly washed with methanol and dried. The general size of sample as recommended by Codex committee for different substrate is as follows.

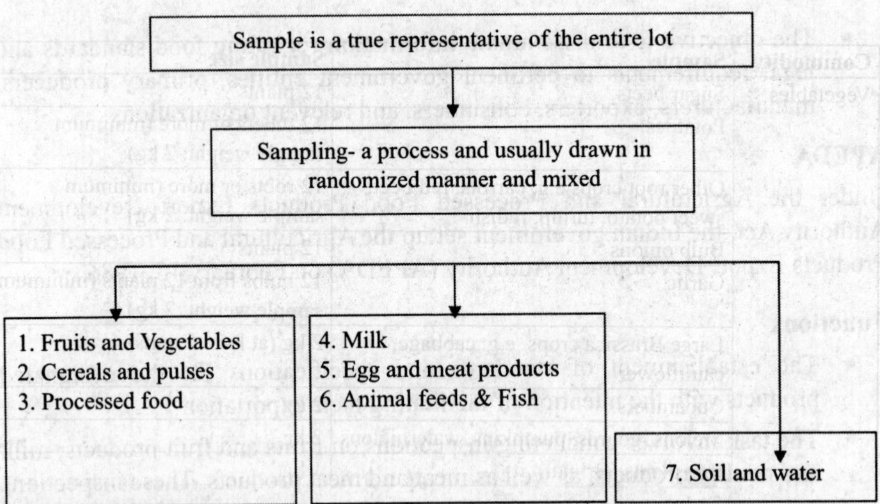

Commodity	Sample	Sample size
Fruits and Vegetables	Small or light foods with unit weight of 25g or less (pea, berries)	1 kg
	Medium-sized foods are typically between 25 and 250 grams per unit (apples, citrus, carrots, potatoes)	1 kg
	Large sized food with unit weight more than 250g (cabbage, cauliflower, melons, cucumber)	2 Kg
Fruits	**Citrus fruits** e.g. orange, lemon, mandarin, grapefruit	A total of 10-12 fruits (5-6 from each vendor)-1 kg
	Apple, guava	1 kg
	Pomegranate	Total of 6-7 fruits (3-4 from each vendor)-1 kg
	Mangoes	1 kg
	Grapes	Minimum of 2 kg
	Bananas (Raw banana)	1 kg
	Pineapples	2-3 fruits (1-2 from each vendor)
	Papaya, grapes	2 kg
	Raspberries and other small berries	A total of 0.5 kg was obtained from 12 different sites or bushes.
	Strawberries, Gooseberries	1 kg total from 12 different places or shrubs
	Miscellaneous small fruits e.g. olives, dates, figs	A total of 1 kilogram was collected from various locations on four trees.
	Tree nuts e.g. walnuts, chestnuts, almonds	A total of 1 kg

Commodity	Sample	Sample size
Vegetables	Sugar beets	12 plants
	Potatoes	12 tubers or more (minimum sample weight: 2 kg)
	Other root crops e.g. carrots, red beet, sweet potato, turnip, radish	12 roots or more (minimum sample weight: 2 kg)
	Bulb onions	12 plants
	Garlic	12 bulbs from 12 plants (minimum sample weight: 2 kg)
	Large Brassica crops e.g. cabbage, cauliflower	2 kg (at least 5 units)
	Cucumbers	2 kg
	Melons, gourds, pumpkins, watermelons	2 kg
	Egg plants (Brinjal)	1 kg
	Sweet corn	12 ears (the sample should weigh at least 2 kg; if necessary, take a higher number of objects to obtain a 2 kg sample)
	Tomatoes	1 kg
	Okra, Capsicum, Green chilli, Peas, moringa/drum sticks, Moringa leaves, Spinach	1 kg
Spices	Coriander seed, Fennel, Cumin, Fenugreek, Dry ginger powder, Cardamom and Black pepper	100 g
	Dry chilli powder, Red chilli dried	100 g
Herbs	Curry leaves (Fresh)	200 g
Plantation	Coconut	A total of 12 nuts
	Tea and dry leaves (Black tea)	200 g
Cereals	**Cereal grains** (wheat, rice, barley, oats, and other small grain cereals, maize (off the cob)	1 kg
	Straw of wheat, rice, barley, oats, and other small grain cereals; maize	0.5 kg
	Maize cobs	12 ears or more (the sample must weigh at least 2 kg)
Pulses	Chickpea, Cowpea, Bengal gram, Green gram, Pigeon pea, Black gram	1 kg
Mushroom	Mushrooms	12 items (at least 0.5 kilogram for the sample)
Processed foods in retail shops	Number of retail units in the lot	Minimum number of retail units to be samples
	1-25	1
	26 – 100	5
	101 – 250	10
	Over 250	15

Commodity	Sample	Sample size
Poultry and Dairy products	Milk	One pouch 0.5 L
	Producers	2 litres
	Pouched milk	3 x 500 ml pouches of similar nature
	Cheese, Butter, Cream	0.5 kg
	Eggs and Egg products	
	Producers (unpacked eggs)	6 dozen eggs
	Packed eggs	1 dozen eggs (12 units)
	Animal feeds	(10 locations × 1 kg) portions to comprise 1 sample
Non-Vegetarian food	Dry Fish	50 g
	Wet Fish	500 g
	Meat	500 g
Soil	Smaller size fields (3×3 grid with 9 total sample)	0.5-1.0 kg (15 cm deep and 3 to 5 cm in diameter are to be taken)
	Medium size fields (4×4 grid with 16 total sample)	
	Larger size fields (5×5 grid with 16 total sample)	
Water		Surface water (pond, river, canal etc) 2.5 L in a glass bottle

Note: Water and soil samples should be quickly refrigerated to 4°C before being transported to the lab.

2. Sample Preparation

Storage of Samples

Though, it is advised that samples collected from field be processed (extracted) within 2-4 hours, but sometime it may not be possible. In such cases, sample should be stored in refrigeration or deep fridge (0 to - 10°C), so that the level of residues presumably remains unchanged and truly represent the level of residues at the time of sampling.

However, samples should not be stored for more than 2-3 days because even under such a low temperature some pesticides tend to degrade. The degradation is faster in chopped, homogenized samples. If samples are required to be stored for longer period then for accurate results, a fortified control sample with known level of pesticide is stored along with the field sample and analyzed simultaneously. The analysis of fortified sample will show the loss of pesticide during storage. The residue level in field sample is corrected for the loss during storage.

3. Extraction

Extraction is the transfer of toxic substances from treated, bulky biological material to a solvent. The extraction technique should be designed to remove pesticides from the matrix quantitatively (with high efficiency) and without causing chemical changes to the pesticide.

Extraction Methods

a. **Liquid-liquid extraction (LLE):** Separation of compounds based on their relative solubility in two immiscible liquids, typically water and an organic solvent such as hexane, ether, chloroform, ethyl acetate, etc.

b. **Soxhlet Extraction (SE):** Franz von Soxhlet devised the Soxhlet extraction in 1879 for the extraction of lipids from solid materials

c. **Super-critical fluid extraction (SFE):** The Supercritical Fluid Extraction (SFE) method involves passing carbon dioxide gas through the matrix at supercritical temperatures (31°C) and pressures (74 atm) (liquefied carbon dioxide) to extract pesticides and then transferring the pesticides to an analytical equipment for quantification

d. **Solid Phase Extraction (SPE):** The solid phase extraction approach is based on the device's selective retention of the analyte. In QuERChERS method commonly the dispersive solid ohase extraction (d-SPE) technique is followed for the extraction of the sample matrix as it is a selective, robust and eliminates complex liquid extraction mrthods while extending the range of recovered pesticides. While performing dSPE, an aliquot of the sample extract is added to a centrifuge tube containing a relatively small amount of the SPE sorbent (PSA) and the mixture is vortexed and centrifuged to facilitate the better clean-up process.

e. **Solid Phase Micro-Extraction (SPME):** In this technique, an extractant droplet or fiber wrapped with the extractant is added to the solution to be extracted before the analytical device is moved.

f. **Accelerated Solvent Extraction (ASE) or Pressurized Liquid Extraction (PLE):** The passing of the solvent through the matrix at room temperature or pressure, the analyte can be extracted in a shorter amount of time (often less than 20 minutes) and with less solvent. This extraction method used for pesticides in soil, chlorinated and organo-phosphorous insecticides and herbicides.

g. **Microwave - Assisted Solvent Extraction (MASE):** To extract the analyte from the matrix, the technique employs microwave energy and a suitable solvent, water is frequently used in this procedure. For example, pesticides can be extracted from meat, eggs, and dairy products, soil and sand, dust, water, and plants.

4. Clean-up

Cleanup is a step or group of steps in the analytical procedure that removes the majority of possibly interfering co-extractives using physical or chemical means. Sugars, lipids, organic acids, sterols, proteins, pigments, and excess water are removed with this process. To remove chorophyll or colour pigments from matrix generally they add graphitized carbon but excess it may affect the residue analysis. To avoid that **ChloroFiltr®** is a new polymeric sorbent available exclusively from UCT. It is designed to replace graphitized carbon black (GCB) for efficient removal of chlorophyll without loss of planar analytes.

The classification of the clean up techniques employed for removal of co extractive is given below

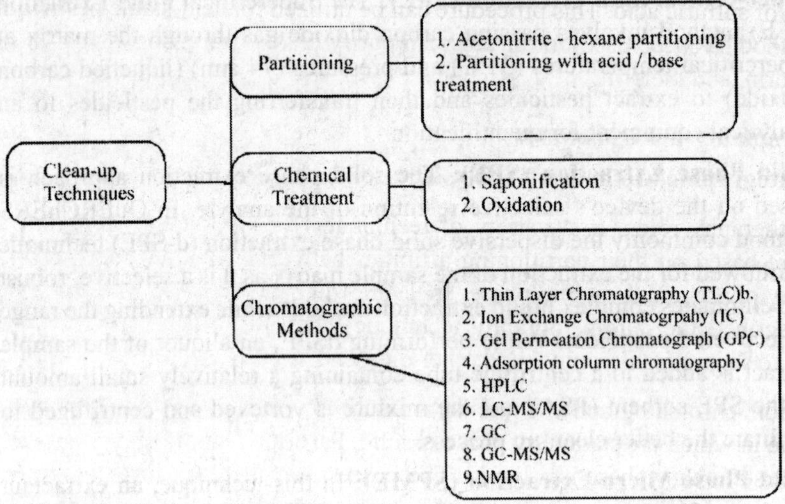

4A. Partitioning

Co-extractives from the extract are eliminated using this procedure by partitioning residues between two immiscible solvents. Due to difference in polarity of pesticide and co-extractives, one of them gets partitioned in non-polar solvent whereas other remains in polar solvent.

4A.1 Acetonitrile – hexane partitioning

Partitioning using acetonitrile and hexane is used to remove oil and fat from extracts. The extract is dried, and the residues are dissolved in acetonitrile (25ml) and partitioned three times with hexane (15ml). This method is used to clean extracts of oil seeds, milk, butter, among other materials.

4A.2. Partitioning with acid / base treatment

This method can be applied to insecticides that are either acidic or basic in nature. The solvent extract is concentrated to dryness. The residues are dissolved in dilute aqueous acid / base solution (acidic solution when pesticides is basic and alkaline solution when pesticide is acidic). This technique cannot be used for pesticides with neutral property.

4B. Chemical treatment

4B.1. Saponification: This method is used to remove fats and oils from extracts. The treatment with an alkaline aqueous solution saponifies or hydrolyzes the fat and oil.

4B.2. Oxidation: In this process, the co-extractables are oxidized with copious amounts of sulfuric acid. This procedure can be utilized for acid-stable pesticides. This method has been used, for instance, to purify milk extracts containing HCH, DDT, aldrin, and dieldrin.

4C. Chromatographic methods

Chromatographic Methods of Analysis

Chromatography refers to a variety of techniques for separating substances in a mixture based on their partitioning affinities between two phases, typically a stationary phase and a mobile phase. In 1906, Michael Tswett, a biologist, first time described the chromatography technique where he used petroleum ether solution to separate chlorophyll and other pigments from plant extracts.

Adsorption chromatography is the term used to describe a chromatographic technique in which the stationary phase is solid. Partition chromatography is used when the stationary phase is a liquid. The process is called gas chromatography when the mobile phase is a gas, and liquid chromatography when the mobile phase is a liquid. Gas-liquid chromatography refers to a separation technique in which the mobile phase is a gas and the stationary phase is a liquid. It is possible to have paper chromatography, with paper serving as the stationary phase, when liquid is the mobile phase.

4C.1. Thin Layer Chromatography (TLC)

A liquid solvent (mobile phase) and a glass plate covered with a thin coating of adsorbent (solid phase) are used to separate dried liquid samples.

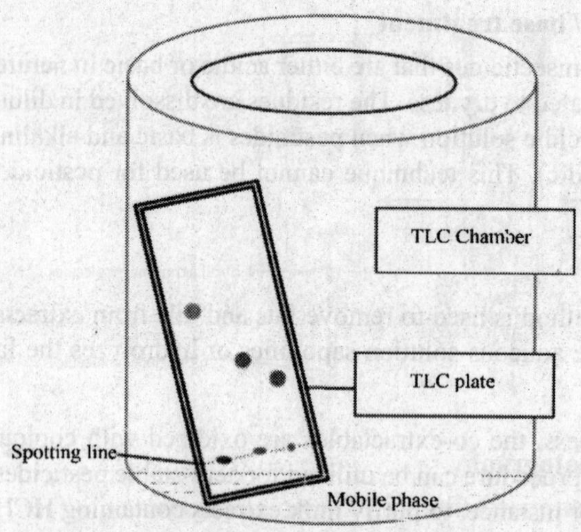

4C.2. Ion Exchange Chromatography (IC)

Chromatography based on differences in the ion-exchange affinities of anions and cations covalently attached to the stationary phase, typically resin.

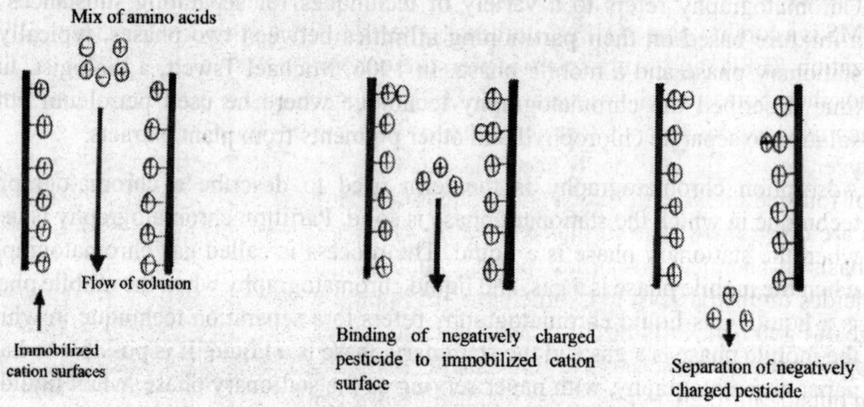

4C.3. Gel Permeation Chromatography (GPC) or Size-Exclusion Chromatography (SEC)

When the separation is based on the size (weight) of the molecules, the bigger molecules can't get through the holes in the gel and move faster because of this. On the other hand, the smaller molecules enter the gel beads and are left behind, so they come out slowly. Gels like Sephadex, Bio-gel, Sepharose, Agarose gel, Styragel, etc. are used that are available in market.

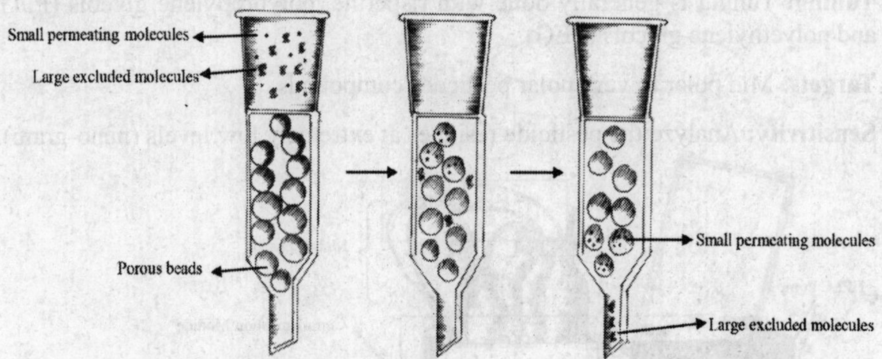

4C.4. Adsorption column chromatography

Chromatography based on the adsorption affinities of sample components to the surface of an active solid. Adsorbents include silica gel, alumina, charcoal powder, and calcium hydroxyapatite.

4C.5. HPLC: refer to pesticide formulation analysis chapter

4C.6. Liquid Chromatography and Mass Spectroscopy (LC-MS/MS)

LC-MS is a combination of HPLC with mass spectrometry (MS) using Electrospray Ionization (ESI) and Atmospheric pressure Chemical Ionization (APCI) or Atmospheric pressure photoionization (APPI), an atmospheric pressure interface is then used to transfer ions (open iqnization) under the high vacuum of the mass analyzer. In MS, the commonly used mass analyzer is quadrupole (Q1, Q2, Q3), a set of four parallel metallic rods and the mass to charge ratios (m/z) of gas phase ions are determined. The former is used for physical separation of pesticides on the basis of chemical properties (polarity) and the later used for quantitation of pesticides on the basis of mass to charge (m/z) ratio with a detection limit in the range of Pg to Fg level. In LC-MS/MS, electron spray is normally nebulized using nitrogen, which even acts as a drying gas for solvent. LC-MS/Ms needs Argon gas from outside for collision purposes inside the mass analyzer.

LCMS/MS can analyze about 200-250 pesticides in a single analysis. In LCMS/MS the mobile phase pH is maintained by adding **formic acid or ammonium acetate or ammonium formate. Selective Ion Monitoring (SIM-mode)** is a technique used in **Liquid Chromatography-Mass Spectrometry (LC-MS)** to selectively detect and quantify specific ions of interest from a complex mixture. It is particularly useful for improving the sensitivity and specificity of the analysis.

Detectors: Electron multiplier (Dynode), Multi-channel plate (MCP) ion, Photo multiplied tube, Florescence detector, UV detector and Photodiode Array (PDA).

Tuning: Tuning is generally done with risperine, polypropylene glycols (PPG), and polyethylene glycols (PEG).

Targets: Mid polar to very polar pesticdes compounds

Sensitivity: Analyze the pesticide residues at extremely low levels (nano-gram).

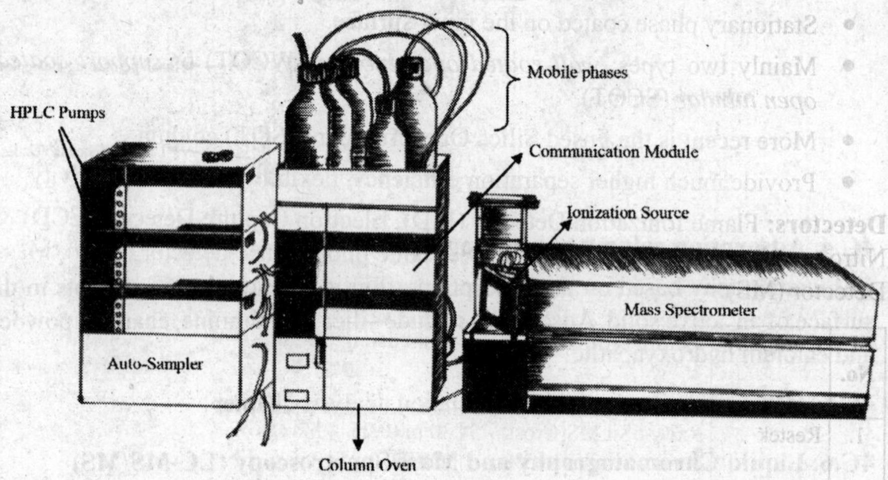

SI.No.	Company	Colum type	Dimension	Name of the suitable equipment
1.	Shimadzu	Shim-pack XR-ODS III	2.0 mm I.D. × 150 mm	LCMS/MS

4C.7. Gas Liquid Chromatography (GLC) or Gas Chromatography (GC): refer to pesticide formulation analysis chapter

4C.8.Gas Chromatography and Mass Spectroscopy (GC-MS/MS)

It is a technique for analyzing samples that combines the benefits of gas-liquid chromatography (GLC) and mass spectrometry (MS). An analytical chemist can do both qualitative and quantitative analyses of a solution containing many chemicals by combining the two methods. Mass spectrometers' sensitivity is based on the m/z ratio of chemicals, which is measured by the instrument. GC-MS can only analyze compounds that are volatile and heat stable. In the case of GC-MS/Ms, the ionization is electron impact ionization (EI) (Closed ionization); the collision gas is Argon. Biopolymers are detected in GC-MS using MALDI (Matrix Assisted Laser Desorption/Icnization) ionizing techniques. The Helium (Purity-99.999 %) is a carrier gas.

Generally, capillary columns are used in pesticide residue analysis.

Characteristics of Capillary columns

- Thin fused-silica.
- Typically, 10-100 m in length and 250 mm inner-diameter.
- Stationary phase coated on the inner surface.
- Mainly two types; *wall-coated open tubular* (WCOT) or *support coated open tubular* (SCOT).
- More recent is the Fused Silica Open Tubular (FSOT) column.
- Provide much higher separation efficiency, flexibility and low reactivity.

Detectors: Flame Ionization Detector (FID), Electron Capture Detector (ECD) & Nitrogen phosphorous Detector (NPD), Flame photometric detector (FPD), Mass Detector (MS).

SI. No.	Company	Colum type	Dimension	Details	Name of the suitable equipment
1.	Restek	Rxi®-5Sil MS(Cross bond®, Similar to 5% Diphenyl/95% Dimethyl polysiloxane)	30 m (0.25 mmID, 0.25 µm df	Max Prog Temp 350°C, Min bleed at 320°C	GCMS/MS

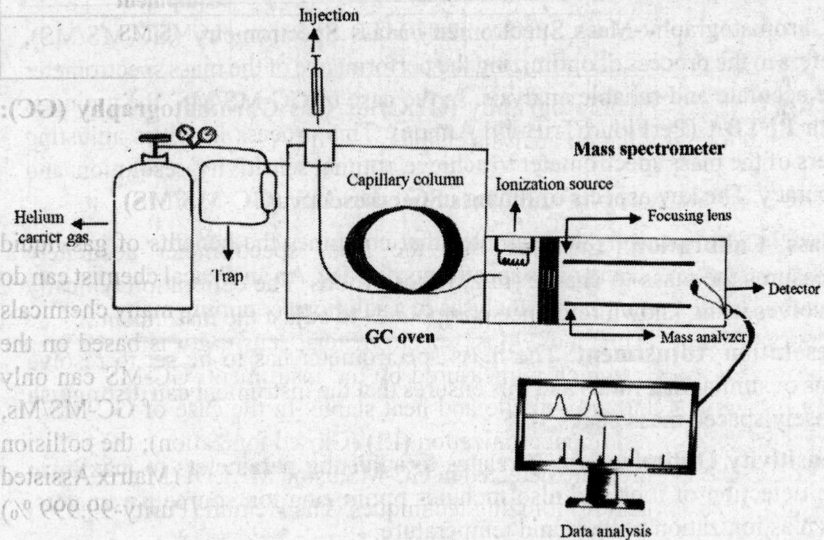

Comparison between GC and GC-MS/MS

- Retention time is the main key for identification in GC whereas m/z is the main key in GCMS.

- GC identifies the components with reference to standard while GCMS does with reference to library spectrum
- GC requires different detectors for different types of components in trace level whereas different types of samples can be analyzed in the same run using GCMS.
- Retention time and elution pattern of a mixture of components is relative to column temperature, nature of carrier gas as well as carrier gas flow rate (mL) in GC where as in GCMS retention time has little role.

Comparison between GC-MS/MS and LC-MS/MS

The advantage of LC-MS is that it can detect pesticide residues at very low levels (nanograms), that is not possible with GC-MS and LC-MS is to be used when the GC faces trouble finding pesticides that elute close together. Pesticides can be quantified and found using LC-MS that are hard or impossible to study with GC-MS. LC-MS can detect both the original parent compound and its metabolites.

Type of samples considered for detection through LC-MS/MS or GC-MS/MS

LC-MS/MS or HPLC	GC-MS/MS or GC-FID
Non-volatile compounds	Volatile compounds
Thermally unstable compounds	Thermally stable compounds
Mid polar to polar compounds	Non polar to mid polar compounds
In LC-MS/MS no library present	In GC-MS/MS have their own library

In Gas Chromatography-Mass Spectrometry/Mass Spectrometry (GC-MS/MS), tuning refers to the process of optimizing the performance of the mass spectrometer to ensure accurate and reliable analysis. In the case of GC-MS/MS, the tuning is done with **PFTBA** (PerFlouroTriButyl Amine). This process involves adjusting parameters of the mass spectrometer to achieve optimal sensitivity, resolution, and mass accuracy. The key aspects of tuning in GC-MS/MS are

1. **Mass Calibration**: Ensuring that the mass spectrometer accurately measures the mass-to-charge (m/z) ratios of ions. The calibration typically involves using known reference compounds to adjust the instrument.
2. **Resolution Adjustment**: The mass spectrometer has to be set to resolve ions of similar m/z ratios and this ensures that the instrument can distinguish closely spaced mass peaks.
3. **Sensitivity Optimization**: It relates to adjusting parameters to maximize the detection of ions and also includes optimizing ion source parameters, such as ionization voltage and temperature.
4. **Ion Source Tuning**: The ion source conditions (electron energy in Electron Ionization (EI), reagent gas flow in Chemical Ionization (CI)) are adjusted to produce ions efficiently.

5. **Quadrupole Tuning**: For quadrupole mass spectrometers, the voltages are adjusted that are applied to the quadrupole rods to optimize the transmission and filtering of ions.
6. **Detector Optimization**: Setting the detector parameters (e.g., gain, voltage) to ensure efficient ion detection and signal amplification.
7. **Checking Instrument Performance**: Running performance checks using standard compounds to ensure that the instrument meets the required specifications for accuracy, precision, and sensitivity.

Modes of Mass Analysis both LC-MS/MS and GC-MS/MS

Mass spectrometers are typically used in one of the three modes: full scan or selected ion monitoring (SIM) and multiple reaction monitoring (MRM)

Full Scan MS

In full scan mode, a target range of mass fragments is determined and programmed into the instrument's method and a typical range might be m/z 50 to m/z 400.

Selected Ion Monitoring (SIM)

In SIM mode, specific ion fragments are entered into the instrument method, and only those fragments are detected by the mass spectrometer.

Multiple Reaction Monitoring (MRM)

In MS/MS mode, an ion from the mass spectrum is selected for fragmentation while all other masses are discarded. Multiple Reaction Monitoring (MRM) is a technique used in mass spectrometry, particularly in tandem mass spectrometry (MS/MS), to selectively and sensitively detect and quantify specific compounds within a complex mixture.

In MRM, specific precursor ions (parent ions) are selected and fragmented into product ions (daughter ions).

Precursor ion (Parent Ion): The parent ion, also known as the precursor ion, is the ion that is initially selected and isolated in the first stage of mass spectrometry.

Fragmented ion (Daughter Ion): The daughter ion, also known as the product ion, is the ion that results from the fragmentation of the parent ion during the second stage of mass spectrometry.

4C.9. Nuclear Magnetic Resonance Spectroscopy (NMR)

In 1946, Edward Purcell at Harvard and Felix Bloch at Stanford were the first to find NMR. It is an analytical chemistry method for detecting a sample's content, purity, and molecular structure for quality control and study. For example, NMR can measure the amount of known chemicals in a mixture.

Principle

Nuclear magnetic resonance (NMR) occurs when certain atom nuclei are placed in a static magnetic field and subsequently subjected to an oscillating magnetic field. Nuclear magnetic resonance (NMR) spectroscopy depends on the distribution of positive charge and the net spin of protons and neutrons, which both have a spin quantum number of 1/2.

Schematic diagram of Nuclear Magnetic Resonance Spectrometry

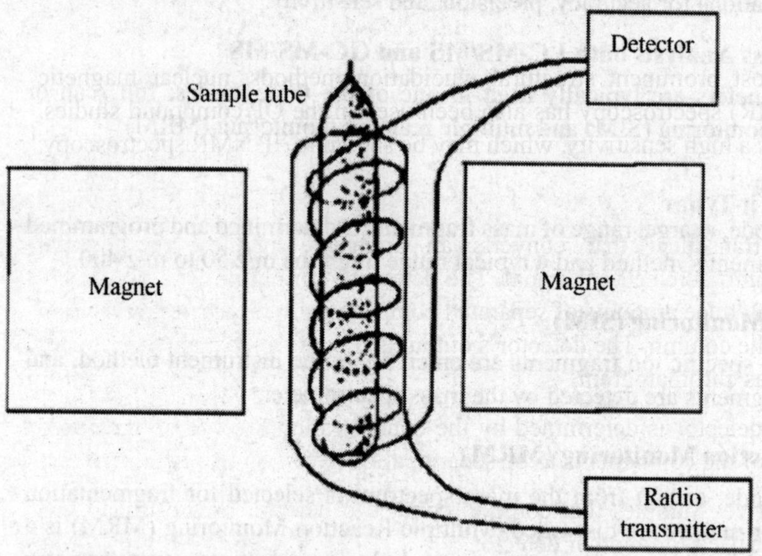

Standard used in NMR

Tetramethyl silane (TMS) Si $(CH_3)_4$ is the NMR standard **(central solvent or internal solvent)** and because of its solubility in organic solvents, TMS is used as an internal standard in nuclear magnetic resonance (NMR) spectroscopy to calibrate chemical shifts.

Solvents used in NMR Spectroscopy

1. Tetrachloromethene, CCL4
2. Deuterochloroform ($CDCl_3$)
3. Deuterium oxide (D_2O)
4. DMSO-d_6
5 Acetic Acid
6. Acetone
7. Acetonitrile
8. Benzene
9. Chloroform
10. Dimethyl Sulfoxide
11. Methanol
12. Methylene Chloride
13. Pyridine

Nuclei	Unpaired protons	Unpaired neutrons	Spin
^{1}H	1	0	½
^{2}H	1	1	1
^{31}P	0	1	½
^{23}Na	2	1	3/2
^{14}N	1	1	1
^{13}C	0	1	½
^{19}F	0	1	½

Application

One of the most prominent structural elucidation methods, nuclear magnetic resonance (NMR) spectroscopy has also been used in the OP compound studies. Phosphorus has a high sensitivity, which may be shown in ^{31}P NMR spectroscopy.

Detectors and it Types

Detector: A transducer that converts an analyte's chemical or physical characteristics into an electrical signal. The detector's primary role is to identify and quantify any trace amounts of separated components in the carrier gas stream that is exiting the column. The detector's output is fed into a machine that creates a trace known as chromatogram.

The choice of detector is determined by the concentration level to be measured and the nature of the components to be separated. Detectors of two types namely,

1. **Destructive-** Samples that come in contact to detector are destroyed and cannot be used for another detector.
2. **Non-destructive-** It is not destroyed after sensing and can be utilized with another detector.

Properties of detectors: Sensitivity, linearity, stability, and selectivity vary depending on the detector.

Sensitivity: The detector response (mV) per unit concentration of analyte (mg/mL)

Linearity: The signal is directly proportional to the amount (or concentration) of analyte within the specified concentration range

Stability: The degree to which the signal output, given a constant input, remains constant over time, is one of the paramount importance properties of detector.

Universal or selective response

The ability of a detector to react to every component present in a mixture is referred to as a universal response. A selective detector, in contrast, only detects specific components in a sample.

Pesticide formulation analysis is performed using a flame ionization detector (FID) and a thermal conductivity detector (TCD), whereas insecticide residue analysis is performed using an ECD, NPD, FPD, and PDA.

Detectors

a) **Electron capture detector (ECD):** The ECD operates by ionization of the carrier gas from a radioactive source (H^3 or Ni^{63}). It is used in analysis of organochlorine compounds such as DDT, Dieldrin, and Aldrin. It is also useful for detection of **electronegative compounds** in case of GLC. ECD is an extremely sensitive detector that can find compounds in picogram quantities and because of its sensitivity, it may be used to determine residues at the ppb and even ppt level. In the use of ECD coupled with GC, nitrogen gas is used and gas consumption is also very low.

b) **Nitrogen phosphorous detectors (NPD) or Thermionic detector or alkali ionization detector:** This is used to detecting nitrogen or phosphorus containing compounds Eg: OP compounds.

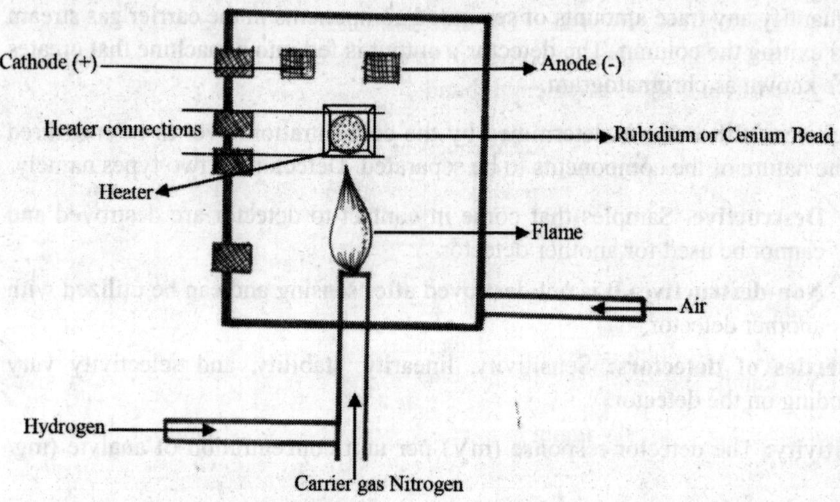

c) **Flame photometric detector (FPD):** Flame photometric detectors (FPD) are used to identify substances that contain sulfur (S) or phosphorus (P) atoms. In order to provide analytical data that is reasonably specific for substances containing these two types of atoms, this device exploits the chemiluminescent reactions of these compounds in a hydrogen/air flame. S_2 stimulated at 394 nm is the emission species for sulfur compounds. HPO stimulated at 510–526 nm serves as the flame's emitter for phosphorus compounds. The FPD-P can detect as little as 0.01 ng of phosphorus, and the FPD-S as little as 0.04 ng of sulfur.

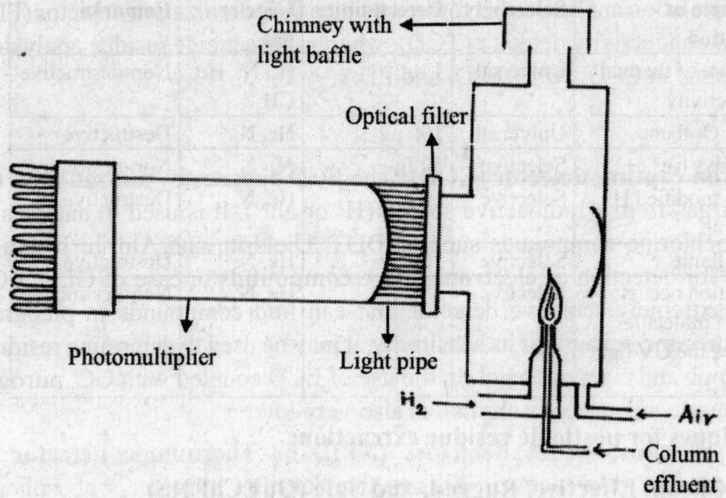

d) **Photodiode Array Detectors (PDA):** The polychromatic light is directed via the flow cell in the PDA detector, where it is diffracted by a grating and finally falls onto a collection of photodiodes that are individually tuned to a particular short wavelength band.

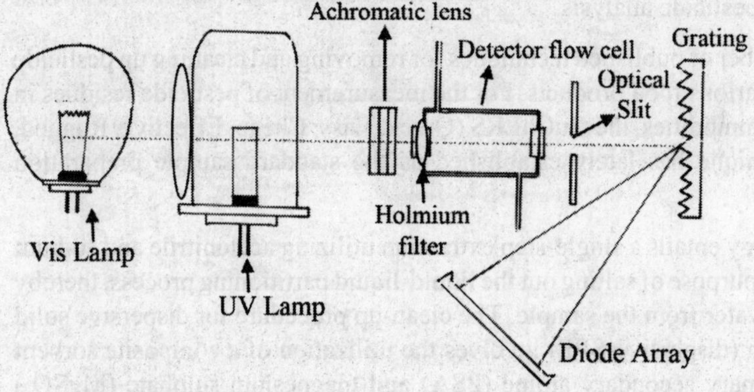

e) Other common detectors such as Hot Wire Detector (HWD) or Thermal conductivity detector (TCD) is a universal detector, Alkali flame ionization detectors (AIFD) is a selective detector, Mass Selective Detectors (MSD) are universal detectors.

Detector	Principle of operation	Selectivity	Detectability	Carrier gas	Remarks
TCD	Measure of thermal conductivity	Universal	1 ng	H_2, N_2, He, CH_4	Non-destructive
FID	N_2 H_2 O_2 flame	Universal	100 pg	He, N_2	Destructive
ECD	N_2 ß rays ($-e^{-1}$ e+)	Selective	50 fg	N_2	Non-destructive
AFID (NPD)	Alkali modified H_2, O_2 flame	Selective	10 pg	He, N_2	Destructive
FPD	H_2,O_2 flame	Selective	100 pg	He, N_2	Destructive
PID	Ionization occurs when a molecule absorbs the UV light energy	Selective	2 pg	He, N_2	Non-destructive

Efficient techniques for pesticide residue extraction:

Quickly, Easy, Cheap, Effective, Rugged, and Safe (QuEChERS)

A variety of pesticides are present in several food samples. These pesticides can be basic, acidic, or neutral, and the sample matrix may include substances viz. lipids and sterols. Cost pressure and rising demand for insecticide residue analysis had made possible to use a multi-residue approach to provide results that are high accurate. In 2003, USDA Anastassiades et al. originally presented the QuEChERS technology for pesticide analysis.

There are a number of published techniques for removing and cleaning up pesticide residues from various food products. For the measurement of pesticide residues in diverse food commodities, the QuChERS (Quick, Easy, Cheap, Effective, Rugged, and Safe) technique has lately established as the standard sample preparation methodology.

This methodology entails a single-step extraction utilizing acetonitrile and sodium chloride for the purpose of salting out the liquid-liquid partitioning process, thereby separating the water from the sample. The clean-up procedure for dispersive solid phase extraction (dispersive-SPE) involves the utilization of a composite sorbent comprising primary secondary amine (PSA) and magnesium sulphate ($MgSO_4$) to effectively eliminate organic acids, excessive water, and other constituents. Ultimately, the residues are analyzed through the utilization of either Gas Chromatography-Mass Spectrometry (GC-MS) or Liquid Chromatography-Mass Spectrometry (LC-MS), contingent depending upon the specific characteristics of the pesticide in question.

The QuChERS has various advantages over the conventional extraction and clean up techniques:

1. It has reduced the cost of sample analysis: Very less sorbent and cartridges are needed thereby reducing the cost for sample analysis.

2. Reduced and less hazardous waste is generated as no chlorinated solvents are used thereby minimizing the load of hazardous waste.
3. Increased scope of pesticide per analysis- The method is very amenable for determination of large number of pesticides in a single run.

Quick **P**olar **P**esticides **(QuPPe) method** used for multiresidue analysis for highly polar pesticides. Eg: Glyphosate, Maleic hydrazide, Gluphosinate, Phosetyl-A1, Ethephon, 2,4-D, Chlormequat, Mepiquat, Paraquat, Phosphonic acid.

Chemicals and their function in insecticide residue analysis

Chemical	Function
Ethyl acetate, Acetonitrile	Organic solvents used for extracting pesticides with the least number of co-extractables
Magnesium Sulphate	removes water from the organic phase
Anhydrous sodium sulphate	Facilitates solvent portioning and improves recovery.
Sodium Chloride	Facilitate phase separation between orhanic solvent & Pesticide
Primary Secondary Amine (PSA)	Sugars and fatty acids, organic acids, lipids, and certain colors are removed using this process
Glassware Class A	Made up of 70 % silica (inert - This inertness is due to the strong Si-O bonds in its structure) and remaining with boron trioxide free from reactions of salts of magnesium and lime, so there is no chance of oxidation
Sonication	Used for degassing purpose in mobile phase

5. Identification and Quantification - This is the most crucial part of analyzing pesticide traces as the goal of residue analysis is to identify and quantify low-level traces of potentially harmful pesticides in a wide number of samples.

Chromatographic techniques used for identification and quantification of pesticide residues are,

1. GLC with numerous detectors (FID, NPD, ECD, etc.)
2. High-performance (Pressure) liquid chromatography (HPLC) with different detectors such as PDA, Fluorescent, etc.
3. GLC combined with a mass detector (GC-MS or GC-MS/MS)
4. HPLC with mass-Detector (LC-MS or LC-MS/MS)

5.1 Detection and quantitation limits of the analytical method

Sample analysis must use the most detailed methods of analysis possible. Most of the time, the Limit of Detection (LOD) and Limit of Quantitation (LOQ) are used to describe how sensitive an analysis method as a whole.

5.1.1 Limit of Detection (LOD)

The lowest concentration can be positively recognized using a specific procedure in a specific matrix. The term "non-detected" (ND) refers to samples that do not contain residues at or above the LOD. But currently, LOD is not considered for determination of pesticide residue analysis.

5.1.2 Limit of Quantitation (LOQ)

The least amount of a pesticide contamination in a sample that can be quantified with reasonable precision and consistency is known as the Limit of Quantitation (LOQ). The term "non-quantifiable" is often used to describe LOQ because samples do not contain residues."

6. Confirmation of results

If the batch's recovery and LOQ measurements are satisfactory, then the negative results (residues below the reporting limit) can be regarded confirmed. Positive results (residues at or above the limit for reporting) normally need more revalidation. Unusually high levels of residues must be found using a combination of techniques, and their amounts must be proven by analyzing at least one more test sample. For confirmation, one can use different combinations of cleaning, derivatization, separation, and detection methods.

The following methods are used for confirmation of residues

- Use of selective detectors, such as ECD, FPD, NPD, DAD, and fluorescence, with GC or LC offers only limited specificity
- In chromatography, the use of dual columns, i.e. a second column with a distinct polarity.
- Utilization of an extremely specific detection system, such as mass spectrometry

Mass spectrometric analysis is regarded as the most precise method of verification. Typically, the term mass spectrometry substantiation refers to "overwhelming evidence."

Mass spectrometric determination usually carried out in conjugation with a chromatographic separation technique to simultaneously provide,

- Retention time
- Mass/charge ratio of ionic species
- Abundance data

For GC-MS procedures, the chromatographic separation should be carried out using capillary columns. For LC-MS procedures, the chromatographic separation can be performed using any suitable LC column. Diagnostic ion chromatograms should have peaks (with minimum 3 data points exceeding, S/N 3:1) of similar retention time, peak shape and response ratio to those obtained from a calibration standard analysed in the same batch. All the peaks with in a spectrum is called as **"full spectrum"**.

Reporting results- Report the exact concentration of pesticide residues in parts per million (ppm).

Calculations in pesticide residues

$$\text{Residues (mg/kg)} = \frac{\text{Sample peakarea} \times \text{Conc. of std. injected} \times \text{std injected } (\mu\text{L}) \times \text{final volume of the sample (mL)}}{\text{Std peakarea} \times \text{weight of sample analyzed} \times \text{samle injected } (\mu\text{L})}$$

$$\text{Wt. of the sample analyzed } (\text{g}) = \frac{\text{Sample weight } (\text{g}) \times \text{Aliquot taken } (\text{mL})}{\text{Volume of extracting solvent } (\text{mL})}$$

Preparation of CRM (Certified Reference material)

1. Calculate the final concentration of the primary stock solution using formula

$$\text{Concentration of the stock solution } (\text{ppm}) = \frac{\text{Weight of the CRM } (\text{g}) \text{xlO}^6 \text{x Purity } (\%)}{\text{Volume of the CRM to be prepared } (\text{mL}) \text{ x } 100}$$

To calculate the working standard solution concentration,

$N_1V_1 = N_2V_2$

Where,

N_1 - available concentration (μg mL^{-1}), V_1- volume to be taken from available stock (mL), N_2 - required concentration (μg mL^{-1}), and V_2 - required volume (mL).

Calculation for Dissipation study

$$\text{Dissipation rate } (\%) = \frac{\text{Initial deposit mg/kg) - Residue at given time } (\text{mg/kg})}{\text{initial deposit } (\text{mg/kg})} \times 100$$

Waiting period: Waiting period (Ttol) is defined as the minimum number of day's lapse before the insecticide reaches the tolerance limit.

The waiting periods or preharvest intervals were calculated by the following formula

$$Total(days) = \frac{[a - \log(tol)]}{b}$$

Where,

Ttol -Minimum time (in days) required for the pesticide residue to reach below the tolerance limit.

a. Log of apparent initial deposits obtained in the regression equation (Y = a + bx). tol- Tolerance limit of the insecticide (MRL)

b. Slope of the regression line

Half-life (RL_{50}): The time in days required to reduce the pesticide residues to half of its initial deposits. Mathematically, it is

$$T_{1\text{-}2}\ (\text{days}) = \frac{e}{b} = \frac{0.301}{b} (e = \log_2 = 0.301)$$

$$SP_{xy} = \sum xy - \frac{(\sum x)(\sum y)}{n}$$

$$SSx = \sum x^2 - (\sum x)^2$$

$$SP_{xy} = \sum xy - \frac{(\sum x)(\sum y)}{n}$$

$$b = \frac{SPxy}{SSx}$$

Where,

n - Number of observation

x - Number of days

y - Log [residues (mg kg^{-1}) × 10^2]

Risk assessment calculation

Based on contamination level, the risk to the general public was calculated using mathematical formulae (Darko and Akoto, 2008).

EDI-Estimated Daily Intake

$$EDI = \frac{C \times CR}{BW}$$

Where,

C -Concentration of the pesticide (mg kg^{-1})

CR- Consumption rate (g day^{-1})

BW- Average body weight (Kg) (60 Kg for Adults)

Hazard index was calculated by

$$HI = \frac{EDI}{ADI}$$

Where,

EDI- Estimated daily intake (mg kg^{-1} body weight)

ADI- Acceptable daily intake (mg kg^{-1} BW d^{-1})

If HI is more than one it is not safe while less than 1, it is considered safe

Dilution factor (F): The dilution factor is the ratio by which a solution is diluted, expressed as the volume of the final solution divided by the volume of the transferred solution or weight transferred. It is used to describe how much a solution or solid material has been diluted and is calculated using the formula:

For liquid dilution factor (F) = $\frac{\text{Final volume}}{\text{Transferred volume}}$

For solid dilution factor (F) = $\frac{\text{Final volume}}{\text{Weight transferred}}$

Note: 1.The maximum dilute factor for one attempt is 1000 fold

2. Do not use more than two different diluents, miscible liquids, or solids.

Method validation parameters and acceptance criteria in different matrices (SANTAE/11813/2017 guidelines).

Guidelines	**What/how**	**Criteria**	
Specificity	Response in reagent blank and blank control samples	< 30 % of reporting limit (LOQ)	
Matrix effect	Comparison of response from solvent standards and matrix matched standards	> ± 20 %	
Linearity	Linearity check from five levels	≤ 0.999 (R^2)	
LOQ	Lowest spike level meeting the method performance criteria for trueness and precision	≤ MRL	
Precision (Repeatability-Intraday)	Repeatability RSD, for each spike level tested	≤ 20 %	
Precision (Reproducibility-Interday)	Within laboratory reproducibility, derived from on-going method validation/verification	≤ 20 %	
Trueness (Recovery)	Average recovery for each spike level tested	Concentration	% Recovery
		> 1µ/kg ≤ 0.01 mg/kg	60-120
		> 0.01 mg/kg ≤ 0.1 mg/kg	70-120
		> 0.1 mg/kg ≤ 1 mg/kg	70-110
		> 1 mg/kg	70-110
Retention time	-	(± 0.1mm)	
Robustness/Ruggedness	Average recovery and Precision (Reproducibility-Interday) derived from on-going method validation/verification	Precision (Reproducibility-Interday)-≤ 20 % Trueness-70-120 %	
Ions	Check compliance with identification requirements for Mass Spectrometry techniques	Minimum 2 product ions	

Pesticide Standard Weight units

In the field of residue analysis, it is common practice to employ weights that are typically smaller than milligrams (mg) and following explanations are provided below

1. Milligram (mg) 1 mg = 10^{-3} g or 1 g = 1000 mg
2. Microgram (μg) 1 μg = 10^{-6} g or 1 g = 1000000 μg
3. Nanogram (ng) 1 ng = 10^{-9} g or 1 g = 1000,000,000 ng
4. Picogram (pg) 1 pg = 10^{-12} g or 1 g = 1000,000,000,000 pg
5. Femtogram (fg) 1 fg = 10^{-15}g or 1 g = 1000000000000000 fg
6. Attogram (ag) 1 og = 10^{-18}g or 1 g = 1000000000000000000 ag
7. Zeptogram (zg) 1 zg = 10^{-21}g or 1 g = 1000000000000000000000 zg

PPM = Parts Per Million

Parts per million (PPM) is a widely employed concentration unit utilized to express exceedingly low concentrations of a solution. A quantity of one gram in a volume of 1000 milliliters corresponds to a concentration of 1000 parts per million (ppm). Similarly, a quantity of one thousandth of a gram (0.001g) in a volume of 1000 milliliters corresponds to a concentration of 1 ppm.

A unit of measurement known as milligram (mg) is equivalent to one thousandth of a gram, while one liter (L) is equal to 1000 milliliters (ml). Consequently, the expression "1 ppm" can be understood as "1 milligram per liter" or "mg/L".

PPM (Parts Per Million) to % (Parts Per Hundred)

Divide the ppm amount by 1,000,000 and multiply by 100 to get %. e.g. :

1 ppm = 1/1,000,000 = 0.000001 = 0.0001%

10 ppm = 10/1,000,000 = 0.00001 = 0.001%

100 ppm = 100/1,000,000 = 0.0001 = 0.01%

200 ppm = 200/1,000,000 = 0.0002 = 0.02%

5000 ppm = 5000/1,000,000 = 0.005 = 0.5%

10,000 ppm = 10000/1,000,000 = 0.01 = 1.0%

20,000 ppm = 20000/1,000,000 = 0.02 = 2.0%

(Parts Per Hundred) % to PPM

Divide the % value by 100 and multiply by 1,000,000 to get ppm. e.g. :

1% =0.01 x 1,000,000 = 10,000 ppm

0.5% =0.0.005 x 1,000,000 = 5,000 ppm

0.1% =0.001 x 1,000,000 = 1,000 ppm

0.01% = 0.0001 x 1,000,000 = 100 ppm

Analytical units

1 mL in 99 Ml	1 %
1 mL in 999 Ml	1 ppm
mg/mL or mg/g	**1000 ppm**
µg/mL or µg/g	1 ppm
mg/kg or mg/L	1 ppm
1 ppm = 1000 ppb	
µg/kg or µg/L	1 ppb
ng/mL or ng/g	1 ppb
µg/kg or µg/L	1 ppb
µg/g	1000 ppb
µg/mL	1000 ppb
mg/kg	1000 ppb
mg/L	1000 ppb
1 ppb = 1000 ppt	
ng/kg or ng/L	1 ppt
pg/Ml	1 ppt
ng/Ml	1000 ppt
ng/g	1000 ppt
µg/kg	1000 ppt
µg/L	1000 ppt

All India Network Project on Pesticide Residues & other contaminants

Considering the aforementioned, the Government of India established an AICRP on Pesticide Residues, which was later renamed as the All India Network Project on Pesticide Residues (AINP-PR), and again renamed as All India Network Project on Pesticide Residues & other contaminants. Previously, AINP concentrated on pesticide residue only, but as per the new rename, they may detect other contaminants in food commodities, such as antibiotics, heavy metals, adulterants, and microbial contamination. This initiative was undertaken by ICAR in 1984-85. The primary objective of the project is to establish protocols that ensure the safe utilization of pesticides by advocating for good agricultural practices. The protocols will be developed based on supervised field trials conducted at multiple locations. The ultimate goal is to ensure that the levels of pesticide residues in food commodities are maintained well below the prescribed maximum residue limits (MRLs) that guarantee safety. The data generated in this manner is utilized for the purpose of "**Fixing MRL's**".

Terminologies used in pesticide residue analysis

1. **Acceptable Daily Intake (ADI)-**Acceptable Daily Intake is the maximum quantity of pesticide which is considered to be safe, if taken daily basis, throughout one's life on the basis of current toxicological information. ADI is usually established at **1/100 of the NOEL**, in order to add an additional margin of safety and expressed in terms of Wt/Wt basis **(mg/kg body weight)**.
2. **Acute Reference Dose (ARfD)-** Acute Reference Dose of a chemical refers to an estimated amount representing the maximum quantity of a substance present in food and/or drinking water that can be consumed within a 24-hour period without posing a significant health risk to the consumer. This estimation is based on all available information at the time of evaluation and is typically expressed in relation to body weight **(JMPR, 2002).**
3. **Maximum Residue Limit (MRL)-**The maximum amount of pesticide residue (including its toxic metabolites) present in or on the agricultural produce at the time of harvest when pesticide is used under supervision, adopting Good Agricultural Practices (GAP). MRL is often mis-interpreted as safety limit instead of trading standard. In India, Maximum residue limits (MRLs) for pesticides are fixed by FASSAI but in world CODEX.

   If contaminated food is 100 g, ADI is 0.1 mg/kg/day, and the weight of a person is 50 kg then what is MRL

   Ans: 50 ppm
4. **Partition coefficient (P)-**The partition coefficient is defined as the ratio of the concentrations of a compound in the two phases of a mixture comprising two immiscible solvents. Eg: The ratio of the concentration of a chemical in liquid phase to the concentration in gaseous phase.
5. **Retention time (Rt)-**The time at which the analyte / analytes elutes from the column and detected by the detector is called the retention time of the analyte. Each peak represents a specific compound or a mixture of compounds with same rate of movement through the column.
6. **Chromatogram-**The coloured graphical representation (symmetric peak) of a compound or analyte can be viewed in the computer after its detection by the detector or while the instrument runs, the computer generates a graph from the signal, called a Chromatogram.
7. **Certified Reference Material (CRM)-**The purest form of a chemical substance or pesticide either in powdered or liquid form which is used for the preparation of calibration standards.
8. **Mobile phase-**The organic solvents such as methanol and acetonitrile as well as acidified water solution used to carry the pesticide standard solution along the analytical columns in liquid chromatography.

9. **Carrier gas-**The inert gases such as helium and nitrogen gases used to carry the pesticide residue standard solutions along the columns in gas chromatography.
10. **Stationary phase (Column)-**Hallow tube and capillary tube-like structures coated with silica, functional groups such as amino (-NH_2) and cyano (-CN) columns used for separation of analytes in liquid chromatography and poly-siloxane columns in gas chromatography, respectively.
11. **Pre harvest Interval (PHI) with holding period/Waiting period-**The pre-harvest interval (PHI) refers to the duration between the application of a pesticide and the subsequent harvest, as prescribed by established agricultural practices. This interval ensures that the concentration of pesticide residue in the harvested crop remains below the maximum residue limit (MRL).

Name of the Insecticide	Crop	Maximum Residue Limit (MRL) in mg/kg	Name of the Insecticide	Crop	Maximum Residue Limit (MRL) in mg/kg
Acephate	Rice	1	Beta Cyfluthrin	Okra	0.01
	Safflower seed	2		Brinjal	0.2
	Cottonseed	2		Cottonseed	0.7
Methamidophos- metabolite of Acephate	Safflower seed	0.1		Soyabean	0.03
	Cottonseed	0.1		Soyabean Oil	0.01*
Acetamiprid	Chilli	2	Bifenthrin	Sugarcane	0.03
	Rice	0.01		Rice	0.05
	Okra	0.1		Apple	0.5
	Cabbage	0.7		Tea	30
	Cotton seed oil	0.1		Cottonseed	0.5
Buprofezin	Cottonseed Oil	0.01	Carbosulfan	Chilli	2
	Chilli	2		Rice	0.2
	Mango	0.1	Cartap Hydrochloride	Rice	0.5
	Grapes	1	Chlorfenapyr	Chilli	0.05
	Okra	0.01*		Cabbage	0.05
	Rice	0.05		Beans	0.01*
Chlorantraniliprole	Bengal Gram	0.03*	Chlorpyriphos	gram	0.01*
	Black gram	0.03*		Black gram	0.01*
	Bitter gourd	0.03*		Coconut	0.01*
	Okra	0.3		Tea	2
	Soybean	0.03*		groundnut	0.01*
	Pigeon pea	0.03*		Food grains	Wheat-0.5, Rice-0.5 and Food grains 0.05
	Tomato	0.03*		Milled food grains	0.01
	Chili	0.03*	Cypermethrin (sum of isomers) (Fat soluble residue)	Fruits	Stawberry-0.03, Plum0.5, Pomefruit-1.0 and other Fruits 0.5

Name of the Insecticide	Crop	Maximum Residue Limit (MRL) in mg/kg
	Brinjal	0.03*
	Rice	0.4
	Cabbage	2
	Sugarcane	0.5
	Cotton	0.3
	Groundnut	0.03*
Chlorantraniliprole	Groundnut oil	0.03*
	Maize	0.03*
Chlothianidin (Chlothianidin and its metabolites Thiazolymethylguanidine (TMG), Thiazolymethylurea (TZMU), Methylnitroguanidine (MNG) TMG)	Sugarcane	0.4
	Cottonseed	0.02
	Cottonseed Oil	0.02
	Rice	05
	Tea	0.7
Chromafenozide	Rice	0.03*
Cyantraniliprole	Grapes	0.01
	Pomegranate seed	Pomegranate fruit-0.8 Pomegranate seed-0.01

Name of the Insecticide	Crop	Maximum Residue Limit (MRL) in mg/kg
	Potatoes and onions	Potato-2.0, Onions 0.01
	Cauliflower and Cabbage	1
	Other vegetables	0.2
	Cottons seed	0.3
	Cotton seed oil (crude)	0.05
	Rice	2
	Cottonseed Oil	0.01
	Wheat grains	2
	Milled wheat grains	0.01
	Brinjal	0.2
	Cabbage	2
	Okra	0.5
	Oil seeds except groundnut	0.2
	Pomegranate juice	0.01
	Chilli	0.05
	Cabbage	2
	Tomato	
	Gherkin	
	Okra	
	Brinjal	
	Cotton seed oil	

Name of the Insecticide	Crop	Maximum Residue Limit (MRL) in mg/kg	Name of the Insecticide	Crop	Maximum Residue Limit (MRL) in mg/kg
Deltamethrin (Decamethrin)	Chilli	0.05	Diafenthiuron and its metabolites (CGA 177960, CGA 14408 and CGA227352)	Cardamom	0.5
	Red gram	0.01		Brinjal	1
	Mango	0.01		Chilli green	0.05
	Tea	5		Chilli red	0.05
	Okra	0.05		Cottonseed Oil	1
	Tomato	0.3		Cabbage	1
	Brinjal	0.3		Citrus	0.2
	Groundnut	0.01*	Diflubenzuron	Cottonseed	0.2
	Cottonseed	0.1		Tea	0.01**
	Food grains	Wheat-2.0 and food grains- 0.3	Dimethoate (residue to be determined as dimethoate and expressed as dimethoate)	Mustard	0.01
	Rice	0.05		Fruits and Vegetables	2
				Chilli	0.5
Dinotefuran	Rice	8	Ethion(Residues to be determined as ethion and its oxygen analogue and expressed as ethion)	Gram	0.01
	Cottonseed Oil	0.05*		Pigeon Pea	0.01
Emamectin Benzoate	Cottonseed	0.02		Soyabean Seed	0.01
	Cottonseed oil	0.02		Tea (dry manufactured)	5
	Okra	0.05		Cucumber and Squash	0.5
	Groundnut oil	0.05		Other Vegetables	1
	Tea	0.01*		Cottonseed	0.5
Ethofenprox (Etofenprox)	Rice	0.01		Food grains	0.03
Fenazaquin	Apple	0.2		Milled food grains 0.01	0.01
	Chilli (green)	0.5		Peaches 1	1
	Okra	0.01		Other fruits	2

Name of the Insecticide	Crop	Maximum Residue Limit (MRL) in mg/kg	Name of the Insecticide	Crop	Maximum Residue Limit (MRL) in mg/kg
	Brinjal	0.01	Fenobucarb (BPMC)	Rice	0.01
	Tomato	0.01	Fenpropathrin	Brinjal	0.2
	Tea	3		Okra	0.5
Fenpyroximate	Chilli	1		Chilli	0.2
	Tea green)	2		Tea(green/black)	2
	Coconut water	0.02		Rice	0.03*
Fenvalerate (Fat soluble residue)	Cauliflower	2	Fipronil	Rice	0.01
	Brinjal	2		Chilli	0.01
	Okra	2		Sugarcane	0.01
	Cottonseed	0.2		Cabbage	0.02
	Cottonseed Oil	0.1		Grapes	0.01*
	Red Gram	0.01**		Wheat	0.01*
	Bengal Gram	0.01**	Fipronil and its metabolites (MB-46513, MB45950, MB-46136)	Onion	0.04
	Groundnut	0.01**	Flonicamid	Rice	0.05*
	Cabbage	0.01**		Groundnut Seed	1
	Tomato	0.01**		Mango	0.2
	Cottonseed Oil	0.02*		Sugarcane	0.1
Flubendiamide	Brinjal	0.1		Okra	2
	Bengal Gram	1.0		Sunflower Seed	0.5
	Cottonseed Oil	1.5		Chilli	0.3
	Rice	0.1		Grapes	1
	Cabbage	4		Tomato	1
	Tomato	2		Cucumber	1
	Pigeon pea	1.0		Cottonseed Oil	0.05

Name of the Insecticide	Crop	Maximum Residue Limit (MRL) in mg/kg	Name of the Insecticide	Crop	Maximum Residue Limit (MRL) in mg/kg
	Black Gram	1.0		Rice	0.05
	Chilli	0.02		Brinjal	0.2
Flubendiamide and its metabolite Des-iodo	Tea	50	Indoxacarb	Soyabean	0.01*
				Soyabean Oil	0.01*
	Soyabean	0.07		Tomato	0.5
	Soyabean Oil	0.07		Chilli	0.01
	Soyabean cake	0.07		01 Pigeon pea	0.1
Hexythiazox	Tea 15	15		Chick Pea	0.2
	Chilli (green)	0.01		Rice	0.05
	0.01 Dried Chilli 0.01	0.01		Soyabean	0.5
	Apple	0.3		Cottonseed	1
Imidacloprid	Citrus (Acid Lime)	1		Cottonseed Oil	0.1
				Cabbage	3
Lambdacyhalothrin	Brinjal	0.2	Lufenuron	Cauliflower	0.1
	Tomato	0.1		Pigeon pea	0.1
	Rice	1		Cottonseed	0.01
	Okra	2		Black Gram	0.02*
	Red Gram	0.05		Chilli	0.05
	Bengal Gram	0.05		Cabbage	0.3
	Chilli Green	0.05	Malathion (Malathion to be determined and expressed as combined residues of malathion and malaoxon)	Food grains	Wheat-10.0, Maize-0.05 and other food grains-4
	Chilli Red	0.01		Milled food grains	1
	Groundnut seed	0.01		Fruits	4
	Onion	0.01		Vegetables	3
	Soyabean	0.01		Dried fruits	8

Name of the Insecticide	Crop	Maximum Residue Limit (MRL) in mg/kg	Name of the Insecticide	Crop	Maximum Residue Limit (MRL) in mg/kg
	Mango	0.2	Novaluron	Chili	0.01
	Grapes	0.05		Chickpea	0.01
	Cottonseed Oil	0.05		Cottonseed	0.01
	Tea	0.05*		Cottonseed Oil	0.01
	Maize	0.01*		Tomato	0.01
Metaflumizone	Cabbage	0.05	Phenthoate	Cabbage	0.01
Milbemectin	Chilli green	0.01		Food grains	0.05
	Chilli red	0.01		Milled food grains	0.01
Monocrotophos	Food grains	0.03	Profenofos	Oilseeds	0.03
	Milled food grains	0.01		Gram	0.01**
	Citrus fruits	0.2		Cottonseed oil	0.05
	Other fruits	1		Soybean	0.01*
	Cottonseed	0.1	Propargite	Brinjal	2.0
	Coffee (Raw beans)	0.1		Chilli	2.0
	Chilli	0.2		Apple	2.0
	Cardamom	0.5		Tea	1.0
	Green Gram	0.01**	Pyridalyl	Cottonseed Oil	0.02
	Pigeon Pea	0.01**		Cabbage	0.02
	Coconut	0.01**		Okra	0.02
Thiamethoxam	Rice	0.02		Chilli	0.02
	Okra	0.5			
	Cottonseed Oil	0.01			

Revised maximum residue limits (MRLs) of insecticides

S.No	Name of pesticide	Maximum residue limit(MRL) (mg/kg)
1	Emamectin benzoate	0.06
2	Fenpyroximate	6.0
3	Quinalphos.	0.7

16

Plant Protection Appliances

Chemicals are commonly employed to manage plant diseases, pests, and weeds, but they are only effective if administered promptly once an infestation has been discovered. These chemicals must be sprayed, dusted, or misted over the ground and plants. Since the chemicals are expensive, it is important to have tools for applying them evenly and effectively. Crop Protection Products (CPP) like herbicides, pesticides, and fungicides work best when they are used at the right time during the crop's most productive times. This problem is getting worse because there aren't enough people to do the work. In this situation, the only choice that makes sense is to automate the application process. Most of the time, poisons are put on with dusters or sprayers. Dusting is the easier way to apply chemicals. It works best with movable equipment and usually only needs simple tools, but it is less effective than spraying because the dust doesn't stick around as long. High volume spraying is usually efficient and reliable, but it is more expensive than low volume spraying, which fixes some of the problems with the first two methods while keeping the things that work well about them.

The selection of appropriate equipment for CPP (crop protection products) application is a critical concern that must be addressed in order to achieve efficient pest and weed control. The choice of equipment is contingent upon its intended application and the specific requirements for effective pest management. According to studies, the effectiveness of CPP implementation accounts for 70% of its success.

Sprayers break liquid into small droplets and distribute them evenly over the protected surface. To prevent dangerous or unproductive insecticide overuse, it regulates insecticide application. Proper application requires a sprayer that delivers large droplets that are able to wet surfaces readily. Air currents have a tendency to redirect exceedingly small droplets, measuring fewer than 100 microns in size, resulting in their wastage. It is advisable to implement a systematic approach in the treatment of crops, preferably in the form of frequent swaths. It is possible to obtain uniform application with a boom using constant output of the machine and uniform forward movement of the machine.

Classification of Spraying Techniques

Spraying methods are defined by liquid volume per unit of land as high volume (HV), low volume (LV), or ultra low volume (ULV). With the arrival of new insecticides, the trend is to utilize less carrier or diluent liquid. Different spraying applications require different droplet sizes. Controlling insects, pests, and diseases requires fine droplets, while herbicides require larger droplets. Deposition on target region improves with more fine droplets created by device. Size impacts droplet drift and penetration distance to target. To prevent drift, cover the plant or target area, and increase penetration, a compromise is needed. Ideal droplet sizes are below:

Optimum Droplet Sizes for Different Targets

Target group	Droplet size (Microns)
Flying insects (drift)	10-15
Crawling and sucking insect (drift)	30-50
Plant surfaces (limited drift)	60-150
Soil application (no drift) as in case of herbicide application	250-500

Volume type	Spray fluid (l/acre)	Droplet size (μ)	Area covered/day	Sprayer type
High volume Spraying (HVS)	200-400	150	2.5 acre	Knapsack, Rockery sprayer Hydraulic sprayer
Low volume spraying (LVS)	40-60	70-150	5-6 acre	Power sprayer, Mist blower
Ultra low volume spraying (ULS)	2-5	20-70	20 acre	ULV sprayer, Electrodyne sprayer

Size of spray droplet and their targets in crop protection

Target	Optimum Sizes of spray droplet (μ)
Flying insect	10-50
Insects on surfaces	30-150
Plant diseases	30-150
Weeds	100-300

The spray is also characterized on the basis of droplet size referred as Vmd (Volume median diameter)

Type of spray	Droplet size (μ)
Coarse spray	>400
Medium spray	210-400
Fine spray	101-200
Mist spray	51-100
Aerosol/Fog	1-50
Smoke	0.01-1
Vapour	0.001-0.1

A sprayer atomizes spray fluid into minute droplets and forces them out. Sprayers have a pump, power source, tank, agitator, distribution system, pressure gauge, pressure regulator, valves, filters, pressure chamber, hose, spray lance, cut-off device, boom, and nozzle.

1. Pump

Any liquid that comes out of a spray tip must be broken up into tiny droplets first. The pump helps create the pressure needed for this, and it is also needed to make the energy needed to atomize the spray fluid. It is a sprayer's most important part and expenisive.

Types of Pumps

a. **Air Compression or Pneumatic pumps:** The function of these pumps is to generate air pressure within a sealed tank that holds spray liquids, facilitating the movement of the liquid via a nozzle where it undergoes atomization. Primarily employed in compression sprayers. In this scenario, the force exerted by the pump is applied to the spray fluid, while the pump itself does not physically interact with the spray fluid.

b. **Hydraulic or Positive Displacement Pump (Plunger, rotary and centrifugal pump):** These pumps take in a specific volume of spray liquid and force it under pressure through the delivery system. The pressure produced by the pumps varies. This pump takes a specific volume of liquid from the intake and transfers it without allowing any escape to the outlet.

2. Source of Power

In order to operationalize spray pumps, it is a prerequisite. Depending on the source of power, it can be either:

- Manual
- Traction
- Motor
- Tractor and air craft engines

3. Spray Tank

A sprayer may have a built-in or separate tank for spray liquid. The tank must be large enough to avoid frequent refills but light enough to carry. The tank's big entrance has a strainer and cap for liquid filling. Small apertures make tank filling and cleaning difficult.

4. Agitator

The agitator might be mechanical or hydraulic to ensure liquid spray uniformity. Metal fans, rods, etc. are mechanical agitators. Hydraulic agitators are pipes with multiple side holes and a closed end that are inserted in the tank and fed spray liquid with a pump. Liquid jets erupt from these holes, agitating the liquid completely. Power sprayers with large tanks are more likely to use hydraulic agitation, which isn't very effective but is more convenient, and sprayers without agitators should be avoided when applying CPP emulsion and suspension.

5. Distribution System: Distribution system includes

A. Nozzle

B. Spray lance

C. Spray boom

5. A Nozzle

Spray nozzles are designed to disperse pressurized spray liquid into fine droplets that can then be applied to the intended target. It atomizes the liquid and disperses the droplets like a spray. What it entails is

a) **Body** : a piece of brass with internal threads, which are called female nozzles if inside and male nozzles if outside, but always outside on the opposite end.

b) **Cap**: The strainer, orifice plate, washer, and swirl plate are secured by a body mounted nut. The nozzle tip is its main part. It controls spray flow and distribution. Many spray tips are intended for specific applications.

c) **Swirl plate**: The nozzle has a swirl plate that is carefully drilled to give a clear spray pattern. The swirl plate is made up of brass.

d) **Washer (sealer):** They come in different widths so that the depth of the swirl chamber can be changed and spray fluid doesn't leak out.

e) **Strainer:** A strainer is built into the end of the needle. The holes in the screen are small so that bigger particles can't get in.

When choosing the nozzle, it is important to take into considerations

- Droplet size
- Spray pattern
- Rate of application

Before the spray lance, other important parts in the sprayer are:

Pressure gauge: In most cases, it is connected to the pipe line near the nozzle.

Valves: A spray fluid's direction of flow is controlled by them.

Filter: Most of the time, this is placed between the tank and the pump unit, between the pump and the spray lance, or inside the spray lance. It's function is to protect the pump from wear, keep the valves from stopping working, and keep the nozzles from getting clogged.

Pressure chamber: It is used in sprayers that use hydraulic pumps, which keeps the pressure from changing and makes the pouring more even.

Hose: The spray lance is attached to one end, while the sprayer is attached to the other.

5. B. Spray lance

Sprayer nozzles are linked to variable-design brass spray lances. Its brass or steel rod length ranges from 35 to 90 cm, and it is attached to a sprayer's delivery hose pipe and has a replacement nozzle. At its nozzle, a CPP spray lance forms a goose neck at the hose end and has a trigger to control liquid flow. We can attach a detachable plastic shield to the spray lance to avoid chemical drift. It may bend 120□C from a goose neck to spray under leaves.

Cut-off valve: Its purpose is to turn off the liquid. This can be controlled by a knob or by a spring (trigger cut-off). There are three types: (a) wheel cut-off valve with strainer. (b) Strainer-equipped trigger cut-off valve (c) Trigger cut-off valve with no strainer.

5. C. Spray bar or Boom

It has a horizontal pipe with two or more nozzles 50 cm apart and a boom length of 1-15 m. Short booms with 2-3 nozzles are used with manual sprayers, while tractor sprayers use longer ones.

Spray boom is preferable over spray lance because it covers more in the field. Anyone can get (i) a uniform spray, (ii) a targeted spray, or (iii) a band spray by adjusting the boom's width. The number of nozzles increases with the boom.

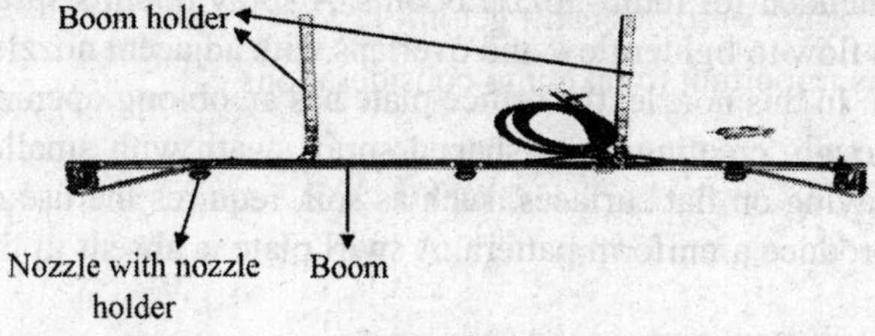

Droplet size

In CPP application, droplet size reduces ambient contamination. A medium or coarse spray is needed to limit drift, regardless of volume.

Droplet size affects drift and coverage. Droplets from CPP nozzles vary greatly in size. Large droplets, which reduce spray drift, may not cover well, while extremely small droplets lack velocity and drift more in windy situations. The distance droplets travel from a nozzle depends on the liquid flow rate (orifice size), liquid pressure, and physical changes to the nozzle design and operation. Additionally, evaporation and drift may occur if the droplet is too small.

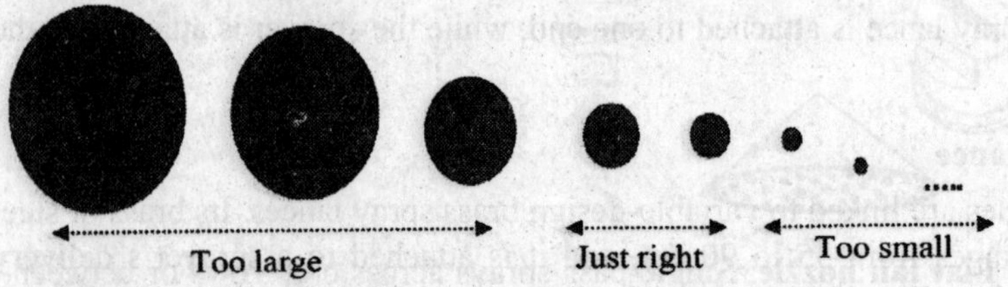

5. D. Nozzle selection

It is crucial to use a spray nozzle that is appropriate for the task at hand. The nozzle controls how much spray is applied, how evenly it is distributed, how much of the surface is covered, and how much of the spray drifts. There are many different types of nozzles available, but only a few are typically utilized for spraying crop protection agents. These include extended range flat-fans, flood jets, and others. Hydraulic energy used in both flood fan and flat fan nozzles. Common spray nozzle types include the following eight:

1. Fan spray
2. Cone spray
3. Flood jet spray
4. Variable cone spray
5. Multiple nozzle spray

5.D.1. Fan spray nozzles

Flat Fan nozzle tips are intended for multi-nozzle booms. A spray boom's spray pattern is tapered from full flow to lighter flow and overlaps with adjacent nozzles to create a uniform pattern. In this nozzle, the orifice plate has an oblong opening that forces spray fluid through, creating a fan-shaped spray swath with smaller droplets in the center. Spraying on flat surfaces, such as soil, requires the use of more than one nozzle to produce a uniform pattern. A swirl plate is absent in the fan spray nozzle.

5.D.1.a. Flat fan nozzles

The boom's flat fan nozzles are placed so that neighboring spray tips overlap by 30%. Boom height and spray tip angle determine spacing. The boom may be spaced further apart on larger angle tips with a wider spray pattern.

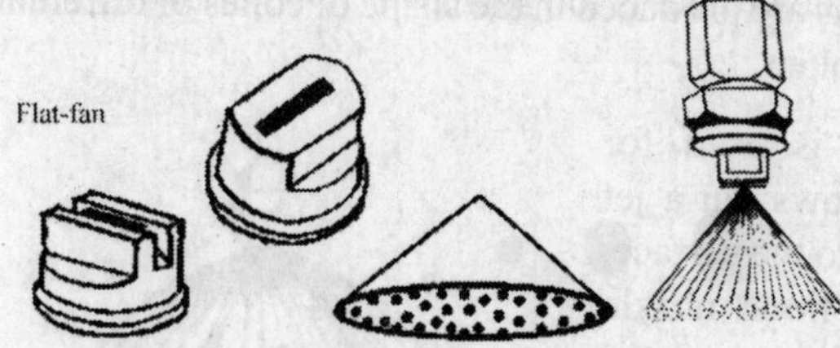

5.D.1.b. Even Flat fan nozzle: Single-pass sprays across crop rows or between rows are possible using Flat fan nozzle tips. Even fan spray tips are not designed for multiple nozzle booms since the spray pattern is full flow.

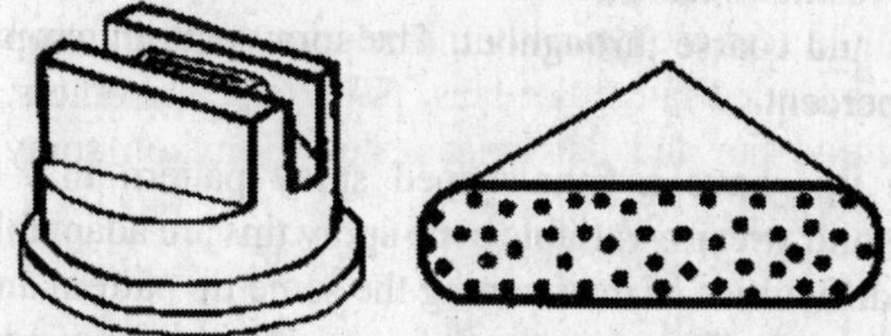

5.D.2. Cone spray nozzles: There are two types of cones: hollow and solid

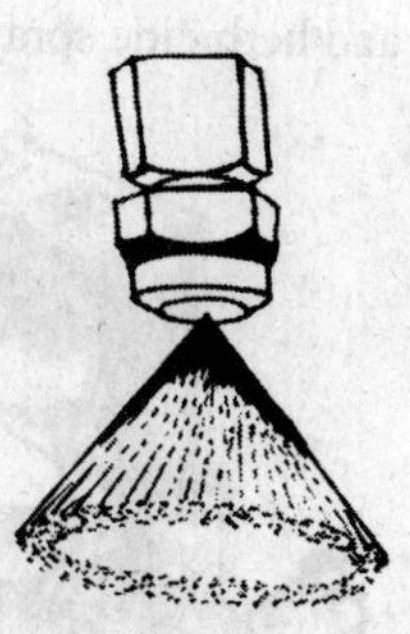

5.D.2.a) Hollow cone: Spray fluid is forced through a slot in the swirl plate to produce a hollow cone shape, which is common for insecticide and fungicide spraying. As the spray approaches the target from different angles, the coverage area increases and the tips produce a fine spray concentrated on the outside edge of the pattern.

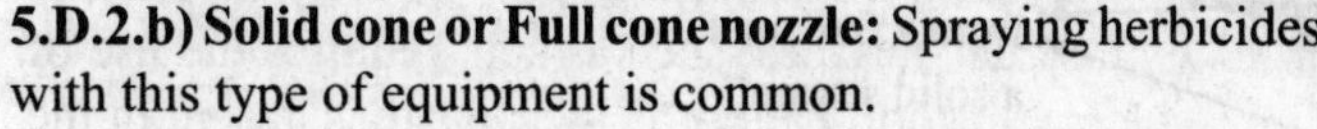

5.D.2.b) Solid cone or Full cone nozzle: Spraying herbicides with this type of equipment is common.

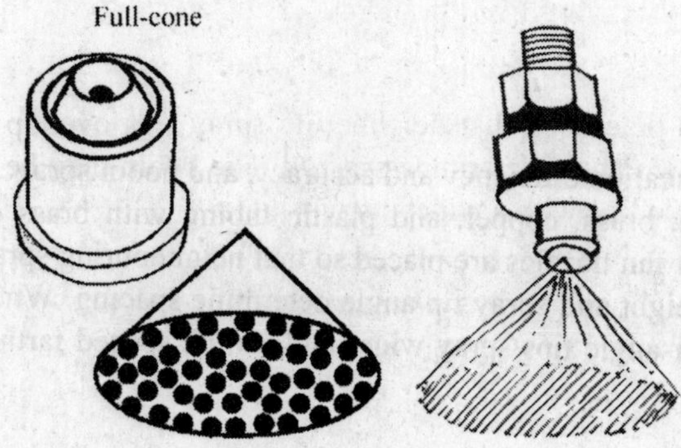

5.D.2.c) Adjustable nozzle: Sprays are produced in the shape of cones of different angles and can also be solid or broken.

5.D.3. Flood jet nozzle: This is used for herbicide spraying because it throws out a jet of coarser droplets at right angles to the surface, minimizing chemical drift. The flood nozzle tip has a wide spray pattern at low pressure, making it popular with knapsack sprayers. Ideal for defoliants and herbicides. The spray pattern tapers from center to edge, although not as consistently as the flat fan. Using this nozzle in a "swinging" manner across a field would usually yield poor application results because the spray is heavy at the edges and coarse throughout. The spray pattern can be improved by overlapping fifty percent.

5.D.4. Variable Cone nozzle tips have a cone-shaped spray pattern that is adjustable from a fine mist to a solid stream. Variable cone spray tips are adaptable due to their design. Due to the difficulties of maintaining the same tip pattern and flow, calibrating these nozzles is tough. These work well for insecticide, fungicide, and herbicide sprays but not most.

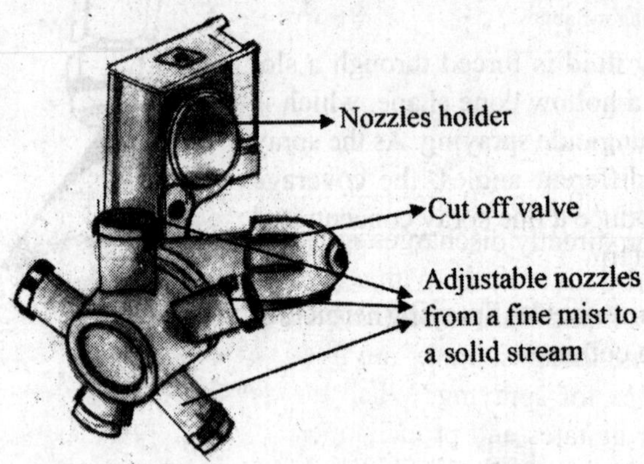

5.D.5. Multiple nozzle booms

Multiple nozzles improve application efficiency and accuracy, and boom sprayers are made of aluminium, steel, brass, copper, and plastic tubing with brass or plastic fittings. The boom's flat fan nozzles are placed so that neighbouring spray tips overlap by 30%. Boom height and spray tip angle determine spacing. When height remains constant, larger angle tips spray wider and can be spaced farther apart on the boom.

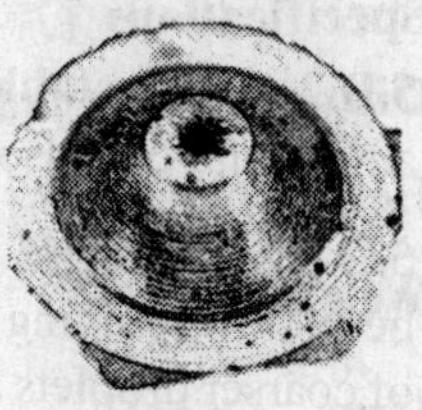

Worn tips

Do not use worn or damaged nozzle tips, which cannot regulate spray pattern. With spray concentrated beneath worn tips, production increases. Distorted spray tips over- and under-apply. Spray tips and patterns should be examined regularly and changed if worn or damaged.

Types of Sprayers

1. Hand syringe

Features

It is a cycle pump with a single operating cylinder. The spray fluid is sucked into the cylinder and delivered during the pressure stroke, after which it is discharged through a nozzle. It's practical if you need to manage a limited space.

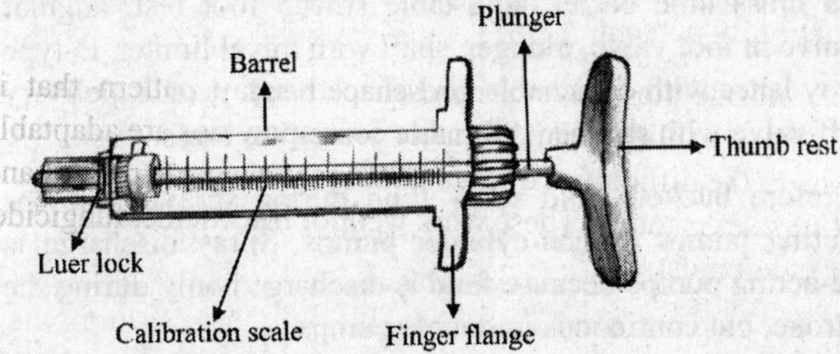

2. Hand sprayer

Features

The plunger pump pressurizes a 0.5 to 3.0 liter tank with a built-in pump. During the pressure stroke, the piston directly discharges spray fluid through the delivery tube. The sprayer has a short delivery tube with a cone nozzle and the air pump is inside the tank. The spray liquid flow is controlled by a spring-actuated lever on the top of the tank in some variants. Tanks are filled to three-fourths capacity and pressurized by air pumps for spraying. When the trigger or shut off valve is activated, compressed air agitates and propels out the spray liquid. With this sprayer, suspension chemicals are usually ineffective. Spraying wettable powders requires continuous shaking to prevent chemical settling. Once charged, the spray nozzle is pointed towards the target. It is made of brass and can be operated by one person.

It is fitted with mist spray nozzle with gooseneck bend. The pump assembly is made of brass and operated by one person. Generally, it is used to spray small kitchen gardens.

Specifications

Diameter of the tank (mm): 130

Height (mm): 210

Weight (kg): 1.2

Uses

Suitable for small nurseries, rose gardens, and kitchen gardens etc.,

3. Bucket pump sprayer/Stirrup pump sprayer

Features

Small farmers and vegetable growers use the stirrup pump sprayer for mosquito control because of its simplicity, low cost, and ease of use. The pump is always submerged in a bucket of spray liquid, hence the name bucket sprayer. Double action pump, seamless brass tube barrel, adjustable stirrup foot rest, angular delivery spout, relief valve in foot valve, plunger, shaft with travel limiter, D-type handle, brass balls, spray lance with detachable gooseneck bend, nozzle, delivery hose, and trigger cut-off valve with strainer. All major pieces are brass.

There is no tank, therefore buckets hold spray fluid during spraying. It has two-cylinder double-acting pumps or one-cylinder pumps. Spray discharge is discontinuous in single-acting pumps because fluid is discharged only during the downward compress stroke, but continuous in double pumps.

The barrel with pump is put in a bucket of spray liquid, and one person operates the pump by placing his foot on the foot stirrup and moving the D-type handle. Another person points the spray lance at the target. A flat fan spray nozzle is typical for sprayers. The sprayer has a 5 m delivery pipe and 0.3 ha/day field capacity.

The sprayers include a brass pump, stirrups, hoses, lances, and nozzles.

Uses

It is used to spray orchards, nurseries, flower crops, and vegetable gardens etc.

3.1. Single barrel stirrup pump

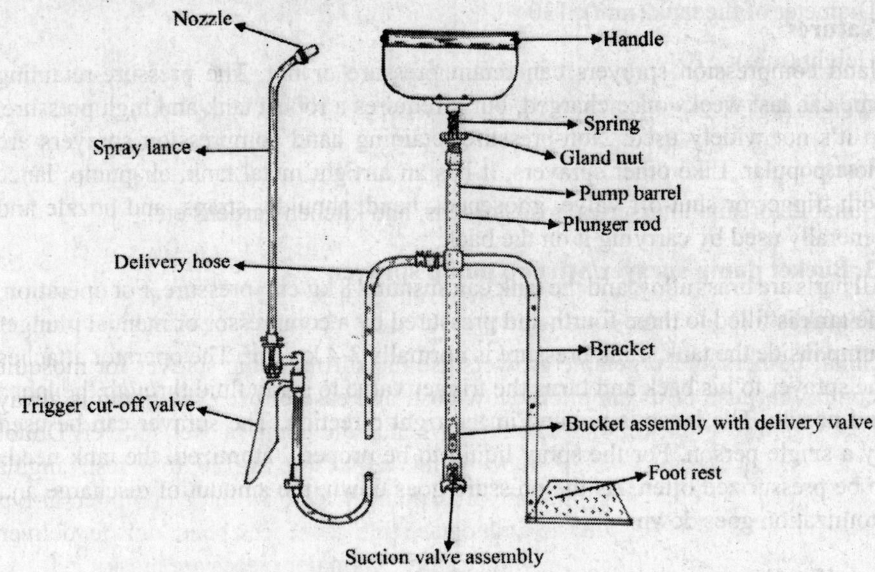

3.2. Double barrel stirrup pump

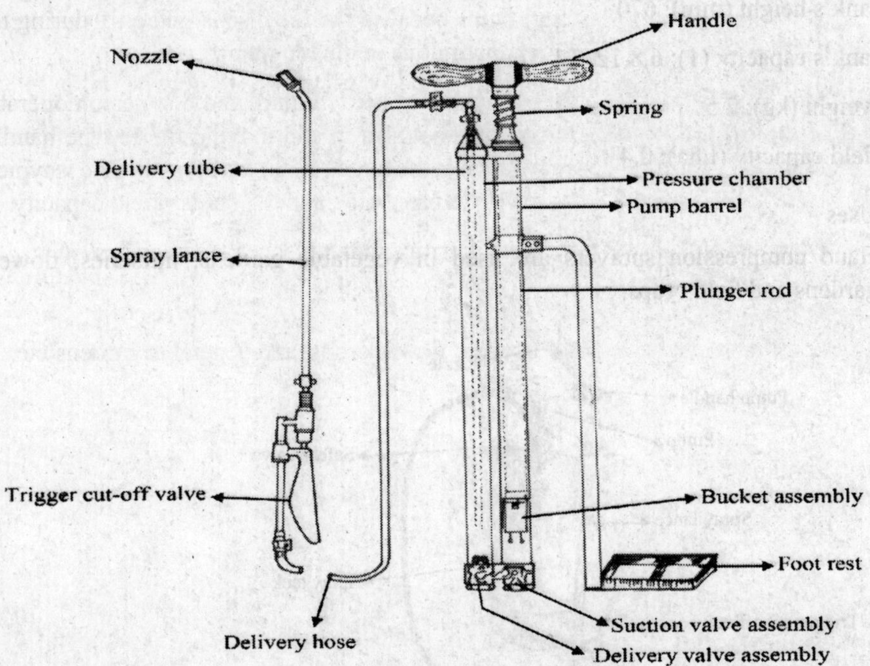

4. Hand compression sprayer

Features

Hand compression sprayers can retain pressure or not. The pressure-retaining type can last weeks once charged, but it requires a robust tank and high pressure, so it's not widely used. Non-pressure-retaining hand compression sprayers are most popular. Like other sprayers, it has an airtight metal tank, air pump, lance with trigger or shut-off valve, gooseneck bend, shoulder straps, and nozzle and generally used by carrying it on the back.

All parts are brass alloy, and the tank can sustain 18 kg/cm^2 pressure. For operation, the tank is filled to three fourth and pressured by a compressor or manual plunger pump inside the tank. Tank pressure is normally 3-4 kg/cm^2. The operator attaches the sprayer to his back and turns the trigger valve to spray fluid through the lance and nozzle. The lance is pointed in the right direction. The sprayer can be used by a single person. For the spray liquid to be properly atomized, the tank needs to be pressurized often. As the pressure goes down, the amount of discharge and atomization goes down.

Specifications

tank's diameter (mm): 200

tank's height (mm): 670

tank's capacity (1): 6,8,12, 14, 16

weight (kg): 7.5

field capacity (l/ha): 0.4

Uses

Hand compression sprayers are used in vegetable gardens, nurseries, flower gardens and field crops.

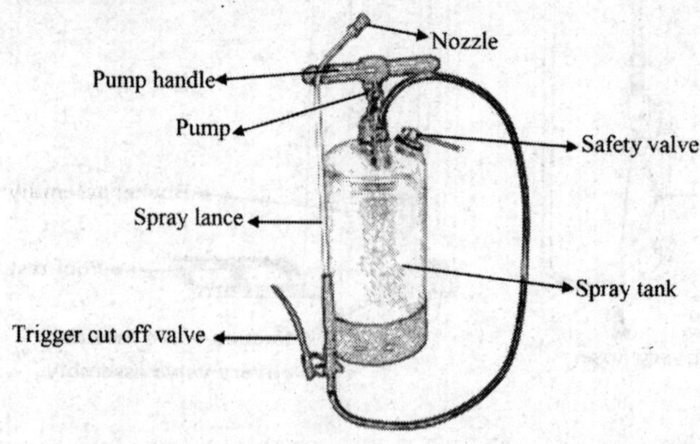

5. Rocker sprayer

Features

The rocker sprayer is a high-pressure sprayer with a long lever that can be used with either one or two lances. The whole sprayer is set up on a wooden board, which is held in place by the operator's foot. The sprayer is made up of a single or double-acting piston pump that creates high pressure, an air chamber, a spray lance with a shut-off valve and strainer, a 5 m suction line with a strainer, and a delivery line.

Most of the parts are made of brass metal. The lance has a gooseneck bend and a tip, and it can be anywhere from 60 to 90 cm long. The pump is turned on and off by moving a long lever back and forth. This makes the inlet pipe, which is buried in the spray liquid, draw liquid into the pump. The other person holds the lance and guides the chemical spray to the target. If two lances are used, it may take at least three people to do the shooting. With a high jet spray gun or a bamboo dart, the chemical can be shot up to 10 meters in the air.

Specifications

Length (mm): 760

Width (mm): 150

Height (mm) : 1230

Weight (kg): 6.0

Pressure developed (kg/cm^2): 10

Field capacity (ha/day): 1.0 ha (with single lance) and 1.5-2 ha (with 2 lances)

Uses

To spray on tall trees such as coconuts, areca nuts, sugarcane, rubber plantations, orchards, vineyards, vegetable gardens, and flowers etc.

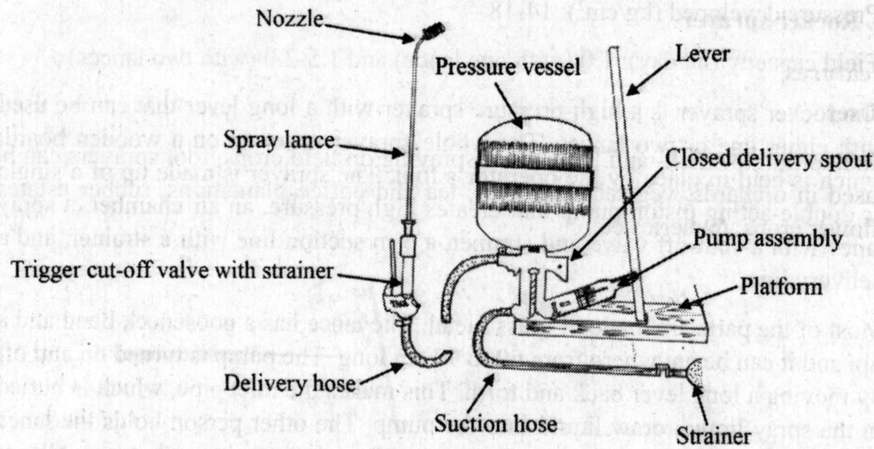

6. Foot sprayer

Features

Foot sprayers are great for multifunctional spraying. It works like a rocker sprayer. Foot-operated pump, suction hose with strainer, delivery hose, spray lance with shut-off pistol valve, gooseneck bend, and adjustable nozzles make up the sprayer. Foot pump duster is also known as plunger type.

The steel frame the pump barrel is attached to makes it quite stable when set down on the ground. After each pumping stroke, the foot lever is retracted to its original position by two powerful springs. Since the sprayer doesn't come with its own tank, you'll need a separate container to hold the spray liquid and keep the strainer of the suction hose submerged while operating. It can accommodate up to two discharge lines where we can expands its utility and performance in the field.

The plunger pump, a positive displacement pump, generates high pressure to spray liquid farther with a boom. The pump barrel, lance, and spray nozzle are brass alloy. The inlet pipe is installed in the storage container, and one person continually runs the pump by foot lever and holds the sprayer at the top by U-type fixture while the other guides the lance to the target area. Spray tall trees up to 10 m with a high jet or bamboo lance.

Specifications

Length (mm): 440

Width (mm): 170

Height (mm): 980

Weight (kg): 10.0

Pressure developed (kg/cm^2): 14-18

Field capacity (ha/day): 1.0 (with one lance) and 1.5-2.0 (with two lances)

Uses

In addition to small- and large-scale spraying on field crops, foot sprayers can be used in orchards, vegetable gardens, tea and coffee plantations, rubber estates, flower crops, nurseries etc.

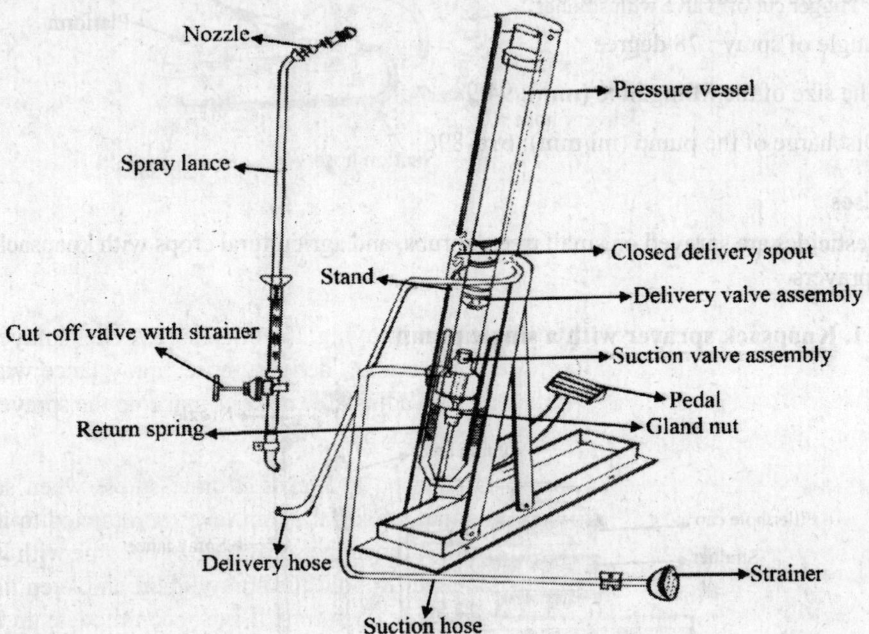

7. Knapsack sprayer

Features

Pumps and air chambers are permanently built into the 9.0-22.5-liter tanks of knapsack sprayers. So the user can pump and spray with one hand, the handle of the pump goes over the shoulder or under the arm. Pressure stays the same because the pump works constantly.

Specifications

No. of person required to operate : One person

Capacity of the tank (L): 9.0-22.5

Inner diameter of pump cylinders (mm): 39-42

Pump cylinder piston number : One

Capacity of pressure chamber (ml): 572-660

Volume displacement (ml): 87.24

Delivery spout type : Threaded

Diameter of the cut-off valve passage (mm): 5

Length of the Lance (mm): 725

The type of nozzle : Hollow cone

Angle of spray : 78 degree

The size of the filling hole (mm): 94.9

Discharge of the pump (ml/min): 610-896

Uses

Pesticides are sprayed on small trees, shrubs, and agricultural crops with knapsack sprayers.

7.1. Knapsack sprayer with a single pump

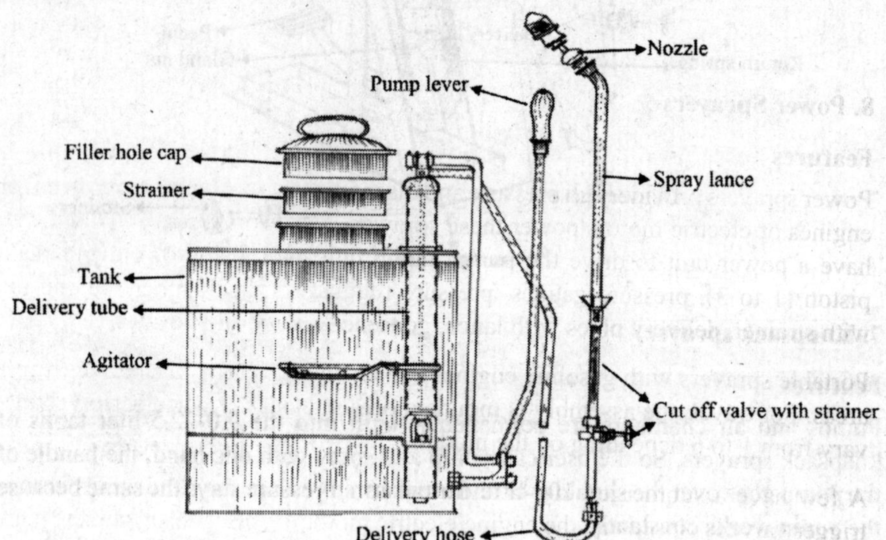

7.2. Compression knapsack sprayer

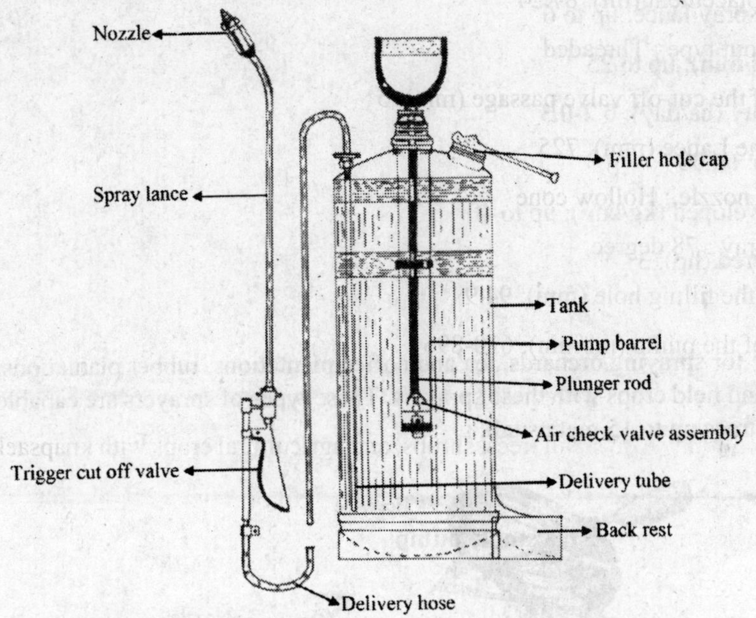

8. Power Sprayers

Features

Power sprayers provide high pressure and discharge to cover wider areas. Auxiliary engines or electric motors power these sprayers. Most sprayers are hydraulic and have a power unit to drive the pump, pump unit with piston or plunger pump, piston (1 to 3), pressure gauges, pressure regulators, air chamber, suction pipe with strainer, delivery pipes with lance, gooseneck bend, and nozzles.

Portable sprayers with gasoline engines can be conveniently transported to spray areas. For travel, the assembly is mounted on a stretcher or wheelbarrow. Lances vary from 1 to 6 depending on the model.

A few have a submerged 100-litre fibreglass storage tank. The lance's shut off trigger valve is closed and the engine/electric motor is started to actuate the pump, which draws spray liquid from the tank, imparts pressure energy, and sends it to the delivery line/lines. The operator then directs the lance toward the target and opens the trigger/shut off valve. An alternative way to deliver spray liquid to long distances/high heights is to use a bamboo lance.

Specifications

Length (mm): 740

Width (mm): 610

Height (mm): 990

Number of spray lance: up to 6

Discharge (l/min): up to 25

Field capacity (ha/day): 0.2-0.3

Weight (kg): 66.00

Pressure developed (kg/cm^2): up to 40

Power required (hp): 3

Uses

It is suitable for spraying orchards, tea and coffee plantations, rubber plantations, vineyards, and field crops with these sprayers. These types of sprayers are capable of spraying trees up to 15 meters tall.

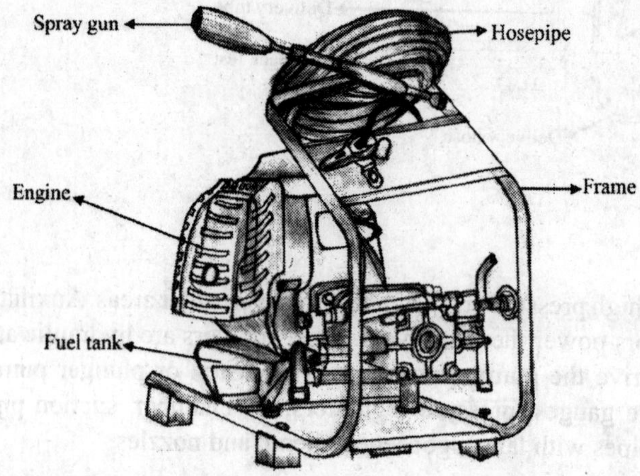

9. Motorised Knapsack Mistblower

Features

The motorized backpack mist blower has a 35-cc 2-stroke petrol/kerosene engine with a centrifugal fan. A vertical centrifugal fan produces a high-velocity air stream that is channelled through a 90-degree elbow to a flexible (plastic) discharge hose with a divergent output. A common frame holds the spray tank, fuel compartment, and engine-fan unit, which fits behind the operator. Tanks are made of polyethylene material. The spray liquid stays in the air stream due to gravity and suction at the nozzle tip.

A roller pump is included in some models of the sprayers to pump the spray liquid into the discharge hose. A tall tree spraying attachment is also available on some models. The spray liquid is placed in the tank, and the engine is started by cranking the rope. As the engine rotates, the fan produces a high-velocity air stream. Flow rate is adjusted by gradually opening the spray liquid control valve. In order to discharge the hose, the operator directs it to the target. After falling into the air stream, spray liquid gets sheared and creates a mist when it comes into contact with the atmosphere.

Specifications

Length (mm): 460

Width (mm): 220

Height (mm): 670

Weight (kg): 11.00

Air velocity at the outlet (m/s): 65-75

Horizontal range (m): 10-15

Vertical range (m): 6-8

Power (hp): 1.2

Fan output (m^3/min): 7

Field capacity (ha/day): 3

Uses

The sprayer is used for spraying orchards, coffee estates, and tall crops.

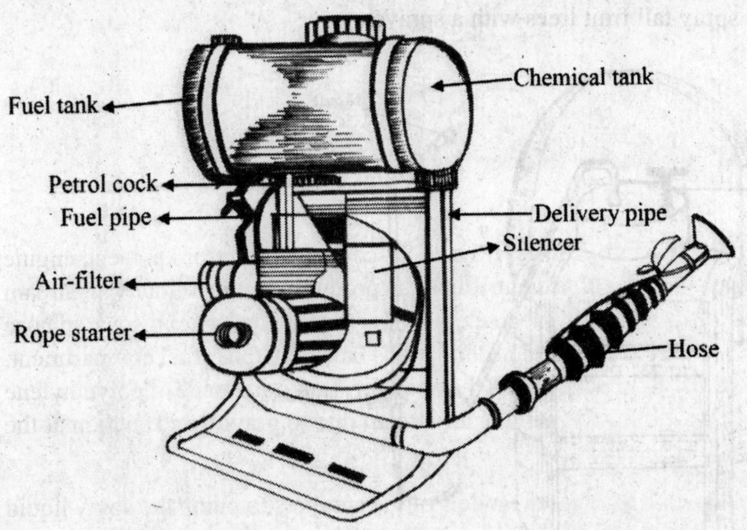

10. Coconut and palm tree Sprayer

Features

The tree sprayer is suitable for spraying tall orchard fruit plants. It has a 4-stroke petrol/kerosene engine to drive the fan, a centrifugal fan that produces high volume and velocity, a micronizer nozzle that produces uniform and fine spray liquid droplets of 150-200 microns, a plastic tank for spray liquid storage, a rotary pump to draw spray liquid from the tank and feed it to the nozzle, and a fibre glass casing.

Two people can easily transport the sprayer to the spraying location thanks to the sprayer's stretcher-type frame, which holds all the necessary components and keeps them together. To use, you'll need to fill the tank with spray liquid, crank the engine to turn on the fan and the rotary pump, and then set the flow rate by opening the control valve. The sprayer is positioned under the tree and then moved manually around it.

Specifications

Power (hp): 3

Amount of fan output (m^3/h): 1000

Capacity of the tank (L): 10, 90

Range of vertical dimensions (m): 20-25

Field capacity (ha/day): 3-4

Uses

It is used to spray tall fruit trees with a sprayer.

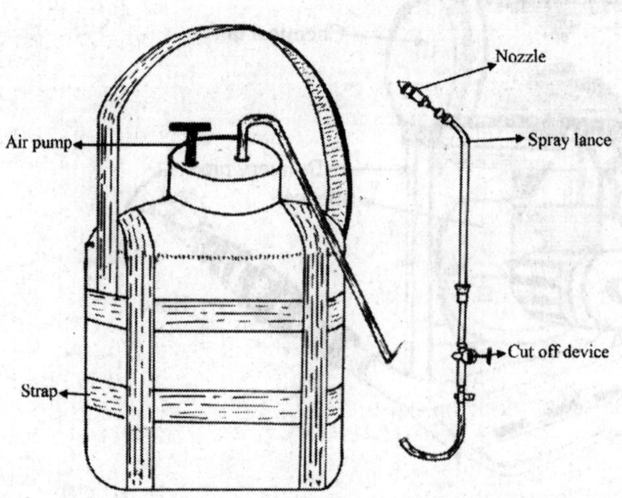

11. Knapsack Power Sprayer

Features

The pump is horizontally gear-driven. Powerful and steady under pressure, the oil-soaking crank box lubricates well. Heat-treated and wear-resistant pistons in a double-cylinder pump increase efficiency. Special materials make 'V' packing durable. Electronic ignition makes the engine easy to start and take care of. Its engine has a lot of power for how much it weighs. It runs on gasoline and makes one power stroke for every 1800 turns of the crank. The oil valve controls the pressure, and the spraying pressure can be changed easily up to 30 kg/cm^2. Its frame is strong, and the materials are good and easy to maintain.

Specifications

Length (mm): 650

Width (mm): 450

Height (mm): 570

Weight (kg): 9.8

Number of gangs: 2.0

Plunger: 1 No. (Double acting)

Cylinder: 2No.s

Diameter of plunger (mm): 16

Stroke (mm): 8

Spraying volume (liters): 4.8-5.2

Pressure (kg/cm^2): 20-25

Uses

The sprayer can be used to spray pesticides and fungicides on rice, fruits, and vegetables etc.

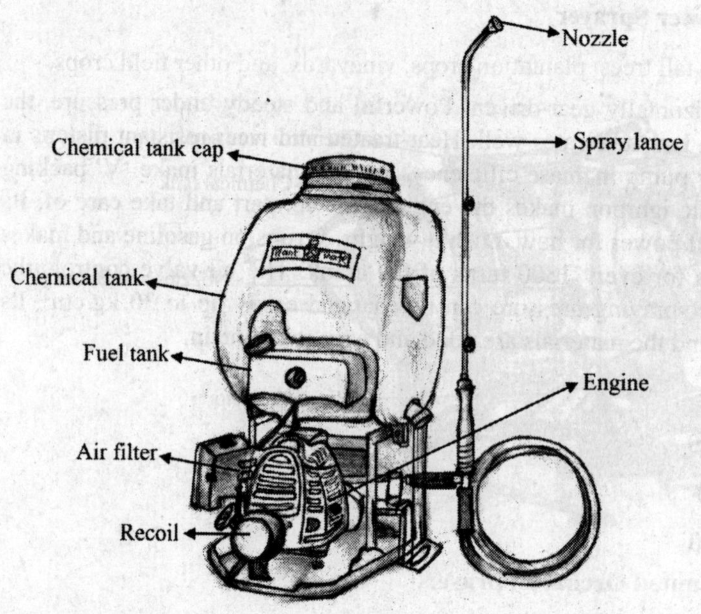

12. Blower Sprayer

Features

These tractor-mounted or trailed air carrier sprayers spray orchards. The sprayer can be high, low, or very low volume depending on the need. High-pressure piston pump atomises spray solution, axial or centrifugal fan produces air, interchangeable and orientable spray heads, high-capacity spray tank, and other control accessories make up the sprayer. All sub-assemblies are installed on the frame and can be coupled to the tractor's 3-point linkage or trailer. Radially placed spray heads cover the trees' 2-rows. Spray tall trees using upward-facing spray heads. Spray heads can be customized for each tree. Spray volume and air stream velocity and volume can be changed as needed. Operation involves filling the tank with spray liquid, driving the fan and pump with tractor power, and sending the liquid to the spray head, which atomises it and collides droplets with the air stream to convey them to the target.

Specifications:

Length (mm): 1360

Width (mm): 970

Height (mm): 2050

Length of propeller shaft (mm): 600

Uses

It is used for spraying tall trees, plantation crops, vineyards, and other field crops.

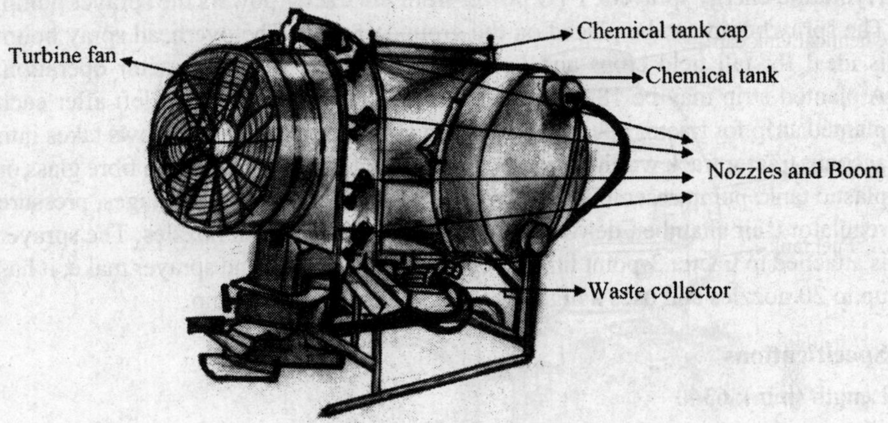

13. Power Tiller Mounted Orchard Sprayer

Features

It has a HTP (horizontal triplex piston) pump, trailed chassis with transport wheels, chemical tank with hydraulic agitation, cut off device, and boom with turbo nozzles. Its turbo nozzles operate at 9-18 kg/cm^2. It produces 100-150 micron droplets. Boom orientation depends on plant size and row spacing. Behind the operator are spray booms.

Specifications

Power source (hp): 12-14, power tiller

Discharge (L/min): 36

Speed (rpm): 950

Pressure (kg/cm^2): 30-35

Drive: Clutch pulley

Tank capacity (L): 200

Number of nozzles: 6

Nozzle spacing (mm): 325

Uses

It is used for spraying foliage in orchard crops like pomegranates, oranges, limes and grapes.

14. Tractor Mounted Sprayer

Features

Hydraulic energy sprayers. PTO power from the tractor powers the sprayer pump. The spray boom can be placed on the ground or aloft. The overhead spray boom is ideal for tall field crops and leaves a 2.5-m-wide strip for tractor operation. A planted strip may be 18-20 m wide, and a fallow strip must be left after each planted strip for tractor operation. Planting ground spray booms in rows takes into account tractor track width. It works for tiny crops. The sprayer has a fibre glass or plastic tank, pump assembly, suction pipe with strainer, pressure gauges, pressure regulators, air chamber, delivery pipe, and spray boom with nozzles. The sprayer is attached to tractor 3-point links. Depending on the crop and sprayer make, it has up to 20 nozzles and uses a high-pressure, high-discharge pump.

Specifications

Length (mm): 6340

Width (mm): 1290

Height (mm): 1570

Tank capacity (L): 400

Weight (kg): 150

Field capacity (ha/day): 8 (with 14 nozzles)

Uses

It is used for spraying vegetable gardens, flower gardens, vineyards, and tall field crops *viz.* sugarcane, maize, cotton, sorghum, millets.

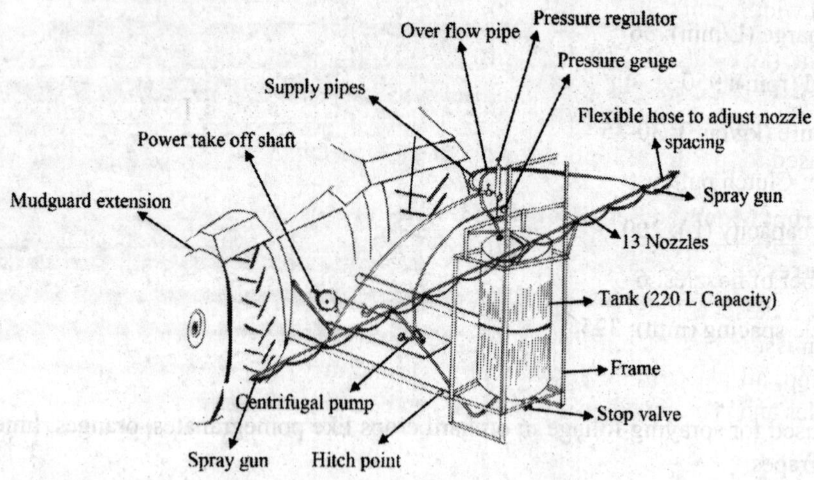

15. Self Propelled Light Weight Boom Sprayer

Features

The operator manipulates the 5 hp diesel engine by means of the handle. A lightweight power tiller and a spraying unit make up the machine. The engine drives the spray pump and two skinny pneumatic wheels via a series of gears, chains, and sprockets. There is also a caster wheel at the rear, which acts as a support wheel. Spray booms are mounted to power units via canopy frames.

Specifications

Power required (hp): 4.8, diesel engine

Engine speed (rpm): 2600

Type of pump: Roller type

Ground clearance (mm): 500

Boom height (mm): 600-1300

Number of nozzles: 12

Type of nozzles: Flat fan/ any other

Nozzle spacing (mm): 500

Tank capacity (L): 100

Pump speed (rpm): 1600

Swath (mm): 6300

Wheel diameter (mm): 440

Tread width (mm): 920

Weight, (kg): 208

Uses

It is used for pesticide application on wheat, vegetables and other crops.

16. Front Mounted Self-Propelled Boom Sprayer

Features

It has a diesel engine (5.5 hp), a spray boom with hollow cone nozzles, a power transmission system, a gear box, lugged cage wheels, a chemical spray tank, and a pump, all placed in a chassis with a ground drive system.. This pump has 14 nozzles and is mounted on a stand that can be attached to the frame.

Specifications

Power source (hp): 5.5, diesel engine

Boom length (mm): 6210

Number of nozzles: 14

Type of nozzle: Hollow cone

Nozzle pressure (kg/cm^2): 3

Distance between nozzles (mm): 300-600 (adjustable)

Chemical tank capacity (L): 50

Weight (kg): 140

Uses

It is used for uniform spraying of horticultural crops as well as cotton, maize, and groundnuts.

17. Self Propelled High Clearance Sprayer

Features

It has a chassis with 1200-mm ground clearance, four wheels, a 20 HP diesel engine, transmission, water tank, operator seat, spray pump, and 18-nozzle boom. Boom width is 10.80 m and nozzles are 675 mm apart. The boom height can be changed from 31.5 cm to 168.5 cm for different crops and folded for transport. There are two narrow wheels at the front (20 cm) that are driven, while there are two steering wheels at the back. For mechanical protection, crop branches are deflected by fenders in front of the drive wheels. Two rows of cotton crop are under the machine chassis on the 135 cm wheel track. First and second gears are for field operation, while third and fourth gears are for road transport. The machine has four forward speeds and one backward speed. The field and road speed may reach 5 and 25 km/h, respectively.

Specifications

Length (mm): 4250

Width (mm): 1950

Height (mm): 2650

Crops for which machine is suitable: All row crops except paddy

Number of gears: four forward and one reverse

Ground clearance of machine (mm): 1200

Length of boom (m): 10.80

Number of nozzles: 18

Nozzles spacing (cm): 67.5 (fixed)

Width of coverage (m): 13.50

Tank capacity (litres): 1000

Maximum field speed (km/h): 5

Maximum road speed (km/h): 25

Weight (kg): 1900

Power required (hp): 20, diesel engine

Uses

It is also suitable for spraying sunflowers, wheat, and other crops with the machine.

18. Aeroblast Sprayer

Features

The equipment has a 400-litre tank, pump, fan, control valve, filling unit, spout adjustable handle, and spraying nozzles to release pesticide solution into the centrifugal blower's air blast. Very small chemical particles are dispersed by the air blast across the area to one side of the tractor. The primary blast from the primary spout treats the bulk of the swath, while the secondary blast from the auxiliary nozzles treats the swath region closest to the tractor. Sprayer is mounted on tractor 3-point linkage and operated by tractor PTO. It is possible to adjust the direction and width of the air outlet.

Specifications

Power required (hp): 35 or above

Type of power (rpm): Tractor power take-off (pto) at 540

Crops suitability: Cotton, sugarcanes, sunflower and horticulture crops

Type of blower: Centrifugal

Type of pump: Piston

Flow rate (L/min): 120

Spray rate (L/ha): 100-400

Spray swath (m): 15 (without wind)

Fan speed (rpm): 3660

Working speed (km/h): 2-6

Tank capacity (L): 400

Machine weight (kg): 230

Uses

It is useful for spraying horticultural trees and crops such as cotton and sunflowers etc.

19. Battery powered low volume knapsack spinning disc sprayer

Features

A plastic tank, grooved spinning disc, fractional horsepower DC motor to spin the disc at fast speed, 12 volt power supply (dry battery or lead acid battery), and light handle with spray head are the sprayer's components. It is possible to adjust the angle of the spray head by adjusting the handle. For operation, the tank is filled with minimum-diluted spray liquid, which flows as drops at the spinning disc. The motor spins the disc at high speed, and the spray liquid travels in its grooves and fragments into very thin droplets when it leaves the disc's periphery due to centrifugal force.

Specifications

Dimensions (l x w x h) (mm): 1810 X 660 X 370

Weight (kg): 7

Tank capacity (L): 10

Number of spinning discs: One

Power source: 6 volts rechargeable battery

Uses

Paddy, cotton, groundnuts, pulses, and vegetables can all be sprayed with it. It saves 30% effort, operational time, and 15% expense over manual spraying. It increases efficiency by 27% over hand spraying.

20. Ultra Low Volume (ULV) Sprayer

Features

Droplets can be applied to the correct size with controlled droplet application sprayers (CDA). They can be used with very low dilution or no dilution at all because of their low volume application features.

ULV sprayers are usually hand-held and can be installed to the tractor boom to improve field capacity. A 1-2-litre plastic container, grooved spinning disc, fractional horsepower DC motor to spin the disc at high speed, 12 volt power supply source (air dry battery or lead acid battery), and light handle with spray head attachment make up the sprayer unit.

The spray head angle is adjustable on the handle. The spinning disc contains fine radial grooves in high-quality plastic. To operate, the container is filled with minimum-dilution spray liquid, which flows as drops in the centre of a spinning disc attached to a high-speed motor. The liquid spray moves along the disc's grooves and breaks up into tiny droplets as it exits the disc's edge, where centrifugal force scatters them. It is possible to spray for up to 15 hours on a single battery charge. There is also fan-assisted ULV sprayers that work well in greenhouses and greenhouse glasshouses. In India, there is currently no ULV sprayer pesticide formulation.

Specifications

Length (mm): 1950

Width (mm): 250

Height (mm): 280

Weight (kg): 8.5

Droplet size (micron): 35-100

Application rate (L/h): 1-5

Field capacity (ha/day): 2-3.

Uses

It is suitable for spraying nurseries, vegetable crops, vineyards, and other fields.

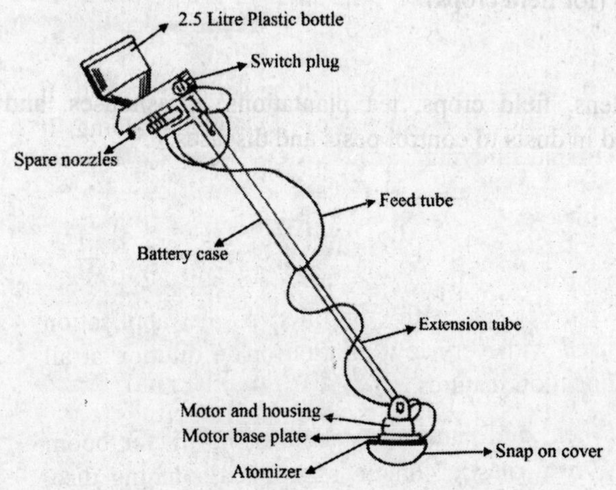

21. Hand Rotary Duster

Features

The hand rotary duster comes in shoulder and belly variants. Farmers use this duster often. The duster has a hopper, fan/blower, rigid/flexible discharge pipe, reduction gearbox, rotating handle, shoulder straps, and metering system. Rotary type duster is also known as crank duster, fan type duster, plunger type duster.

Either plastic or aluminium is used for the hopper. To ensure a longer life, the hopper is made from mild steel sheet coated with anticorrosive material. The gearbox-connected mechanical agitator in the hopper mixes the chemical and prevents outlet clogs. Below the hopper outflow, an adjustable orifice plate controls application rate. Through the gearbox, the fan/blower is rotated with the handle. The hopper is loaded 1/2 to 3/4 full for operation. Adjustable straps secure this to the shoulder/belly. Rotating the handle directs the spoon-shaped deflector-equipped discharge pipe towards the target. Chemical dust/powder falls from the hopper into the discharge pipe with a blower-generated air stream. The cloud of dust particles from the discharge pipe lands on the plant's leaves, stems, and other sections. Bellow/belly mounted duster works on the blacksmith's principle.

Specifications

Length (mm): 280

Width (mm): 330

Height (mm): 330

Weight (kg): 3.5

Hopper capacity (L): 5

Field capacity (ha/ day): 0.6 (for field crops).

Uses

In nursery, vegetable gardens, field crops, tea plantations, glasshouses, and godowns, chemicals are used in dusts to control pests and diseases.

21.A. Shoulder mounted rotary duster

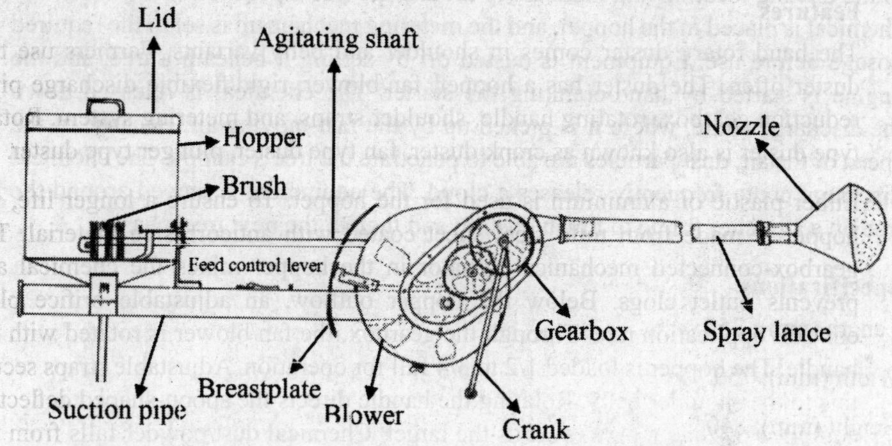

21.B. Belly mounted/bellow rotary duster

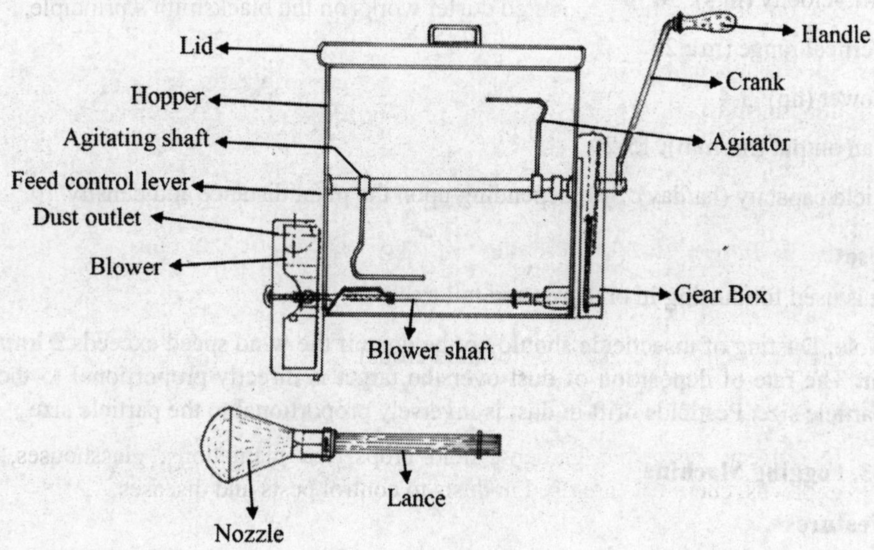

22. Tree Duster

Features

The orchards and tall trees can be dusted using a tree duster. The tree duster comprises of a gas or gas/kerosene-powered engine, a blower/fan directly attached to the engine to provide a strong stream of air, a hopper to hold the pesticide, a discharge chute and several control attachments. All the parts are put together and attached to the stretcher. The equipment is light enough for two people to carry to the dusting location and manoeuvre around the tree's perimeter with ease. The chemical is placed in the hopper, and the metering mechanism is set to the required dosage before use. Equipment is dusted off by setting it beneath a tree, and the engine is started by hand-cranking the starter. The chemical is released down the discharge chute, where it is picked up by the fast-moving air. Because of the speed of the air, dust particles are able to penetrate the tree's canopy. The chemical discharge chute frequently releases a cloud. The equipment is moved around the tree on a stretcher frame to finish dusting and then to the next tree's basin.

Specifications

Length (mm): 750

Width (mm): 750

Height (mm): 840

Weight (kg): 45

Air velocity (m/s): 70

Vertical range (m): 25

Power (hp): 3-4

Fan output (m^3/min): 15-20

Field capacity (ha/day): 3-4, depending upon the plant distance and density.

Uses

It is used for dusting in orchards and tall trees.

Note: Dusting of insecticide should not be done if the wind speed exceeds **5 km/hr.** The rate of deposition of dust over the target is directly proportional to the particle size. Pesticide drift in dust is inversely proportional to the particle size.

23. Fogging Machine

Features

Fog is defined as an atmospheric condition in which droplets smaller than 15 μm occupy a volume of air to the point where visibility is impaired. Fog is typically manufactured using **thermal fogging** machines, which work by injecting hot gases produced by burning fuel with a pesticide dissolved in oil of appropriate flash point. When the vaporised liquid meets cool, dry air, it condenses back into liquid. A fuel tank/reservoir, formulation tank, pump, fogging nozzle, fogging coil, water pump, and other controls make up the fogging machine. These spraying machines are ideal for usage in greenhouses and other enclosed environments. In order to keep the fog low to the ground, the fogging process is performed first thing in the morning.

Specifications

Length (mm): 350-980

Width (mm): 150-295

Height (mm): 150-300

Formulation tank capacity (L): 2.0 - 2.2

Type of fuel: liquefied petroleum gas/ petroleum base

Fuel consumption (L/ha): 0.5-1.5

Weight (kg): 4.0-6.1

Uses

Fogging machines are used to vaporize pesticides in the form of fog to kill flying insects.

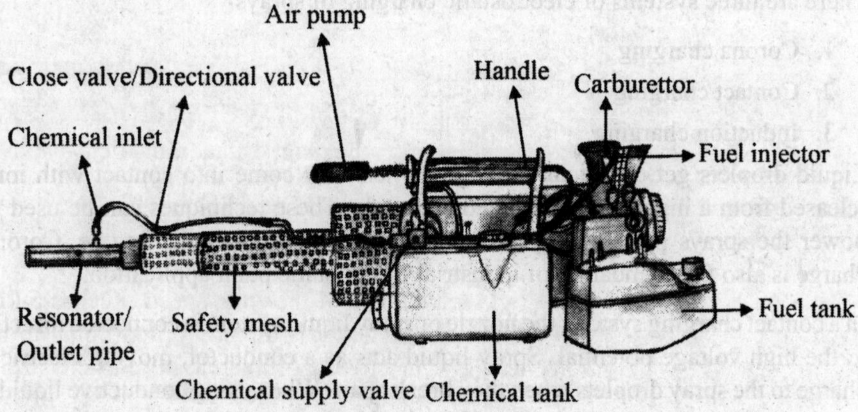

24. Electrodyne sprayer (EDS) or Electrostatic sprayer

An Electrodyne sprayer is a completely new way to apply pesticides with controlled droplets. Electricity is used to form droplets and propel them towards target plants. It consists of a spray stick and a unique combination of a bottle and nozzle (called a BOZZLE). There are two batteries and a generator in the spray stick. A readyily formulated pesticide is contained in the bottle for immediate application to crops. The advantages of this type of system are as follows:

1. Droplets attracted to target crop by charge
2. Leaves are covered even on their ventral surfaces

3. Non-target drift is minimal
4. It is very easy to use
5. For spraying, no water is required
6. Temperature and wind have no effect on spray
7. An energy and labor minimum is necessary

Among the limitations are the high cost and the lack of suitable pesticide formulations.

Power requirement: 13 to 24 KV or 6 V DC (4 torch cells)

Application volume: 0.5 to 1.0 L/ha

Spray particle size: 50 µm

Principle

Spraying pesticides electrostatically reduces the volume of application considerably and improves pesticide deposit quality. Liquid atomization is accomplished through electrostatic forces.

There are three systems of electrostatic charging of sprays

1. Corona charging
2. Contact charging
3. Induction charging

Liquid droplets get electrically charged when they come into contact with ions released from a high-voltage pointed electrode. These techniques can be used to power the sprays produced by a rotating nozzle or a hydraulic nozzle. Corona charge is also the foundation of industrial electrostatic paint application.

In a contact charging system, the nozzle or spray liquid system is connected directly to the high voltage potential. Spray liquid acts as a conductor, moving electrical charge to the spray droplets when they break apart. When using conductive liquids, this setup performs admirably. The entire setup requires top-notch insulation..

Spray droplets are charged using an electric field in an induction charging device. The conductive liquid and the charging electrodes in this system require reliable insulation.

The spray cloud is formed when the negatively charged droplets leave the nozzle and repel each other due to their comparable charge. These negatively charged drops are easily deposited on the upper and lower surfaces of vegetation (both of which are positively charged).

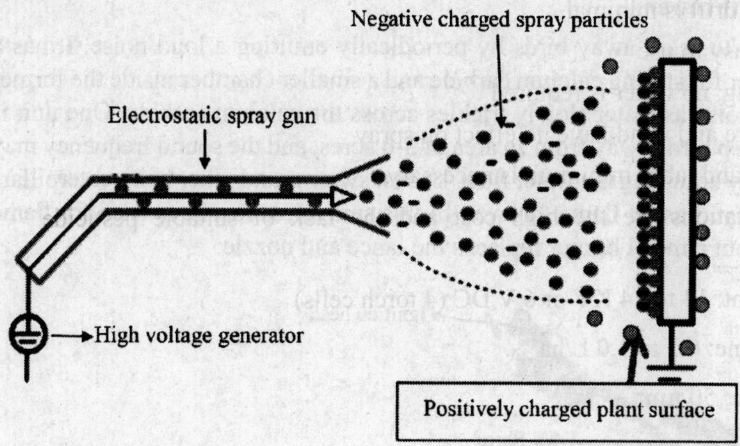

25. Other appliances

25.1. Soil injecting gun

It is used for fumigating the soil at different depths to control the nematodes and soil insects. It consists of a tank, a pump barrel, plunger assembly, injector nozzle, thrust handle and an injection handle. By holding with thrust handle, the equipment is thrust into the soil, till the nozzle rod gets into the soil completely and the injection needle is pressed to release the calculated quantity of liquid fumigant.

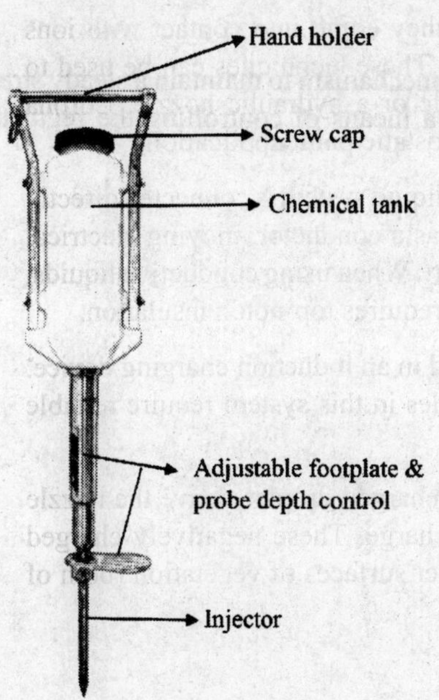

25.2. Flame thrower

It's designed to scare away birds by periodically emitting a loud noise. It has a large chamber for storing calcium carbide and a smaller chamber inside the former for making noise as water slowly trickles across the calcium carbide. One unit is enough to keep birds away from an area of 3-6 acres, and the sound frequency may be changed by adjusting the water flow. Locust swarms and other hairy caterpillars can be destroyed using flame thrower. In this sprayer, kerosene is used for flame production, but a metal burner replaces the lance and nozzle.

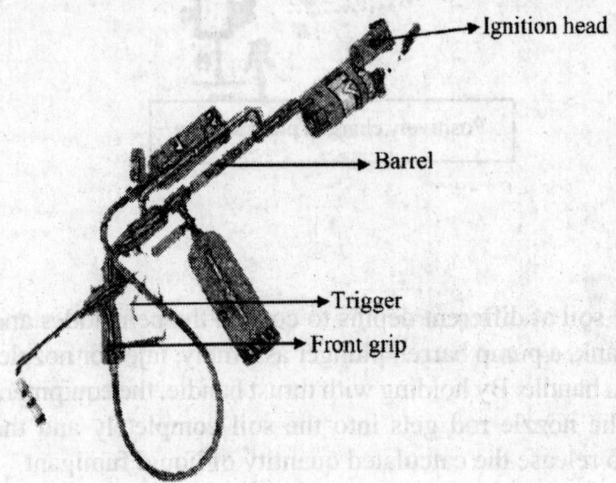

25.3. Granular applicator electrodyne

A hopper to store the granules, a regulating mechanism to maintain a steady stream of granules to the distribution point, and a means of controlling the regulating mechanism are all necessary parts.

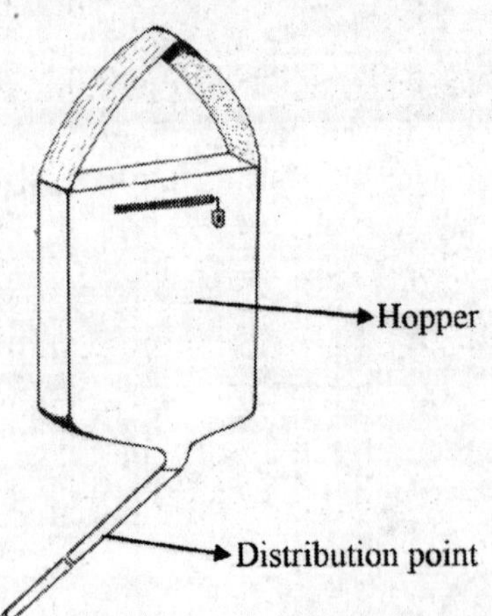

25.4. Drones

Drones are more formally known as Unmanned Aerial Vehicles (UAV's) or Unmanned Aircraft Systems. Essentially a drone is a flying robot that can be remotely controlled or fly autonomously using software controlled flight plans in its embedded systems,that work in conjunction with onboard sensors and a Global Positioning System (GPS).

Types of Drones: Drones are classified in several ways based on their weight, structure, altitude etc. But most importantly drones are classified into four main types.

They are: Multirotor Drones

Fixed wing Drones

Single Rotor Drones

Fixed wing hybrid VTOL Drones (By James Rennie)

Agriculture Drones

Agriculture Drones are defined as Unmanned Aircraft Systems (UAS) that is used for agricultural purposes such as crop mapping, field analysis, irrigation, pesticide spraying & crop monitoring.

Types of Agriculture Drones

The most commonly used Agriculture drones are fixed wing drones and rotary wing drones. Fixed wing drones are typically used for tasks such as crop mapping & field analysis. Rotary wing drones are used for tasks such as pesticide spraying and crop monitoring. Rotary wing drones are classified into Helicopter, Tricopter (3 propellers), Quadcopter (4 propellers), Hexacopter (6 propellers), Octocopter (8 propellers).

Difference between Quadcopter and Hexacopter

Most commonly used agriculture drones are Quadcopter & Hexacopter.

Stability: Hexacopter is more stable in air than Quadcopter because of its six propellers.

Motor failure: If any of the motor fails in Hexacopter, the remaining motors supports the drone and gives safe landing. If anyone of the motor fails in Quadcopter it leads to crashing.

Payload: Hexacopter lifts more weight than the Quadcopter, because of its six motors that gives more thirst.

Cost and Maintenance: Hexacopter is costlier because of its heavy building and extra parts and maintenance is heavier. Quadcopter is cheaper compared to Hexacopter.

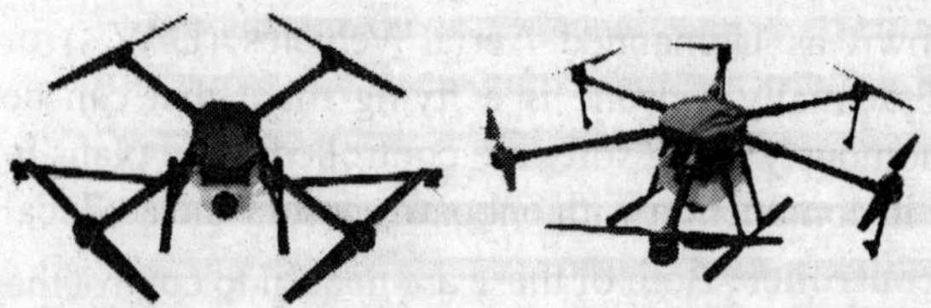

Quadcopter Hexacopter

Parts of Agriculture Drone

Agriculture drone is complex and consists of several parts.

Landing Gear: It is the bottom part of the drone which gives safe landing to drone.

Water tank and water pump: Water tank is used to store the water. Water pump is used to pump water from tank to nozzles.

Mainframe and Arms: The top most part is mainframe, which consists of flight controller and remote controller and all the wirings of drone. Arms are the parts, which supports propulsion motors and are extended from mainframe horizontally.

Propulsion motors: These motors provide thirst to lift the drone.

Propellers: Propellers are the blades that are connected to the propulsion motors. Propellers along with motors provide thirst.

Flight controller and Remote controller: Both of these are helpful to control the drone entirely.

GPS module: It is helpful to locate the area and to travel the drone straight.

Antennae: These are helpful to connect to the remote and to send signals to drone.

Camera: It is situated in the front to look for the any obstacles and for monitoring.

Accelerometer: It is used to know the speed of the drone.

Altimeter: It is used to know the height at which drone is travelling.

Sensors: Collision Avoidance Sensor: To avoid the drone to hit obstacles in middle. Liquid sensor: It is used to measure the amount of liquid flow.

Battery: A power source of the drone.

Nozzles and pipes: Pipes are used to connect from water tank to nozzles. Nozzles are the parts from where liquid is delivered.

Advantages of Agriculture Drone

Uniformity: Spraying will be done uniformly all over the field

Time: Great reduction in spraying time, takes approximately 6-7 minutes per acre.

Spraying: spray liquid will be discharged up to the bottom of the plant.

Safety: Due to non-involvement of man power, no much health issues.

Spraying Cost: Spraying cost will be reduced almost to half.

Disadvantages of Agriculture drones

Skills: It requires some basic knowledge and skills to operate drone.

Batteries: Batteries should be replaced after a acre of spraying.

Weather: Extreme weather conditions make drones difficult to perform.

Obstacles: If there are any obstacles in the middle of spraying area, the area around the obstacle will be left unsprayed.

Maintenance: Maintenance of drones is difficult and very high, because few parts need regular change.

Battery: Batteries last for only 300-400 acres.

Propellers: These need to be changed for 300 acres because of the chemical reaction with pesticide as they lead to breakage.

Registration & License of Drones

Director General of Civil Aviation (DGCA), Government of India controls and regulates the drone manufacturing and working in India. Digital sky is the platform for registration and licensing of drones. All drones except those in nano category must be registered and issued a Unique Identification Number (UIN). Drone pilots must need a pilot license to operate drone and they should maintain a direct line of sight with drone while operating. Drones cannot be flown in areas specified as No fly zones.

General information about plant protection appliances

Spray Applications

It is crucial to apply pesticides consistently. Crop damage, ineffective pest control, and damage to the following crop are some issues that can arise from improper use of residual pesticides.

Single Lance

A single-nozzle lance can apply pesticides, however spraying a straight swath with proper overlap is problematic. Walking with the single nozzle lance swung side to side will result in considerable under and over application. Choose multi-nozzle booms to avoid such mistakes.

Multi nozzle boom

When using a boom sprayer rather than a single flood jet nozzle, spray uniformity will be increased. With a three or four nozzle boom, the pass width may be smaller, but herbicide distribution will be very standardized. For spray swaths of three metres or greater, use a backpack with appropriate nozzles and adjust the spray pressure to produce enough output.

Only the middle third of the spray pattern of a single flat fan nozzle can be expected to have a full rate application. The tapering shape of adjacent nozzles ensures a uniform spray distribution when two or more are spaced so that they overlap by 30%. The outside border of the spray will be lightly applied on boom ends without nozzles. The most frequent spray tip angles are 800 and 1100. This is the side-to-side spray pattern angle. When using a boom sprayer, height is very important. If the boom is too low to the ground, the spray nozzles won't meet, so the spray will come out in bands with little or no spray in between. To find out how wide a spray spread is on a boom with more than one nozzle, multiply the distance between the nozzles by the number of nozzles. The spray swath width can be determined by using the boom at the appropriate height when making multiple passes across the field.

For example: 4 nozzles × 50 cm spacing = 200cm swath width.

Nozzle Height

Spray nozzles and booms need to be set sufficiently high above the target to ensure even spray output. Double overlap or double coverage reduces the effect of boom height or it changes by doubling the heights or halving the distance between the nozzles. If in doubt, set the boom too high rather than too low.

Factors affecting the efficacy of the Crop protection Products (CPP)

In order for CPPs to be effective, many factors must be considered. The equipment can be calibrated and operated properly to achieve improved accuracy. In order to deposit plant protection products accurately on their targets, agricultural CPP equipment is necessary. To minimize contamination of the operator and losses to the environment, the aim is to deliver the amount of chemical necessary to control the pest. It is estimated that just 1% of CPPs used worldwide truly achieve their intended purpose, making CPP application equipment and methodology, the two most crucial variables in maximising CPP efficacy and human safety. Better application technology can solve overdosing, uneven application, accidental CPP diffusion into the environment, and human exposure hazards. Hence, a good understanding of the causal agent, CPP method, equipment, and operator is necessary.

Three factors influence the efficacy of a CPP in any application technique:

1. An average level of deposit (dosage) represents the total amount of toxicant (active ingredient) applied per unit area of treatment
2. A chemical (active ingredient) deposit may cover the leaf entirely in runoff (high volume) spray, but it may be unevenly distributed.
3. Wetting agents reduce droplet size and boost spreads, while low-volatility carriers prevent small droplets (low and very low volumes) from evaporating and improve distribution.

Sprayer Calibration

A spray pattern and droplet size that covers the target area evenly without runoff is achieved by sprayer calibration. Therefore, calibration should be considered

1. **Target Area**- A large area will require a higher quantity (for a large area, a higher quantity would be needed)
2. **Droplet size**- Droplets with fine particles cover a large area with less volume, reducing runoff, but they can also cause more drift and evaporation losses
3. **Nozzle size and spacing**- After determining spray volume and droplet size, the boom nozzle size and spacing should be based on crop height.
4. **Nozzle capacity** -Nozzle capacity, a manufacturer's rating, is directly proportional to sprayer output at constant pressure and speed. Increased nozzle capacity boosts output. Knapsack sprayers with numerous nozzle booms may need to limit nozzle capacity to avoid overloading the pump. The typical nozzle size is 700, 800, or 900 ml/minute. Companies make smaller nozzles; however, they may not be widely available.
5. **Speed** – Spray application is inversely proportional to boom output and as you walk faster, you apply less spray
6. **Pressure** - Pressure improves sprayer production, but four times more pressure is needed to double nozzle output. Variable pressure affects output. There is a specific pressure rating for the nozzle angle. When the pressure goes below the recommended minimum pressure for a nozzle, the spray angle will get smaller. Low pressure nozzles, which have spray directions that stay same at low pressures, are made, but they may not be available everywhere.

Calibrating a multiple boom nozzle

Step1

- A straight boom should be aligned
- The nozzle interspacing should be equal, usually 50 cm
- A boom usually measures 1.5 meters in length

Step 2

- Make sure the nozzles are producing equal output by tying containers of equal size to the nozzles and observing the output.
- If the nozzle outputs are different, adjust the nozzles accordingly.

Step 3

- Pour water into the knapsack sprayer from a bucket.

Step 4

- Make a rectangle measuring 33 m by 3 m, approximately 100m^2 in area
- Spray from one end of the rectangle in a straight line, then turn around and make a second pass so that all the area is sprayed
- Once the area has been sprayed, measure out the amount of water (A1 ml) left in the knapsack

Step 5

Calculations:

Initial volume of water taken = **A** ml

Water left after spraying 100 m^2 = **A1** ml

Area sprayed = 100 m^2

Volume per unit area = (x – x1)/100 = a_2

Volume required for spraying 1 acre = $a_2 \times 4000 = a_3$

In terms of Knapsack tanks

One tank capacity = b l

No. of tanks required for 1 acre = a_3/b l

No. of tanks required per katta = (a_3/b)/17 (assuming 17 kattas = 1 acre)

Operational Efficiency

Reducing filling times and field stoppages will boost operational efficiency. The biggest problem when spraying is clogged nozzles, demonstrating the significance of efficient mixing, agitation, and filtration.

Filtration

A good filtration system cannot be underestimated. Several filters and filter kits have reduced obstructions. A double system with 100 mesh washable filters for tank filling and in-line screening works best. A second 100-mesh in-line filter can be added between pump and nozzle. Start with fine filters to support nozzle filters. It's best to clean in-line filters with each filling rather than let them dry out.

Filling

When calibrated, concentrate siphon probes reduce the risk of chemical contamination and simplify the mixing of formulae. Various types of containers, including metal and PVC, are commercially available. Some sprayers (for example, computer sprayers) include probes as part of the filling system. Homemade models that fit into the line of the filling pump also work. Contractors usually use different systems for filling concentrates. The spraying unit works and lasts well because it has electrically driven gear pumps and a flow metre built into it.

Flushing

The installation of a flushing system enables the stationary cleansing of pipelines and fittings using clean water. Additionally, it provides a safer method for unclogging and replacing nozzles, as well as stationary calibration of computer-controlled sprays. The utilisation of nozzles greatly facilitates the process of checking, and with regular use, the occurrence of nozzle blockages tends to decrease.

Maintenance and storage of Sprayers

Maintenance of Sprayers

a. Ensure that only clean water is used
b. Ensure the inlet spray is screened
c. Nozzles can be cleaned with metal objects
d. Sprayers should be flushed before use
e. Immediately after each use of a sprayer, clean it thoroughly

Cleaning of Sprayers

In order to remove all residues of CPPs completely after spraying, it must be completely cleaned. It plays an important role in preventing the following:

a. After spraying with different CPPs, crop plants are damaged
b. A negative interaction between the residual CPP and the newly used CPP
c. Wear and tear on sprayer parts due to corrosion

Procedure for Cleaning of Sprayers

a. The screens and boom extensions should be cleaned with kerosene and a small brush
b. The tank should be cleaned with detergent
c. After removing the nozzle from the boom, flood it with full of clean water

Storage of sprayers

a. Avoid sunlight and frost damage when storing the sprayer
b. Keep the sprayer away from children

17

Honeybees: Toxicity and Safety

Since bees are indicator species, their abundance on earth serves as a barometer for environmental conditions and the health of ecosystems. With up to 60,000 honey bees living in a single hive, each with a distinct job, the life of a bee is undoubtedly one of nature's greatest miracles. There is much to be learnt from the exquisitely smart, well-organized lives of bees. Around the world, honey bee hives have been experiencing severe decreases over the past ten years, and local pollinating species are also experiencing significant losses. A growing number of studies point to a particular class of pesticides as the principal cause of the worldwide pollinator disaster, and if our government agencies and certain legislators don't act quickly, the pollinator situation will only get worse.

Rachel Carson written Silent Spring 50 years ago, she expressed her concerns about the use of systemic insecticides like neonicotinoids. We are currently seeing first-hand how her forecasts come true

"The world of systemic insecticides is a weird world, surpassing the imaginings of the brothers Grimm... It is a world where the enchanted forest of the fairy tales has become the poisonous forest in which an insect that chews a leaf or sucks the sap of a plant is doomed. It is a world where a flea bites a dog, and dies because the dog's blood has been made poisonous, where an insect may die from vapors emanating from a plant it has never touched, where a bee may carry poisonous nectar back to its hive and presently produce poisonous honey (Rachel Caron, Silent Spring)".

The situation is made worse by a move away from agroecological practises and integrated pest management (IPM), which to varied degrees reduce the indiscriminate use of harmful chemicals. Industrial agriculture practises rely on pesticides whether they are required or not, adding weight to Carson's dire predictions by releasing an unsustainable amount of chemicals into the environment. According to one report, the trend in pest management over the past 20 years has been from reactive to preventative. The seeds are now frequently treated with fungicides, insecticides, and herbicides prior to planting, and pre-emptive treatments and routine spraying are frequently used to apply chemicals to plants before pest damage has begun. Today, our indicator species show quite clearly how our environment is degrading and the consequences that result from it.

Pesticides are one of the principal strategies used to damage honeybees and other pollinators in the agricultural ecosystem. Honeybees are particularly sensitive to several pesticide groups. The selective application of pesticides is one of the conditions for integrated pest management (IPM). In this case, selecting pesticides that are effective against the desired pests while being less harmful to pollinators is critical and will undoubtedly decrease bee losses. Many pesticides have lately been tested to discover how effective they are against pests and how dangerous they are to honeybees.

The causes of bee poisoning are

- Use of broad spectrum insecticides for example, chlorinated hydrocarbons, synthetic pyrethroids etc.
- Pesticide treatment during the crop's blossoming time.
- Direct application of insecticides to bees foraging on crops.
- Bees collecting contaminated nectar and/or pollen from treated plants.
- Bees feeding polluted food and water sources.
- Worker bees transport pesticide dusts and polluted pollen/nectar to the hive.
- Toxic chemical drift from the application site onto flowers, pollen, and nectar, or over apiaries.
- The use of insecticide formulations such as Dust and EC, which are more dangerous than WP and granules.
- Insect growth regulators may reduce brood production.
- Insecticide formulations that use diesel oil as a carrier.

Routes of pesticide exposure

Bees are primarily exposed to pesticides during their foraging activities, specifically while searching for nectar and pollen. However, the specific pathways of pesticide exposure can vary depending on factors such as the application method of the pesticides and the behavioral and ecological characteristics of different bee species.

One prominent pathway of potential exposure involves direct contact with the pesticide through topical means, wherein bees come into contact with the pesticide while in flight or while visiting flowers within a cultivated area during pesticide application. The aforementioned scenario has the potential to result in immediate harm or fatality to the bee, or alternatively, it may introduce a significant quantity of the pesticide into the colony.

When bees come into contact with treated objects like leaves or flowers after they have been sprayed, indirect topical exposure occurs. The risk to bees significantly decreases once a pesticide has dried or begun to break down due to exposure to sunlight and other environmental variables. The advice on bee safety offered on the pesticide label frequently reflects this fact. However, as some pesticides break down, they can produce secondary metabolites that are similarly dangerous. For the majority of pesticides, the residual toxicity to bees has not been well-documented and varies between chemicals. Rapidly absorbed pesticides into plant tissue may also cause reduced indirect topical exposure.

Bees may be exposed to pesticides if they are spread by air movement from the target crop to nearby blooming weeds, bee nests, or honey bee colonies. By utilising coarse sprays and avoiding applying pesticides in windy conditions, drift can be prevented. Additionally, because water-based and oil-based sprays have differing drift properties, it's crucial to carefully analyse the composition.

Bees can be exposed to pesticides by eating pollen or nectar that has been polluted. Numerous neonicotinoid insecticides, which are systemic pesticides, can be sprayed onto leaves and then locally absorbed into plant tissue. They can also be sprayed on soil, where plant roots will absorb them and the vascular system will disperse them. These systemic insecticides can be quite successful at getting rid of aphids, leafhoppers, and other sucking pests, but if they end up in nectar, pollen, or plant guttural fluid, they can also be harmful to pollinators. Different systemic pesticides have different levels of toxicity for bees. Additionally, due to their mechanism of action or because the pesticide does not dissolve in the mouth, certain pesticides have substantially higher oral toxicity than contact toxicity.

Honey bees typically gather water to control hive temperatures in hot weather, thus contaminated surface water and chemical spills are two more possible exposure sources. Bee exposure to pesticides can be prevented in part by keeping agricultural water sources clean and avoiding chemical spills.

Symptoms of bee poisoning

Different classes of chemicals that bees are poisoned with result in varied symptoms in the bees. The following are the general and most prevalent signs of bee poisoning

- The appearance of a significant number of dead bees around the entrance of beehives or colonies, as well as in fields.
- Dead bees on the top or bottom of frames.
- Bees that are paralysed crawling on surrounding things.
- Bees lose orienting power and may conduct strange communication dances on the horizontal landing board at the hive entrance.
- Bees' legs, wings, and digestive systems cease to function.
- Bees become angry and aggressive, sting excessively, and fail to recognise guard bees.
- The queen may exhibit unusual behaviour and deposit eggs in an irregular manner.

- Bees fighting among themselves.
- Rapid decrease in food storage and brood raising.
- Distended abdomen and regurgitation of stomach contents.
- In the hive, the dead and deserted brood.
- Bees' inability to recognise pollen and nectar.
- A scarcity of foraging bees.
- The colony's bee population has been depleted.
- Adult bee longevity is lowered.

Guidelines to assess the extent of bee poisoning by pesticides

The mortality statistics provided below can be used to estimate the amount of chemical poisoning of bees.

No of dead bees /day at entrance of bee box	Level of poisoning
100	Normal death rate
200-400	Low
500-1000	Medium
More than 1000	High

Effects of bee/pollinator poisoning

Agriculture suffers from three sorts of negative consequences. They are as follows:

- Reduced yield of cross-pollinated crops
- Loss in production of honey
- Bee products contamination

Categories of pesticides on the basis of their toxicity to honeybees

Pesticides are also classified into several groups based on their toxicity to bees. This can aid in the selection of appropriate insecticides. Pesticides are classified into three types based on their toxicity to bees.

Category I: Highly toxic pesticides

These pesticides are extremely hazardous to bees. They should not be used on blossoming crops or weeds. Even after 10 hours, their residual toxicity is generally high. List of very hazardous pesticides that are routinely used.

Carbofuran	Imidacloprid
Chlorpyriphos	Monocrotophos
Clothianidin	Oxydemeton-methyl
Cypermethrin	Permethrin
Deltamethrin	Quinalphos
Dimethoate	Thiamethoxam
Fenvalerate	

Category II: Moderately toxic pesticides

These pesticides should only be used in the late evening, night, or early morning hours when bees are not actively feeding since they are moderately harmful to bees. Usually, their residual toxicity is negligible after 3 hours after spraying. The insecticides are listed below.

Acetamiprid	Malathion
Diazinon	Monocrotophos
Fenitrothion	Oxydemeton-methyl
Fenthion	Thiacloprid

Category III: Relatively non-toxic pesticides

These pesticides are least hazardous to bees and can be used on honeybees at any time with a fair level of safety. With direct application, their toxicity is typically modest.

Bacillus thuringiensis	Ethion
Sabadilla	Nuclear polyhedrosis virus
Endosulfan	Phosalone
Dicofol	Pyrethrum

Reduction of bee poisoning

Honey Beekeeper-Grower Cooperation

Beekeeper-grower cooperation is a key factor in the decrease of honey bee poisoning. It is crucial that the beekeeper and the producer work together and be aware of each other's management concerns since the beekeeper frequently depends on the grower for bee food and the grower depends on the honey bees the beekeeper provides for pollination.

Role of Pesticide Applicator

1. Avoid using pesticides on crops or weeds that are in bloom that are hazardous to bees. Prevent pesticide drift by using low drift spray nozzles and avoiding spraying in windy circumstances.
2. Use pesticides in the late afternoon when bees are not actively searching for food. In general, honey bees won't feed at temperatures below 13 °C. Applying pesticides when abnormally low temperatures are predicted after application is thus not advised since the residues will stay poisonous to bees for a longer period of time.
3. Because there is less product dispersion and fewer crop areas are treated at once, ground treatment of pesticides is often less dangerous than aerial spray.

4. Avoid disposing of pesticides where they might provide a risk of bee poisoning. When pollen is scarce, bees will occasionally gather dry pesticide formulations because the dust particles resemble pollen grains in size. In this case, the bees will bring the dust back to the colony and feed it to the growing brood and adults, perhaps leading to the hive's demise.
5. When using integrated pest management (IPM) programmes for particular crops and insect pests, use low-risk insecticides or formulations that are less harmful to bees. In general, dusts and encapsulated pesticides pose a greater risk to bees than sprays of the same substance. The residual time of wettable powders is frequently longer than that of emulsifiable concentrates. Granular formulations of some pesticides are often the most secure type of insecticide treatment close to honey bee colonies, whereas ultra-low volume formulations of particular pesticides may be significantly more dangerous to bees than ordinary sprays. Use the least toxic recommended pesticide with the shortest residual duration if all recommended pesticides are equally harmful. Be careful to read the pesticide label for details on the toxicity of particular pesticides. Notify beekeepers with apiaries in the area in which insecticides will be applied at least 48 hours before their application.

Reduce drift onto areas where bees are living or foraging. When applying pesticides throughout the season, it is important to take into account the reduction of drift from the target crop to adjacent regions where bees may be feeding or nesting. In order to lessen the risk to bees, it may also be taken into account while designing new farms or fields. Reducing drift that impacts bees and their environment

- Keep the spray on target.
- When driving near hives, turn off the sprayer to avoid chemical drift into open blooms.
- The most critical elements influencing downwind spray deposition are droplet size and wind speed. Use that knowledge to decrease drift.
- If applications cannot be avoided, utilise bigger droplet sizes (more than 150 microns) to decrease drift in low humidity, high temperature, or windy circumstances. This may be accomplished by lowering nozzle pressure, increasing nozzle orifice diameter, and employing a low-drift nozzle.
- Spraying should not be done in windy or cold conditions. Apply only when the wind speed is less than 10 mph. However, keep in mind that extremely still circumstances, such as an air inversion, can also cause microscopic droplets to float. Additionally, under hot, dry circumstances, greater evaporation might result in smaller droplet sizes and a higher danger of drift.

- Place honey bee or bumble bee colonies next to crop fields, but keep them away from places that will be directly sprayed. Some farmers set aside certain parts of their property for bee installation, enabling their beekeeper to build and remove hives quickly. This can also lessen the possibility of machines and personnel disturbing colonies.
- Be mindful of your neighbours who have honey bee colonies.

Select a less toxic formulation: Pesticide formulations include dusts (D), wettable powders (WP), soluble powders (SP), emulsifiable concentrates (EC), solutions (SL), and granular formulations (G). Avoid applying dusts and wettable powders during bloom since they might cling to bee hairs and be accidentally delivered back to the hive or nest, where the residues may end up in the bees' food supply.

Role of grower

1. Use insecticides only when necessary, and follow all pesticide label requirements. When insect pests cause yearly crop loss, start an IPM programme that includes pest monitoring so that pesticides may be sprayed when economic threshold levels are met. This ensures that the pesticide is most effective in killing the bug while minimising non-target effects.
2. Insecticide usage during the day or when crops are in bloom has the potential to kill bees or entire colonies, which will reduce agricultural productivity.
3. Keep in mind that even when the lucerne is not in flower, blossoming weeds like narrow-leaved hawk's beard might contain honey bees that an insecticide treatment could harm.
4. Gain a mutually beneficial awareness of the beekeeper's management concerns regarding insecticide-honey bee interactions and the potential negative effects in order to maintain proper bee pollination of crops.

Role of Beekeeper

1. Recognise the crops that will be cultivated close to your apiaries and educate yourself on the pesticides that may be used on those crops in terms of their bee toxicity.
2. If at all feasible, select apiary locations that are generally separated from frequent pesticide treatments and that are not often vulnerable to chemical drift.
3. Distribute an apiary map to your municipality's office, local aerial applicators, and the provincial apiculturist.
4. If pesticides are to be used, remove colonies from the area. If this is not possible, cover colonies with a tarp, tent-staking the edges and providing a central support. An internal source of fresh water is required to keep the bees cool during confinement and to avoid losses caused by exposure to

tainted water when the tarps are removed. Beekeepers can place damp bags in the entrance of the hive to hinder bee flying for up to 12 hours and give the spray additional time to dry. A 2.5 cm (1 inch) hole on each side of the hive entrance is required so that the bees may exit and ventilate the hive. Remove honey bee colonies as soon as pollination is complete and before any post bloom insecticides are applied.

Safeguarding Other Bee Pollinators

Information on the effects of pesticides on bee pollinators other than honey bees is often scarce. The amount of toxicity may differ depending on the size and behaviour of the particular pollinator, but pesticides that are toxic to honey bees will probably also be poisonous to other bee pollinators.

18

Insecticide Poisoning and Therapy

Insecticide use for agriculture has increased along with food quality and quantity, but self-harm has also increased. Despite a high rate of insecticide poisoning and sometimes usage for homicide, little is known about insecticide management.

Insecticide poisoning kills up to 300,000 people around the world every year. According to hospital admission data, most estimates of insecticide poisoning in the short term have been based on numbers of people poisoned by insecticide s. This represents only a small part of the real number, as it includes only the most dangerous cases. IS 4015 is a standard guide for handling cases of pesticide poisoning.

Classification of general symptoms of insecticide poisoning

Mild poisoning	**Moderate poisoning (any mild symptoms plus any of)**	**Severe poisoning (any mild and moderate symptoms plus any of)**
Headache	Abdominal cramps	Inability to breathe
Dizziness	Vomiting	Chemical burns on skin
Weakness	Diarrhoea	Respiratory distress
Fatigue	Excessive salivation	Loss of reflexes
Nervousness	Constriction in throat and chest	Uncontrollable muscle twitching
Loss of appetite	Abdominal cramps	Unconsciousness
Thirst	Rapid or slow pulse	Convulsions
Nausea	Excessive perspiration	
Irritation of throat and nose	Trembling	
Eye irritation	Muscle incoordination	
Constriction of pupils	Mental confusion	
Blurred vision		
Skin irritation		
Changes in mood		
Loss of weight		

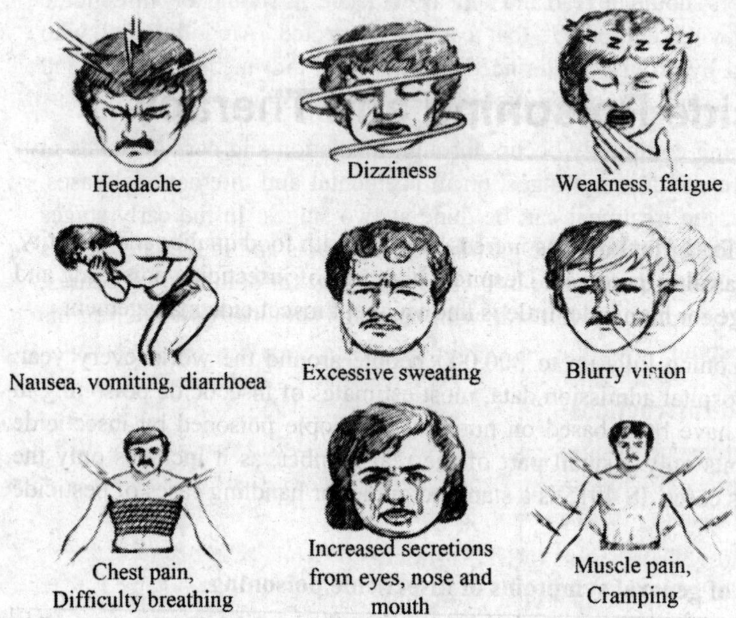

Indications of insecticide poisoning

General: Extremely tired and weak.

Skin: Irritation, a burning sensation, increased perspiration, and discoloration are all possible symptoms.

Eyes: Itching, a burning feeling, wet eyes, impaired or hazy vision, and dilated or enlarged pupils are all symptoms of a bacterial infection.

Digestive system: Excessive salivation, nausea, vomiting, abdominal discomfort, and diarrhoea are all symptoms of a burning feeling in the mouth and throat.

Nervous system: Headache, dizziness, confusion, nervousness, twitching muscles, stumbling walk, slurred speech, fits, and falling asleep are all signs of a stroke.

Respiratory system: a cough that causes pain and tightness in the chest, trouble breathing, and hacking.

It's important to find out more about it

- Did the person work with insecticide s?
- Did contamination occur?
- Which exact product was used?
- What quantity was consumed?
- When was it?

Labels on containers should be read and kept as evidence, as should any insecticide containers or spraying equipment that can be inspected. An individual who has been poisoned by an insecticide needs prompt first aid, medical advice and assistance, and, if feasible, transportation to the nearest hospital.

Insecticide poisoning commonly occurs through inhalation and dermal routes in industrial exposure and through ingestion in accidental and international cases. In poisoned cases, the treatment can be done at two stages: In the early stages when the poison could still be in the gastrointestinal tracts and in the later stages when it has been absorbed and gone into circulation and for both the procedures, the patient should be put on his left side with head and shoulder down during his transport to the hospital.

General First Aid Instructions

- If there is a chance of direct exposure to insecticide s, take precautions by wearing protective clothing and equipment whether providing first aid or removing a casualty from a confined space.
- Material Safety Data Sheets (MSDS) should be kept up-to-date.
- Telephone numbers for emergency response should be readily available
- Prepare a first aid kit that contains the necessary supplies.
- Make sure you always have clean water to drink. In a real situation, you could mix insecticide s with water from a farm pond, an irrigation system, or a watering trough.
- Normally, the first step in preventing absorption of insecticide s is to dilute them and prevent their absorption into the body when exposure has occurred via oral or dermal routes.
- The victim must be immediately brought to fresh air if exposure to inhalation occurs
- If a person is unconscious, don't give anything to them orally
- If a person's breathing has stopped or become impaired, he or she may need artificial respiration

Treatment during early stages- Following measures should be taken when the poison is expected to be in the gut

1. Induce vomiting (emesis) : 1-2 table spoons of table salt in ½ glass warm water or by putting blunt spoon or fingers
2. If emesis (process of vomiting) fails, gastric lavage (stomach-wash) can be given. It involves inserting a large bore rubber tube in the stomach, washing it with water and draining the contents out so a doctor is the right person to undertake lavage.

3. Further cleaning (Purgation/Cleansing) of gut or intestine, a pinch of Mg citrate or $MgSo_4$ is used and to do this, castor oil (1 spoon) can also be taken.
4. A pinch of activated charcoal : It binds the poison by absorbing it
5. Induce artificial respiration if breathing is irregular
6. Decontamination with soap : in case of exposure on skin wash with soap repeatedly
7. Whenever you suspect eye contamination, wash the eyes thoroughly with running water and avoid using chemicals

Treatment during late stages

1. Support all the vital functions like heart beat & respiration
2. Increase excretion by giving divertic agent (medicine given to increase the flow of urine) and the process is called diveresis
3. If possible go for dialysis (purification-removing impurities and purifying the blood)
4. Good nursing and care antidote prescription contains to monitor vital functions
5. Try to obtain the detailed history to find out the exact poison ingested and also by examining the container of the poison
6. Check for the blood pressure and the samples of blood and urine
7. One should not hide the true factor for the fear of police under investigation
8. Provide sufficient ventilation

Toxicological signs and symptoms of poisoning

Insecticide poisonings are usually acute and result from extensive skin contact or ingestion of insecticide s. It is possible to confuse symptoms of insecticide poisoning with symptoms of other illnesses, depending on the kind of insecticide used. The toxic effects of insecticide s are diverse and varied. Three ways are possible for them to enter the human body: via absorption, through the skin or eyes (dermally), via the mouth (orally), or through breathing into the lungs.

Dermal exposure

It causes absorption as soon as a insecticide comes into contact with the skin or eyes. The insecticide continues to be absorbed as long as it is in touch with the skin. Dermal absorption happens at a distinct pace for each body component. The relative absorption rates are calculated by comparing each absorption rate to the forearm absorption rate.

Oral exposure

If swallowed, insecticide s can cause serious illness, severe injury, or even death. The ingestion of insecticide s can be accidental, careless, or purposeful.

Respiratory exposure

It is dangerous because insecticide chemicals are quickly absorbed by the lungs into the bloodstream. Inhaled insecticide s can damage the nose, throat, and lungs. Vapours and tiny insecticide amounts provide the greatest risks. Inhaling insecticide particles, droplets, or vapours exposes the lungs. Inhaling concentrated wettable powders during mixing is dangerous. The risk of respiratory exposure from insecticide spray droplets is relatively low when low pressure application equipment is used, as most droplets are too large to remain airborne and long enough to be inhaled. However, this risk increases when high pressure, ultra-low volume (ULV), or fogging equipment is used. Air currents can carry mists and fogs along with the droplets formed during these operations.

Toxicity of insecticides: A insecticide's toxicity in an organism can generally be classified as:

Acute toxicity: The term "acute toxicity" is used to describe the harmful effects of a insecticide after a single exposure or several exposures in a short period of time. Acute poisonings can cause a wide range of symptoms. If even a small amount of an acutely hazardous insecticide is ingested, it can be fatal. Acute toxicity can be evaluated using the following methods: ingestion, skin contact, and breathing.

Chronic toxicity: Long-term or repetitive low-level hazardous exposures cause chronic toxicity. Chronic exposure can take years to show symptoms. Chronic poisoning can cause the following effects:

- It causes cancer when it is carcinogenic.
- It causes genetic changes by being mutagenic.
- The tendency to cause birth defects as a result of teratogenicity.
- Oncogenicity- Cancer-inducing ability (not necessarily cancer).
- Liver damage- Fibrosis, liver cirrhosis, and jaundice (yellowing of the skin).
- Reproductive disorders- Miscarriage, reduced sperm count, and sterility.
- Nerve damage - An accumulation of organophosphate insecticides leads to depression of cholinesterase.
- Allergenic sensitization- As a result of allergic reactions to insecticide s or chemistry in their formulation, insecticide s may cause allergies.

Similar to acute toxicity, chronic toxicity is also dependent on the dosage administered. To clarify, it should be noted that minimal exposure to chemicals with the capacity to induce prolonged consequences may not result in immediate harm. However, the likelihood of chronic detrimental effects is significantly amplified when such exposures are recurrent due to negligent handling or improper usage. Observable indicators of poisoning can be discerned by external observers, such as the manifestation of symptoms like emesis, diaphoresis, or constricted pupils. Symptoms encompass any alterations in physiological functioning from the baseline state, as reported by those affected by poisoning. These manifestations may encompass sensations of nausea, headaches, diminished physical strength, dizziness, and various other indicators.

A. Poisoning of Organochlorine Insecticides

Organochlorine insecticide s exhibit limited biodegradability and demonstrate long-lasting presence in the environment, leading to their impact on the nervous system as stimulants or convulsants. Organochlorines frequently cause nausea and vomiting shortly after their use.

Anxiety, agitation, hyperactivity, dizziness, headache, disorientation, weakness, skin tingling or pins and needles, and muscle twitching are all symptoms that might appear early on. This is then followed by incoordination, fits resembling epileptic seizures, and eventually unconsciousness. Apprehension, twitching, tremors, bewilderment, and convulsions are all possible reactions to chemical absorption through the skin.

Organochlorine poisoning has no known particular antidote. Insecticide may be washed off skin and hair with soap and water, therefore do this as soon as possible after removing contaminated garments. Helpers should take precautions to avoid contamination with the insecticide, such as wearing chemical-resistant gloves. Ipecac (syrup) and water should be given to a cognizant patient who has ingested the insecticide, or a finger can be inserted down the patient's throat to force the insecticide back up.

B. Poisoning of Organophosphorus Insecticides

Humans are affected by OP insecticide s because they inhibit acetyl cholinesterase, an enzyme necessary for normal nervous system function. Symptoms of OP insecticide exposure appear quickly after exposure and can be life-threatening for those with compromised immune systems, breathing problems, epilepsy, and other conditions. A mild exposure to organophosphate insecticides may result in the following signs and symptoms:

Common symptoms experienced by individuals include headache, fatigue, dizziness, loss of appetite accompanied by nausea, stomach cramps, and diarrhea. Additionally, blurred vision accompanied by excessive tearing, contracted pupils of the eye, excessive sweating and salivation, a slowed heartbeat typically below 50 beats per minute, and rippling of surface muscles just beneath the skin are frequently observed.

These symptoms may mimic flu, heat stroke, heat exhaustion, or upset stomach. Moderate organophosphate and carbamate insecticide poisoning cases have all the symptoms of mild poisonings, plus the victim is unable to walk, complains of chest discomfort and tightness, has pinpoint pupils, twitches, and has involuntary urination and bowel movements. Symptoms of severe poisoning include incontinence, unconsciousness, and seizures.

If the product is eaten, stomach and other gastrointestinal symptoms appears first, but if it is absorbed through the skin, gastric, and respiratory routes, they appear simultaneously. Organophosphate or carbamate poisoning sufferers can get effective antidotes at emergency centers, hospitals, and many physician clinics, but time management is crucial as with all insecticide poisonings. Swallowed insecticides require immediate medical attention, and cutaneous exposure requires removal of contaminated clothing, washing, and medical attention.

Organophosphates mainly acts on two receptors and cause nicotinic and muscarinic effects.

B.1. Nicotinic Receptors

Nicotinic receptors are of two types central (neuronal) and peripheral (neuromuscular). Central nicotinic receptors, also known as NN or N_2, are located in the central nervous system (CNS). They can also be found in the sympathetic and parasympathetic ganglia of the peripheral nervous system (PNS) and the adrenal medulla. Peripheral nicotinic receptors, or NM or N_1, are located at the neuromuscular junctions. The N_1 neuromuscular junction can cause fasciculation and muscular weakness, whereas the N_2 autonomous nervous system is associated with hypertension and tachycardia.

B.2. Muscarinic Receptors

All five subtypes of muscarinic receptors, M_1 to M_5, are distributed throughout the CNS. Postganglionic muscarinic receptors provide parasympathetic innervation to the heart, exocrine glands, and smooth muscles of the internal organs. Sympathetic postganglionic fibers provide innervation to the sweat glands.

Stimulation of each specific receptor yields distinct clinical signs and symptoms, as mentioned below.

- M_2 receptors in the heart: Hypotension and bradycardia
- M_2 and M_3 receptors in the eyes: Miosis
- M_2 and M_3 receptors in the gastrointestinal system: Abdominal cramps, drooling, and salivation
- M_2 and M_3 receptors in the respiratory system: Bronchospasm, bronchorrhea, and rhinorrhea

- M_2 and M_3 receptors in the smooth muscles of internal organs: Abdominal cramps and urinary urgency
- M_1 to M_5 receptors in the CNS: Seizure, anxiety, and agitation

Adverse Effects of Organophosphate

The adverse effects of exposure to organophosphate pesticides can be categorized based on the duration of exposure, as mentioned below.

- Acute effects: Occur within minutes to 24 hours
- Subacute effects: Occur between 24 hours and 2 weeks
- Chronic effects: Extend beyond weeks to years

C. Poisoning of Carbamate Insecticides

As cholinesterase inhibitors (nerve poisons), these insecticide s are low to mildly toxic:

- Mild exposure - constricted pupils, salivation (slobbering), profuse sweating
- Moderate exposure - fatigue, uncoordinated muscles, nausea, vomiting
- Severe exposure - diarrhoea, stomach pain, tightness in the chest.

D. Poisoning of Synthetic Pyrethroids

Synthetic pyrethroids mimic natural pyrethrins. Some pyrethroid insecticides are oral toxins, although most are safe. Few very big dosages can produce incoordination, tremors, salivation, vomiting, diarrhoea, and sound and touch irritation. Renal excretion of most pyrethroid metabolites is rapid. Pyrethroids don't block cholinesterase. Crude pyrethrum causes skin and respiratory allergies. Exposures caused skin irritation and asthma. Although less allergic, purified pyrethrins may still irritate and/or sensitize.

When people are exposed to commercial products, it is important to think about the likely role of other toxicants in the products. The synergists, like piperonyl butoxide, are not very dangerous to people, but the organophosphates or carbamates in the product could be very dangerous. The cholinesterase enzyme is not stopped by pyrethrins on their own. Pyrethroids rarely cause systemic poisoning in humans due to inhalation and cutaneous absorption. Dermal contact might cause stinging, burning, itching, tingling, and numbness.

E. Poisoning of Oxadiazines

Methemoglobinemia is a condition in which an abnormally high level of methemoglobin, a form of hemoglobin, is present in the blood. Hemoglobin normally carries oxygen from the lungs to tissues and organs, but in methemoglobinemia, hemoglobin is altered and cannot effectively release oxygen to body tissues. This results in reduced oxygen delivery, which can cause symptoms such as cyanosis (bluish skin and lips), fatigue, shortness of breath, and, in severe cases, neurological symptoms and even death.

Indoxacarb, an insecticide, can cause methemoglobinemia as a toxic side effect when it is ingested or absorbed in significant amounts. This condition occurs because indoxacarb interferes with hemoglobin's ability to release oxygen, converting it to methemoglobin.

F. Rodenticides Poisoning

F.1. Warfarins

After several doses, these rodenticides disrupt vitamin K-dependent coagulation factors II, VII, IX, and X and damage capillaries, causing rats to haemorrhage. Most cases involve asymptomatic prothrombin time prolongation. Larger doses can cause many bleeding locations. Look for petechial haemorrhages, haematuria, and occult blood in the stools.

F.2. Aluminium phosphide

Northern India uses pellets and powder of aluminium phosphide, a cheap solid fumigant rodenticide, to preserve grain. Nontoxic aluminium phosphate and hypophosphite remain in the grain after spraying. Celphos®, Alphos®, Quickphos®, and Phosfume® are 3g pellets of 57% aluminium phosphide. Unexposed pellets of aluminium phosphide are fatal for adults at less than 500 mg, however air quickly degrades them.

Symptoms of poisoning generally start within 30 minutes, and they include severe pain in the upper stomach, repeated vomiting, diarrhoea, low blood pressure, fast heart rate, and metabolic acidosis. In these people, a bad sign is that they have severe low blood pressure that does not respond to dopamine. Respiratory symptoms include cough, shortness of breath, cyanosis, lung edoema, and oedema, and most patients stay awake until the very end.

F.3. Zinc phosphide

Zinc phosphide smells like rotten fish and is crystalline. It releases phosphine gas when water touches its powder or pellets. Zinc phosphide poisoning has comparable symptoms to aluminium phosphide poisoning but a later onset due to slower phosphine production. Early symptoms include nausea and vomiting, which can occur with a ingestion of 30 mg. Patients complaint about chest tightness and may be excited, irritated, or thirsty. Shock, oliguria, coma, and convulsions are possible.

F.4. Ethylene dibromide (EDB)

Ethylene dibromide, which is also called 1,2-dibromomethane, is a popular insecticide used in India to keep cereals and grains from going bad when they are stored. It is a clear liquid with a strong smell of sweetness.

EDB is absorbed through all pathways in people, easily penetrates clothing, and there is no effective antidote. Small doses of EDB may not be lethal, but exposure to 5-10 ml is frequently fatal. EDB has been identified as a fatal occupational hazard among grain stores and handlers. EDB can cause ulcers, conjunctivitis, gastrointestinal and mucosal irritation, central nervous system irritation, and depression when applied topically. Vomiting, diarrhoea, and throat burning occur shortly after intake of EDB, and these symptoms may linger for 1-3 days. Tremors and central nervous system depression also occur in patients.

Treatment and Therapy to Insecticide Toxicity

These steps should be familiarized by those who use insecticide s frequently:

- Identify insecticide poisoning's signs and symptoms.
- Immediately seek medical attention from a hospital, physician, or poison control center
- Provide medical authorities with information on the insecticide to which the victim has been exposed
- The insecticide label should be present when medical attention is being provided. Information on the label will assist a victim of insecticide poisoning.
- An individual poisoned by OP or carbamate insecticides can be treated with atropine, a medicine that counteracts the muscarinic effects, keeping them alive. For OP poisoning, pralidoxime prevents enzyme/insecticide complex "aging". If aging is prevented, the insecticide molecule dissociates from the enzyme, slowly restoring enzyme activity.

First-aid treatment

If the insecticide has been spilled on the skin or clothing, remove any contaminated clothing as soon as possible and thoroughly wash your skin with soap and water. Scrubbing too hard increases insecticide absorption. Rinse the afflicted area with water, then wash and rinse again. Dry the injured area gently and, if required, cover it in a soft towel or blanket. If you have chemical burns on your skin, cover the affected area with a clean, soft towel. Unless otherwise directed by medical personnel, avoid using ointments, greases, powders, and other drugs.

If the insecticide has entered into the eyes, hold the eyelid open and begin gently cleaning the eye with clean flowing water right away. There should be no chemicals or medicines in the eye wash water. Wash for another 15 minutes. Avoid infecting the other eye if just one is affected. To eliminate debris, flush the eyelids with water. Cover the eye with a clean towel and seek medical assistance right once. If you are wearing contact lenses, remove and discard them before washing your eyes as instructed above. If no water is available, gently wipe the area with towels or paper to absorb the insecticide while avoiding hard rubbing or scrubbing.

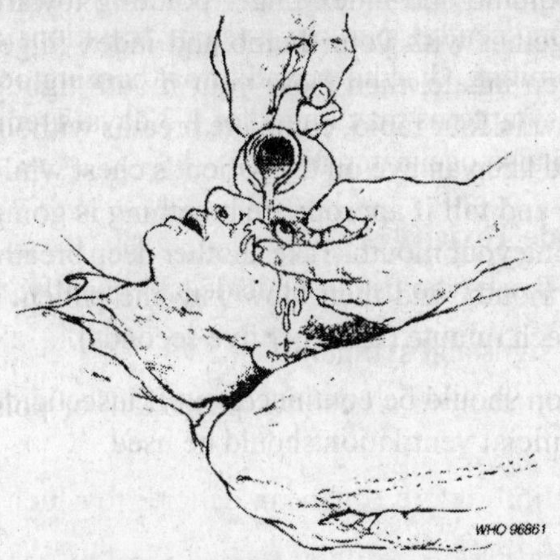

If the insecticide has been inhaled, get the person out into the open air right away. But don't try to help someone who is trapped in a small space unless you are wearing the right safety gear. Tell the person to lie down and take off some of their clothes. Keep the person calm and warm. Watch how the person is breathing and cover their head if they are shaking. Keep your chin up to keep your airways open so you can breathe. If a person stops breathing, they should be given mechanical air.

If the insecticide has been swallowed, Call a doctor right away and tell them the name of the poison and about how much of it was swallowed. If the poison got in your mouth but you didn't swallow it, rinse your mouth out with a lot of water. Most poisonings, including insecticide poisonings, don't call for making someone throw up.

If breathing has stopped: allow for artificial respiration. If no insecticide was consumed, mouth-to-mouth resuscitation may be administered. To maintain the airway open, pull the patient's chin up and tilt the head back with one hand. Place the second hand on the patient's brow, thumb and index finger pointing toward the nose. Pinch the patient's nostrils together with your thumb and index finger to prevent air from escaping. Take a deep inhale, then close your mouth tightly over and around the patient's mouth. Blow in four rapid, complete breaths without allowing the lungs to totally collapse, and keep an eye on the patient's chest while inflating the lungs. The chest should rise and fall if appropriate breathing is going place. Allow the patient to exhale by closing your mouth. Take another deep breath, create a tight seal around the patient's mouth, and then blow into the patient's mouth once again. Repeat 10-12 times each minute (once per five seconds).

If there is still a pulse, artificial respiration should be continued, but if insecticide has been swallowed, another form of artificial ventilation should be used.

Vomiting

Induce vomiting only if the patient has consumed an extremely hazardous insecticide and medical care is not expected soon. Never induce vomiting if the patient has ingested oil spray or items diluted in diesel or kerosene, because the vomited substance might be inhaled and be more hazardous than the intestinal poisoning. The insecticide's toxicity level should be indicated on the product label (skull-and-crossbones symbols). Only if the patient is conscious should vomiting be induced. Sit or stand the individual up if required, and tickle the back of the neck with your finger. Whether or not the patient vomits, offer him or her a drink made of three tablespoons of activated charcoal in half a glass of water. Rep until medical assistance comes.

Caring for the patient

Make the person lie down and rest because organophosphorus and carbamate poisoning gets worse when the person moves. Place the patient on his or her side with the head lower than the body. If the person isn't awake, pull the chin forward and the head back to make sure the mouth is open. Cover the patient with a blanket if he or she feels cold. If the person is sweating too much, use a sponge with cold water to cool him or her down. If the patient throws up on his or her own, make sure he or she doesn't breathe in the vomit. If the person has fits, put something soft between their teeth to keep them from getting hurt. Don't let them smoke or drink booze, and don't give them milk, but you can give them water. If a patient needs more medical care, they should be sent to the nearest medical center.

Antidotes for insecticide poisoning

Antidotes are medicines or other things that are used to stop the effects of poisons. There are medicines that can stop the poisoning affects of the chemicals. But if you don't use them right, they can be even more dangerous than the poison itself. Only a doctor should give or write a prescription for a cure. Some poisons have no known way to treat them. Once a lethal amount has been taken, the consequences can't be undone and the person will die.

The inability to produce enough of the enzyme cholinesterase results in the poisoned person's death because it controls the chemical transmission of nerve impulses. This enzyme in the blood is targeted by both carbamate and organophosphate insecticides, rendering them ineffective. If a doctor knows a patient's baseline cholinesterase level and then does a blood test and finds that it has dropped, it is conclusive evidence that the patient has been overexposed to an organophosphate or carbamate insecticide. One should wait until his cholinesterase level has restored to normal before coming into touch with these chemicals again. Antidotes can be administered medically in extreme circumstances.

Common antidotes

1) Universal antidote:- This mixture contains 7 g of activated charcoal, 3.5 g of magnesium oxide, and 3.5 g of tannic acid in half a glass of warm water. If poisoning is caused by corrosive substances, gastric lavage should be performed.

Charcoal - To bind the insecticide

Tannic acid – neutralizes

Magnesium oxide – To clean the gut

Warm water – Fast reaction

2) Specific antidote: - which varies with the toxicant group

Types of Antidotes

1. **Cholinesterase reactivators**: To counteract the effects of acetylcholinesterase inhibition, these drugs are given to patients. The 2-formyl-1-methylpyridinium ions found in oxime compounds are responsible for their efficacy. Pralidoxime, also known as Protopam or 2-PAM, is a nucleophilic substance that binds to the OP molecule and reactivates phosphorylated AChE. The sooner these antidotes are given to a patient who had muscular weakness or respiratory muscle weakness as a result of known or suspected OP exposure, the better the outcome is likely to be. In addition to 2-PAM, atropine can be given to counteract the effects of the OP toxin on the respiratory center and the muscarinic effects of AChE poisoning.

   With the addition of 2-PAM to the therapy regimen, atropinization symptoms might manifest sooner. Patients who have been poisoned with OP insecticides or nerve agents may experience worsening symptoms from the effects of barbiturates, which are potentiated by AChE inhibitors, antagonistic with neostigmine, pyridostigmine, and edrophonium, morphine, theophylline, aminophylline, succinylcholine, reserpine, and phenothiazines. Rapid injection can result in tachycardia, laryngospasm, and muscle rigidity, pain at the injection site, blurred vision, diplopia, impaired accommodation, dizziness, drowsiness, nausea, tachycardia, hypertension, and hyperventilation, as well as myasthenia crisis in myasthenia gravis patients and muscle rigidity in healthy volunteers. Since 2-PAM is eliminated in urine and can momentarily raise creatinine phosphokinase levels, SGOT and/or SGPT levels rise in 1 of 6 individuals, a loss in renal function raises serum medication levels.

2. **Anticholinergics:** Atropine is an anticholinergic that blocks excessive anticholinesterase activity at muscarinic receptors pharmacologically. Oximes undo nicotinic effects, such as muscular paralysis, and AChE inhibition.
3. **Atropine (Atropair):** In known or suspected OP or carbamate poisonings, this medication is used to treat GI (Gastrointestinal) or pulmonary discomfort. It should be kept up until the bronchoconstriction and bronchorrhea are under control. In extreme situations, high dosages may be necessary for the first 24 hours of therapy, and treatment may be required for 48 hours.
4. **Pralidoxime (2-PAM, Protopam):** As a result of OP poisoning, pralidoxime may be indicated for muscle weakness (especially respiratory weakness). When used early in poisoning, before the OP-AChE bond has aged, it is effective. The treatment may be beneficial in preventing intermediate and delayed neuromuscular and neuropsychiatric OP syndromes.
5. **Diazepam (Valium)**: There is some evidence that diazepam inhibits the activity of GABA at all levels of the central nervous system (eg. limbic and reticular formations). A potentiating effect is produced by phenothiazines, narcotics, barbiturates, MAOIs (monoamine oxidase inhibitors) and other antidepressants. Monitor respiratory depression with high or repeated doses, particularly if you are taking other CNS depressants, have low albumin levels, or have hepatic disease (may increase toxicity).
6. **Lorazepam (Ativan):** The dose of DOC may not exceed 2 mg/min and may be administered IM (Intramuscular) if IV (Intravenous) access is not possible because of its prolonged duration in the central nervous system. It contains benzyl alcohol, which can be toxic to infants in high doses, which may cause respiratory depression in high doses. Take caution if you have renal or hepatic impairments, myasthenia gravis, organic brain syndrome, Parkinson disease, or are unable to clear benzodiazepines (e.g., nicotine or cimetidine).

Methods of Administration of the Antidotes for Different Groups of Insecticides

A. Antidotes for Organochlorines

Assure that the pharynx and trachea have an unobstructed airway by suction if necessary. Anticonvulsants such as diazepam and paraldehyde, as well as soluble barbiturates, may be administered to control convulsions, Phenobarbital upto 0.7 gm per day or Pentobarbital @ 0.25 to 0.5 gm per day or 10% Calcium gluconate to be given intravenously.

It may become necessary to dialysis the kidneys or inject corticosteroids to protect the liver in extreme situations. Administer a low-fat, high-protein, high-carbohydrate, high-calcium diet. Since adrenaline derivatives can induce ventricular fibrillation, they are contraindicated. In the event of one or more convulsions, patients should be observed closely for at least 48 hours.

B. Antidotes for Organophosphates

A contraindication to these medications is morphine, phenothiazines, succinyl chloride, xanthene derivatives, epinephrine and barbiturates.

Glycopyrrolate: Similar to atropine in its peripheral actions, this is a quaternary ammonium antimuscarinic substance. It has a longer duration of action than other antidotes, but it cannot neutralise the poison's effects on the central nervous system since it does not cross the blood-brain barrier.

Atropine sulphate should be injected intravenously every 5-10 minutes until signs of atropinisation appear, such as a dry mouth and dilated pupils. Once the atropinisation is stopped, carefully observe the patient; if signs of poisoning return, it may be necessary to recommence treatment. Antidote therapy with atropine, a muscarinic antagonist, competes with ACh and effectively reverses muscarinic signs of toxicity. The initial dose for adults is 2 to 5 mg, administered intravenously (IV), and 0.05 mg/kg IV for children until they reach the adult dose. If the patient does not respond to the treatment, the dose can be doubled every 3 to 5 minutes until respiratory secretions clear and bronchoconstriction subsides. Notably, atropine does not reverse nicotinic symptoms such as muscle fasciculations and weakness.

Pralidoxime (2-PAM) is recommended as an AChE activator, with the potential to prevent the irreversible inactivation of the enzyme. The process of aging, in which the AChE enzyme is irreversibly inhibited, is a time-sensitive phenomenon that depends on the specific organophosphate agent. Administering 2-PAM before this aging process theoretically restores the enzyme and reverses toxicity. Maintaining full atropinization, administering an oxime and, if possible, a cholinesterase reactivator such as 2-PAM at 1000–2000 mg/kg body weight IM (Intra Muscular) or IV (Intravenous) for adults (25 mg/kg body weight for children) or Toxogonin

(Merck®) at 250 mg for adults (4–8 mg/kg body weight for children) can help with convulsions and anxiety. If necessary, repeat in another one to two hour.

Pralidoxime is administered as a loading dose of 30 mg/kg over 30 minutes, followed by a continuous infusion of 8 mg/kg/h for a maximum of 7 days. Caution should be exercised to avoid rapid administration, potentially resulting in cardiac arrest. To mitigate any exacerbation of muscarinic symptoms, it is advisable to administer atropine before pralidoxime. Controversy surrounds the efficacy of 2-PAM, with a study indicating no clear benefit and potential harm. Pending further research, the World Health Organization advises the combined use of both atropine and pralidoxime in treating organophosphate toxicity.

Management of OP poisoning in simple way

- Airway, breathing, circulation. Start NS infusion with aim for SBP > 80 mm/Hg & urine output > 0.5 ml/kg/hr. Foleys catherization.
- Injection Atropine 1 amp (0.6 mg) and assess every 5 minutes:

No improvement in parameters **Double the dose** (1-2-4-8-16..) **Give until:** 1. Sweating stops (dry axillae)- **Most reliable sign.** 2. Chest auscultation: Clear. (Focal areas of aspiration may persist). 3. Pulase > **80/min** 4. SBP > **80 mm/Hg** 5. Pupils dilated- **delayed sign**. (Very dilated pupils are a sign of atropine toxicity).	**Improvement begins** Similar or smaller dose continue

Start atropine infusion @ 10-20 % of the total dose of atropine given/hour in NS

Reduce the rate by 20 % every 6 hours once the patient is stable Stop

Too little atropine: Cholinergic features reappear after some time Up-titrate the dose and re atropinize if necessary.	**Too much atropine:** 1.Urinary retention 2. Absent bowel sounds 3. Stop infusion for 30-60 minutes, restart the infusion at a lower rate (80 % of the initial).

Next step give Pralidoxime chloride (2-PAM)

- **Loading dose:** 30 mg/kg (2 gm) over 20-30 min in NS.
- **Maintenance dose:** Infusion of 8 mg/kg /hr (500 mg/hr) until atropine is not needed for 12-24 hrs & patient has been stabilized.

Note: Atropine must be administered to patient before PAM to prevent the exacerbation of muscarinic mediated symptoms.

Adequate sedation with **benzodiazepines** to be given.

Amitraz: a mimicker of organophosphate poisoning. Amitraz not OP compound even though it shows organophosphate poisoning symptoms. A person with drowsiness and have constricted pupils, hypotension and bradycardia. Absence of a hyper secretory state, and the presence of hyperglycaemia and hypothermia along with a normal serum cholinesterase level suggested an alternate possibility. Chlordimeform, Amitraz poisoning cases use antidote **Atipamezole.**

C. Antidotes for Carbamates

Carbamates compounds should be treated with atropine. It is not recommended to administer oximes such as 2-PAM, P2S, and toxogonin. Medications such as morphine, phenothiazines, succinyl chloride, xanthene derivative, epinephrine, and barbiturates should also be avoided. It is possible to treat convulsions with diazepam (valium, Roche)

Additionally, **atropine** is used as part of the treatment. Because carbamates have a short duration of action, **PAM** is used only in cases where atropine fails to adequately treat carbamate poisoning.

D. Antidotes for Synthetic Pyrethroids

In severe skin exposure from handling or application, tingling, burning, or numbness may occur, especially on the face. These feelings will fade in a few hours. Treatment is symptomatic. If more is eaten, lavage the stomach. Use activated charcoal and saline cathartic with sodium sulphate, magnesium sulphate, magnesium citrate, sorbitol. Injectable diazepam or barbiturates can control seizures.

E. Antidote for Oxadiazines

The primary treatment for methemoglobinemia is **methylene blue**, which acts as an antidote. Here's how it works: **Methylene Blue**: It helps convert methemoglobin back to hemoglobin by acting as an electron donor, thereby restoring the ability of hemoglobin to carry oxygen. The typical dose for methylene blue is 1-2 mg/kg administered intravenously over 5-10 minutes. However, this treatment is typically reserved for cases where methemoglobin levels exceed 20% or if symptoms are severe.

Ascorbic Acid (Vitamin C): In milder cases, ascorbic acid can be used as a secondary treatment option because it has reducing properties that help convert methemoglobin to hemoglobin, though it acts more slowly than methylene blue.

F. Antidotes for Rodenticides

1. **Warfarins:** Warfarin compounds do not require gastric lavage after one dose. Serial prothrombin time measurements assess toxin toxicity. The patient can be discharged if it remains normal after 24 hours and there is

no clinical bleeding. If prothrombin time is extended, 10 mg of vitamin K1 (phylloquinone or phytonadione) is given intravenously up to 5 times a day. Vitamin K1 is essential. Neither vitamin K3 (menadione) nor vitamin K4 (menadiol) can reverse these anticoagulants.

2. **Aluminium phosphide:** Since there is no antidote for phosphine gas, the victim should be transported to an open location promptly. Otherwise, care for both inhalation and ingestion of aluminium phosphide is supportive. The goal is to sustain life until the lungs and kidneys naturally eliminate phosphine gas.

   Gastric lavage with potassium permanganate at 1:10000 dilution reduces gut toxin absorption. Infuse a lot of saline under central venous pressure or pulmonary artery wedge pressure to treat shock. Most patients need 4–6 L of fluids over 4–6 hours. If no response occurs, dopamine may be given. Administer intravenous sodium bicarbonate for metabolic acidosis. If the tablet is new and has not been exposed to the air, half a tablet of aluminium phosphide can kill.

3. **Zinc phosphide:** The majority of zinc phosphide poisoning treatment is symptomatic.

4. **Ethylene dibromide (EDB):** Multiple doses of activated charcoal may be necessary in cases of EDB poisoning, and stomach lavage is suggested if ingestion occurred within the previous two hours. Administering intravenous fluids to correct intravascular volume, maintaining oxygenation, and correcting acidosis are all crucial parts of providing the patient with the supportive care they need during treatment.

Antidotes for insecticides poisoning

S.No.	Antidote / Medicine	Used in poisoning due to
1.	Common salt (Sodium chloride)	General stomach poisoning
2.	Activated charcoal (7g) in warm water + Magnesium oxide (3.5g) + Tannic acid (3.5g) **(Universal antidote)**	General stomach poisoning
3.	Demulcents (Substances having Soothing Effect) Gelatin (9- 18 g in 570 ml of water) or Flour or milk or butter (or) cream (or) mashed potato (or) Sodium thiosulphate	General stomach poisoning
4.	Calcium gluconate (10%)	Insecticides containing chlorinated compounds (Organochlorides), ethylene dichloride, carbon tetrachloride, mercurial compounds
5.	Phenobarbital (or) Pentobarbital intravenous administration	Insecticides containing chlorinated hydrocarbons cause stomach poisoning
6.	Sodium bicarbonate	Organophosphate poisoning in the stomach

7.	Atropine sulphate (2-4 mg intramuscular / intravenous administration) or 2-PAM (2-Pyridine-Z aldoxime-N-methliodide)	Organophosphate Compounds
8.	Atropine sulphate (2-4 mg intramuscular / intravenous administration)	Carbamates
9.	Phenobarbital, activated charcoal and saline cathartic with sodium sulphate	Synthetic pyrethroids
10.	Potassium permanganate	Nicotine, Zinc phosphide
11.	Vitamin K_1 K_2 and K_3	Warfarin, Zinc phosphide
13.	Epinephrine	Methyl bromide
14.	Methyl nitrite ampule	Cyanides

If severe pesticide poisoning occurs, doctors may use hemoperfusion

Hemoperfusion

Hemoperfusion is a medical treatment method used to remove toxic substances, pesticides, or metabolic waste products from the blood. It involves passing blood through an external device containing an adsorbent material, typically activated charcoal or a resin that binds and removes harmful substances. This process is particularly effective in cases of poisoning, drug overdose, or when the kidneys and liver are unable to filter out toxins efficiently.

Key steps in hemoperfusion

1. **Blood extraction:** Blood is drawn from the patient and passed through an adsorbent filter in an extracorporeal circuit.
2. **Adsorption:** The toxic substances bind to the adsorbent material (e.g., activated charcoal or resin) within the hemoperfusion cartridge.
3. **Filtered blood return:** The cleaned blood is then returned to the patient's circulation.

Common applications

Overdose of poisoning: Especially for substances that are not easily dialyzable, such as barbiturates, theophylline, or organophosphate poisons. Hemoperfusion is often used in combination with hemodialysis in certain critical care settings, enhancing the removal of both large and small molecules of pesticides.

19

Pesticide Safety

Insecticides possess toxicity towards both pest organisms and human beings. Nevertheless, the potential harm to humans and non-target animal species can be mitigated by implementing appropriate safety measures. The ingestion or prolonged dermal exposure to most insecticide s can result in deleterious effects, whether intentional or accidental. Inhalation of insecticide particles may occur during the spraying process, and there is an associated risk of contamination of drinking water, food, or soil. The issue of safety is consistently a concern when utilizing insecticide s. Exposure to insecticide concentrates or vapour drift has the potential to cause harm to applicators, bystanders, and the surrounding environment. It is imperative to exercise specific precautions when engaging in the activities of transportation, storage, and handling. Individuals who are engaged in the handling and application of insecticide s are required to possess knowledge and adhere to appropriate safety protocols in order to mitigate potential risks. The foundation of ensuring insecticide safety lies in the selection of an appropriate product. The preservation of safety is of utmost significance in the various stages of insecticide management, including storage, transportation, mixing, and loading. The safe execution of equipment cleanup and maintenance is imperative, as it necessitates the proper disposal of unwanted insecticide s and empty insecticide containers. Regular cleaning and maintenance of spray equipment is essential in order to mitigate the occurrence of leaks. Furthermore, individuals who are involved in insecticide application should undergo comprehensive training to ensure their adeptness in handling these substances safely.

Specific precautions to be taken while use of insecticides

1) Selecting and Buying Insecticides

- The insecticide has been officially registered for its intended use.
- The insecticide may be applied using the appropriate equipment that is readily accessible.
- Adequate personal protective equipment (PPE) is readily available.
- The label clearly indicates that the insecticide has been approved for the intended use.

- The insecticide can be safely utilized within the specific site conditions, with minimal impact on non-target organisms, human health, and the environment.
- The insecticide can be incorporated into an integrated pest management program.
- The required quantity of the product has been accurately calculated.
- The restrictions specified on the label are well understood.

2) Label

Packaging and labelling of insecticide s must adhere to World Health Organization guidelines (1). Contents, precautions, and first aid procedures in case of ingestion or contamination should all be clearly labelled in both English and the local language. Insecticide s must be stored in their original packaging at all times. Follow all advised safety procedures, including the use of protective gear.

Instructions and Warnings on the Label:

The insecticide label provides details regarding the requisite personal protective equipment (PPE) for the safe handling of a insecticide. The aforementioned information is situated within the precautionary statement section on the secondary panel of the label. It is imperative to consistently adhere to the instructions provided on product labels.

It's possible that a label won't mention every item of PPE that's required. It might simply state the protection that is required. Labels may contain words that alert consumers to the possibility of a risk (for example, "avoid breathing dust or fumes," "avoid skin contact," or "keep product out of the eyes") in the event that the hazard actually occurs. Consider each of these points before settling on the protective gear that will be worn.

The Material Safety Data Sheet (MSDS) or a insecticide company representative can provide additional information regarding personal protective equipment (PPE) for a specific insecticide. Other sources of PPE-related information include:

- Insecticide pamphlets
- Safety equipment vendors
- Applicator handbooks
- Training personnel

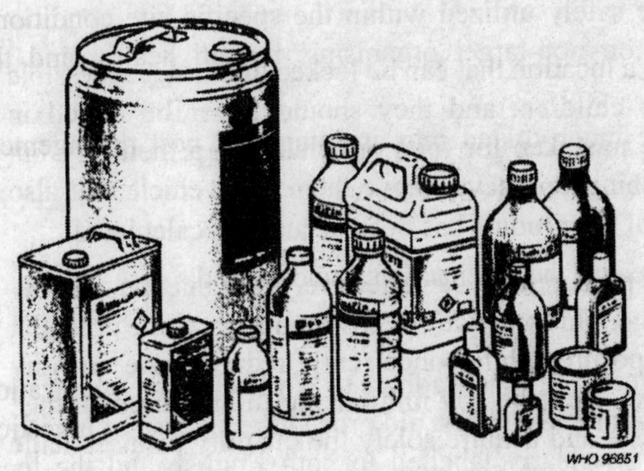

Types of insecticide containers

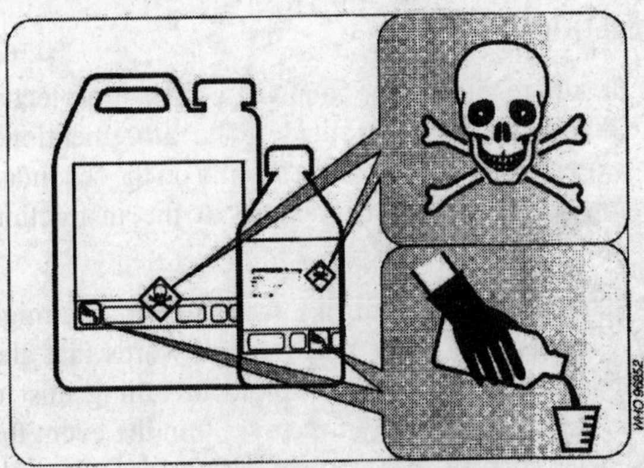

Look for warning symbols, pictograms and colour coding on labels

Difference between label and leaflet

Aspect	Label	Leaflet
Placement	Attached directly to insecticide container	Accompanies the insecticide product
Contents	Primary source of detailed information	Supplemental information
Information	Brand name and product details	Additional application procedures
	Active ingredients and concentrations	Safety precautions and protective equipment
	Directions for use (rates, timing, methods	Environmental considerations
	Precautionary statements and hazards	Information on hazards and emergencies
	Storage and disposal instructions	Regulatory compliance information

3) Storage and transport

Insecticide s should be stored in a location that can be locked and is not accessible to unauthorized individuals or children, and they should never be stored in a location where they could be mistaken for food or drink. Keep them dry but away from fires and direct sunshine, and never move them in a vehicle that also transports food.

Strategic planning of insecticide procurement can effectively mitigate the quantity and duration of insecticide storage. This practice effectively mitigates the potential for human or environmental exposure, while concurrently reducing the quantity of insecticide present on the premises, thereby minimizing the required storage capacity. It is advisable to procure and acquire solely the quantity of insecticide that can be effectively utilized within a brief timeframe, or for a single application.

Keep insecticide s out of reach to children

A storage facility should be:

- Set apart from the workplace, housing quarters, and animal enclosures
- Keep a safe distance from water bodies, wells, and ditches
- Keep away from regions with porous soil and those that are prone to flooding.
- Away from locations frequented by the general public, children, and animals
- Ideally situated 50 meters from residences, hospitals, schools, and occupied structures.
- It is reachable by road for rescue workers

Securely packing insecticide containers will prevent spills and breaks during transit. Be cautious when transporting liquid insecticide s, as a spill could contaminate the rest of the load and the vehicle itself. Also, make sure the containers are packed safely to lessen the likelihood of a leak. In the event of a spill, it is important to clean the vehicle and dispose of or return the contaminated items to the manufacturer.

Never bring insecticide s into a vehicle's passenger area, and never allow humans or animals to ride alongside chemicals (even in the rear of a truck). Do not combine the transfer of insecticide s with consumables for humans or animals, feed for livestock, fertilizer, clothing, or other household items. Spills can lead to poisoning or vehicle pollution, and toxic vapors might be emitted. Do not use a wooden truck bed to transport insecticide s since the insecticide s will soak into the wood and may contaminate subsequent cargoes. Put your insecticide s in a sturdy plastic or metal container. Alternatively, you might use a waterproof sheet to protect the insecticide canisters. Protective gear and spill clean-up equipment (such as a shovel and chemical neutralizer) should be kept in the vehicle at all times, and any storage boxes or bulk insecticide containers being transported should be fastened securely to the vehicle.

4) Disposal

The safest way to get rid of any leftover insecticide suspension is to pour it into a hole dug in the ground or a pit latrine. Some insecticides, including pyrethroids, are particularly hazardous to fish and should not be disposed of where they could end up in drinking water, washing water, fishponds, or rivers. If you live in a mountainous location, dig your hole on the lower side of the hills and at least 100 meters away from any streams, wells, or homes. Fill the pit with water from hand washings and spray washings, bury any insecticide storage containers, boxes, or bottles you find, and cover the hole as soon as possible. Burning cardboard, paper, and clean plastic containers is acceptable in areas remote from homes and water supplies.

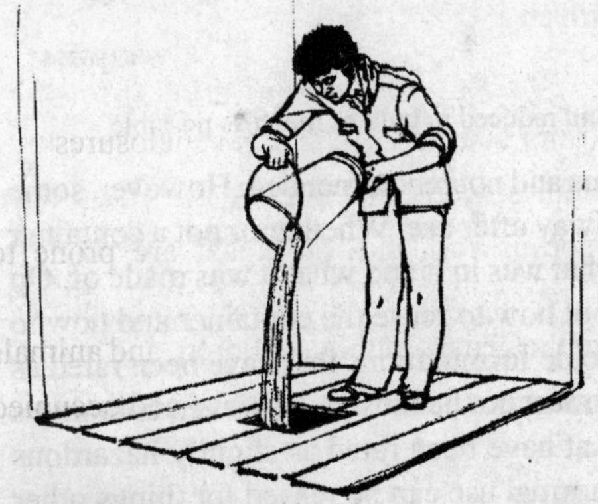

Surplus insecticide solution can be disposed of safely by pouring it into a pit latrine or a specially dug hole in the ground.

Clean paper and cardboard and cleaned plastic containers (not PVC) may be burnt

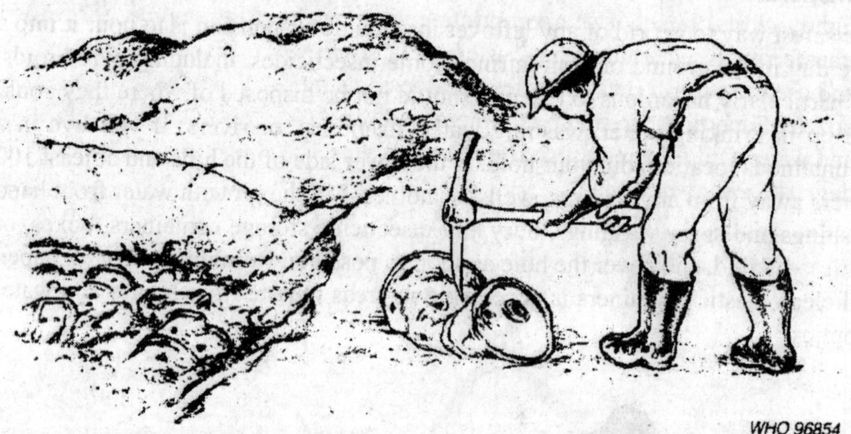

Packages to be buried must be made unusable and reduced in bulk as much as possible

Reusing insecticide containers is dangerous and not recommended. However, some containers may be too valuable to throw away after use. Whether or not a container can be cleaned and used again relies on what was in it and what it was made of. On the label, there should be information about how to reuse the container and how to clean it. Containers that have held insecticide formulations that have been rated as highly hazardous or extremely hazardous must not be reused. However, containers that have held insecticide formulations that have been rated as slightly hazardous or unlikely to present an acute hazard in normal use can be reused for things other than storing food, drink, or animal feed. No matter how they are made, containers made of materials like polyethylene that particularly absorb insecticide s shouldn't be used again if they have held insecticide s whose active ingredient is rated as moderately, highly, or extremely dangerous. As soon as an insecticide bottle is empty, it should be rinsed with water, filled to the top, and left alone for 24 hours. Then they should be dumped out, and the process should be done twice more. Pyrethroid suspensions can be dumped on dry ground, where they are quickly taken and broken down. This doesn't harm the environment. Insect pests such as ants and cockroaches can be killed with extra fluid. Place it under kitchen sinks or in parts of the house where bugs are hiding.

You can briefly stop insects from breeding by pouring the solution in and around toilets and other places where insects like to live. After being made, pyrethroid solutions that are used to treat mosquito nets and other materials can be used for a few days. The solution can also be used to treat sleeping mats or string pillows to keep mosquitoes from biting from below. There are ways to treat beds where bedbugs are a problem.

5) General hygiene

When using insecticide s, do not eat, drink, or smoke, and always keep food in tightly closed boxes. Use appropriate equipment for weighing out, mixing, and transferring insecticides. Do not use bare hands to stir liquids or scoop insecticide, and use the pump's pressure release valve or a soft probe to clear blockages in the nozzle. Each time the pump is replenished, wash your hands and face with soap and water. Only eat and drink after cleaning your hands and face. At the end of the day, take a shower or bath.

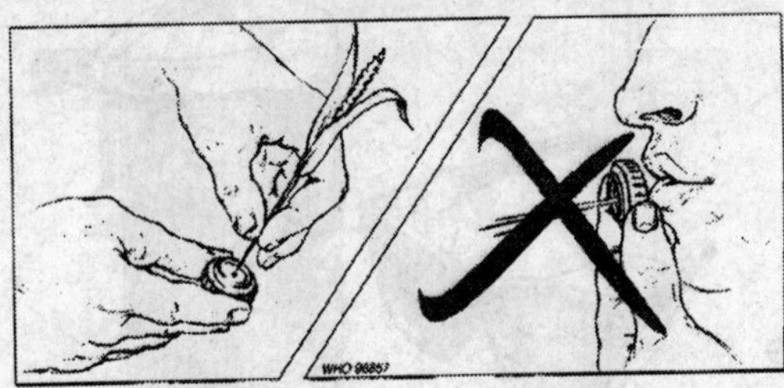

Clean blocked nozzles with soft probe

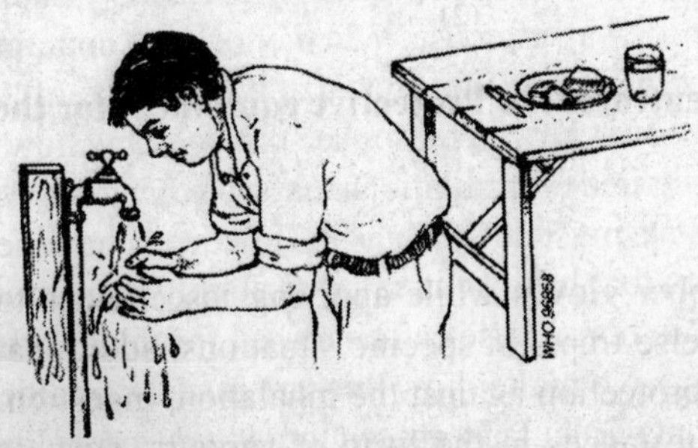

Wash the hands and face before eating or drinking

6) Protective clothing

Proper protective clothing should be ensured at the time of spraying to keep safe from the insecticide s from different insecticide poisonings and ill effects. Following are some of the ways to keep safe.

7) Spraying indoors

Spray workers should wear overalls or long-sleeved shirts and pants, a broad-brimmed hat, turban, or other headgear, sturdy shoes or boots, and no sandals. A disposable paper mask, surgical mask, or clean cotton should be used to cover the mouth and nose. Cotton clothes should cover the body without openings and be easy to wash and dry. Insecticide s should be sprayed at cooler hours in hot and humid settings since extra protective clothes may be uncomfortable.

8) Mixing

When mixing and putting insecticide s in bags, specific safety measures must be taken. Gloves, an apron, and eye protection such a face shield or goggles are advised to be used in addition to the safety gear mentioned above. These face shields offer comprehensive facial protection and offer enhanced comfort during wear. It is advisable to ensure that the mouth and nose are adequately covered, as per the prescribed guidelines for indoor spraying. Additionally, it is important to exercise caution and avoid from making contact between the gloves used for insecticide handling and any part of the body.

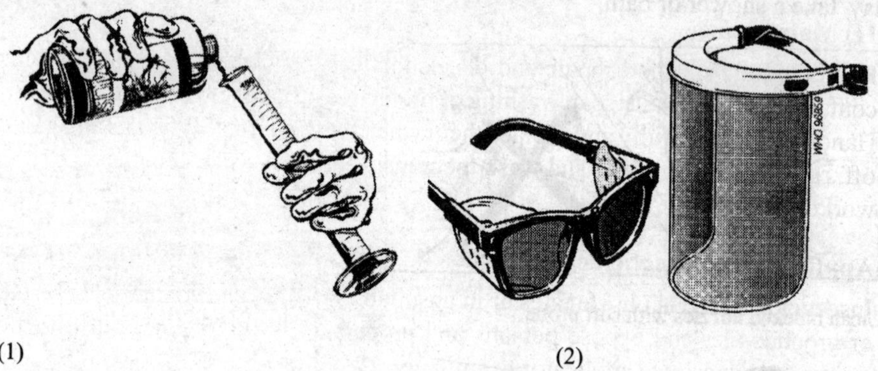

(1) (2)

(1) Wear gloves when handling concentrates (1); Protective equipment for the eyes and face (2)

9) Impregnation of fabrics

It is recommended to utilise long rubber gloves while applying insecticides to mosquito nets, clothing, screens, or tsetse traps. In specific situations, additional safeguards may be necessary, such as protection against the inhalation, ingestion, or contact with potentially harmful substances in the form of vapours, dust, or spray. The inclusion of supplementary safety equipment, such as aprons, boots, face masks, dungarees and helmets, must to be clearly specified on the product label.

10) Personal Protective Equipment (PPE)

The utilisation of personal protection equipment (PPE) serves to mitigate the applicator's potential exposure to insecticide s. In order to provide optimal protection, the selection of personal protective equipment (PPE) should be guided by the information provided on the insecticide label. The durability of this equipment should be sufficient to withstand the demanding conditions of labour and prolonged exposure to insecticide s. It is imperative for a insecticide applicator to possess knowledge and proficiency in the proper fitting, utilisation, cleansing, and upkeep of all personal protective equipment (PPE). It is imperative that this equipment be exclusively utilised for its designated purpose and not be employed for any alternative tasks.

Personal protective equipment (PPE) required for a certain insecticide is determined by the danger associated with its handling. Among the risk factors are

- Insecticide properties
- Type of exposure
- Length of exposure
- Application method

11) Maintenance

Keep clothing in good repair and check for tears or damaged areas that could contaminate skin. Daily soap washing of protective gear should be done separately. Handle gloves carefully and replace them when they wear out. When taking gloves off, rinse them with water and wash them inside and out at the conclusion of each workday.

Application of insecticide s on farm

Insecticide s used at unsafe rates or in unsuitable weather can contaminate surface or groundwater and expose persons and property. Failure to handle insecticides safely can expose the applicator to spillage.

Before and during a insecticide application, applicators should practice the following:

- Read the label
- Wear proper personal protective equipment (PPE)
- Use clean water
- Prevent contamination

- Mix and apply safely
- Never work alone
- Calibrate equipment
- Plan the insecticide application
- Apply granular insecticide s correctly
- Repair equipment malfunctions safely

Overall general precautions

- Safe insecticide handling practices include:
- Reading and following label information and directions
- Wearing clean personal protective equipment (PPE)
- Removing contact lenses before handling insecticide s
- Washing before eating, drinking, smoking, or using the toilet
- Not having food or smoking products on your body when handling insecticide s
- Never eating, drinking or smoking when handling insecticide s
- Immediately, washing any spilled insecticide off an affected person and removing contaminated clothing
- Showering and washing hair and cleaning under the fingernails at the end of each day

Insecticide Drift

Insecticide drift has the potential to cause damage to crops, landscape ornamentals, and lawns that are not the intended targets. People's health can be put at risk, and the environment (including water, wildlife, and aquatic habitat, among other things) can be negatively affected.

To reduce insecticide drift

- Use the lowest possible application rate.
- Deliver the spray as close to the target area as possible.
- Select the slowest speed possible for motorized application equipment.
- Use nozzles that reduce fine droplets (e.g., large nozzle orifice, low pressure).
- Do not apply when wind speeds exceed provincial regulatory requirements or the label maximum.
- Do not apply insecticide s when the temperature exceeds 25^0 C as the release of vapours by insecticide s increases as the temperature rises.

Container Cleaning and Disposal

When a container is emptied, it should be cleaned, since this will eliminate any insecticide residues before they have a chance to dry. When emptying a insecticide container:

- For liquids, drain the insecticide into the spray tank or mixing tank until no drips are visible.
- For solids, gently shake the bag into the tank or hopper until no loose insecticide remains.
- Triple rinse or pressure rinse metal, plastic, or glass containers, unless otherwise indicated on the label.
- Gently rinse bags if possible, unless otherwise indicated on the label.

Insecticide Disposal

Disposal of Concentrated Insecticide

When you plan your purchases of insecticide s, you may reduce the amount of extra insecticide concentrates that are left over after an application or use season, and you can analyse records of previous applications. Before purchasing more insecticide, use what you already have on hand, and check with the maker of the insecticide or a local supplier to make sure that older stocks are not rendered ineffective by time. Utilising insecticide concentrates in accordance with the advice on the label is the most secure method to dispose of them; nevertheless, if this is not an option for you, unopened containers may sometimes be returned to the manufacturer or a local dealer. Applicators also have the option of contacting the provincial organisation responsible for regulating insecticide s to inquire about the correct disposal of unused insecticide s.

Disposal of Surplus Tank Mixture

Applicators can avoid large amounts of surplus tank mixture before mixing by:

- Accurately measuring the area to be treated
- Confirming application rates
- Calibrating application equipment

If there is tank mix remaining at the end of an application, use it according to label directions on another area that requires the same insecticide.

Re-entry into applied field

If they enter treated areas too soon after an application without PPE, humans risk insecticide exposure, which can be as high as during application.

If insecticide labels do not list re-entry times, use provincial rules or wait until liquid insecticide s dry to minimise exposure. Observe the re-entry period before entering a treated area without suitable PPE. If you need to re-enter before the time has passed, wear proper PPE and post signs warning others.

Important points for consideration while using insecticide s

A) Calculate the right dose of insecticide based on insecticide formulations: Always calculate the proper insecticide, herbicide, or fungicide dose before applying or buying insecticide s. This would prevent farmers from buying extra volume of insecticide s. This book describes several insecticide dose calculation methods with examples in an earlier chapter.

B) Insecticide's shelf life: A insecticide's shelf life is the length of time it may be used or stored before losing quality or becoming unusable even if properly stored. Each insecticide has a shelf life, which is listed on its label. The expired insecticide should never be used for pest management since it may lose its activity and become phytotoxic to the crop owing to chemical changes.

C) Waiting period for insecticide s: The "waiting period" is the amount of time that must pass between the last treatment of an insecticide and when the crop can be picked and eaten. It varies by insecticide, crop, and location depending on residual toxicity and environmental factors. Insecticide leaflets usually include crop information. Always wait 15 days before using generally advised insecticides.

D) **Follow the compatibility chart:** Compatibility means if two chemicals can be mixed without affecting toxicity. Farmers sometimes blend insecticide and fungicide or insecticide with fertilisers like urea to manage insects and illnesses. The leaflet's suggestions should be checked for chemical compatibility to minimise crop damage from incompatible combination products' phytotoxic effects.

E) **Insecticide application equipments:** Insecticide s can be applied with a wide variety of tools, some of which are specific to the pest or the treatment area. For effective pest management, it is crucial to use the appropriate insecticide application equipment. Hand sprayers and atomizers (for smaller kitchen gardens), hand compression and knapsack sprayers, hand-carried dusters and granule applicators (for small and marginal farming), motorised knapsack mist blowers and various power sprayers, dusters (for large-scale pest control or spraying on taller crops/trees), ultra low volume or controlled-droplet applicators (for specially designed Keep a separate sprayer for herbicides and label it "WEEDS ONLY" to avoid damaging plants if the same sprayer is used for other insecticide s or fertilisers. If it's not possible, wash the sprayer three times with clean water to remove chemical insecticide residue. Proper nozzles are crucial to bio-efficacy results. Farmers often use the same sprayer nozzle for insecticides, fungicides, and herbicides. This causes poor pest control by under-covering the intended area with insecticide. Jet nozzles are best for herbicides and hollow cone nozzles for insecticides and fungicides.

F) **Minimizing Environmental Contamination:** If the pest is not evenly distributed, use spot treatments and avoid insecticide applications in the entire crop. Spot treatments call for the insecticide to be mixed per the directions on the label and then applied directly to the afflicted area. Some herbicides should be applied using wick or shielded applicators to prevent contamination of the surrounding area. You can also treat tree trunks to get rid of particular insects and set traps for rats and ants.

Refere[illegible]

Abivardi [illegible]

Ambrus, A. [illegible]

Annis, G.D. [illegible]

Banerjee [illegible]

Bartlett [illegible]

Bryant [illegible]

Cordova [illegible]

Dass [illegible]

David [illegible]

David [illegible]

Dhaliwal [illegible]

Dhaliwal [illegible]

Dhingra [illegible]

Dhingra [illegible]

Douglas [illegible]

Dudwal [illegible]

References

Abivardi C, Weber DC, and Dorn S. 1998. Effects of azinphos-methyl and pyrifenox on reproductive performance of Cydia pomonella L. (Lepidoptera: Tortricidae) at recommended rates and lower concentrations. Annals Appl Biol 132:19-33.

Ambrus, A., & Yang, Y. Z. (2015). Global harmonization of maximum residue limits for pesticides. Journal of agricultural and food chemistry, 64(1), 30-35.

Annis, G.D., Barnette, W.E., McCann, S.F., Wing, K.D., 1992. Arthropicidal carboxanilides. International Patent WO 9211248.

Banerjee, T., & Agyani, M. K. (2019). Risk assessment and persistence evaluation of Spiromesifen 240 SC on cucumber in India. International Journal of Postharvest Technology and Innovation, 6(1), 11-25.

Bartlett and Ewart, 1951, Insecticide-induced outbreaks of walnut brown soft scale Coccus hesperidium. J. Econ. Entomol. 43: 819-831.

Bryan E. (2017). Denham. Probit Analysis, Wiley.

Chelliah, S. and Heinrichs, E. A, 1980, Factors affecting insecticide induced resurgence of the brown planthopper, Nilaparvata lugens (Stal) on rice. Environ. Entomol. 9: 773-777.

Cordova, D., Benner, E. A., Sacher, M. D., Rauh, J. J., Sopa, J. S., Lahm, G. P., ... & Tao, Y. (2006). Anthranilic diamides: a new class of insecticides with a novel mode of action, ryanodine receptor activation. Pesticide Biochemistry and Physiology, 84(3), 196-214.

Cordova, D., Benner, E. A., Sacher, M. D., Rauh, J. J., Sopa, J. S., Lahm, G. P., ... & Tao, Y. (2006). Anthranilic diamides: a new class of insecticides with a novel mode of action, ryanodine receptor activation. Pesticide Biochemistry and Physiology, 84(3), 196-214.

Cutler GC, Ramanaidu K, Astatkie T, and Isman MB. 2009. Green peach aphid, Myzus persicae (Hemiptera: Aphididae), reproduction during exposure to sublethal concentrations of imidacloprid and azadirachtin. Pest Manag Sci 65:205-209.

Das, S. (2004). Nuclear magnetic resonance spectroscopy. Resonance, 9(1), 34-49.

David, H and Easwaramoorthy, S., 1987, Resurgence of sucking pests in sugarcane. In: Resurgence of sucking pests-Proc. Nat. Symp. (Ed.) S. Jayaraj, Tamil Nasu Agric. Univ., Coimbatore pp: 29-38.

David Lea. (2003). Agricultural and Mineral Commodities Year Book", Europa Publications (A member of the Taylor & Francis Group).

Dhaliwal, G. S. and Arora, R., 2006, Integrated pest management: concepts and approaches. Entomol. 48, 157-161.

Dhooria, S., Behera, D., & Agarwal, R. (2015). Amitraz: a mimicker of organophosphate poisoning. Case Reports, 2015, bcr2015210296.

Dileep Kumar A D, Narasimha Reddy, D. (2021). The Insecticide Management Bill 2020. Current science 121(3): 348-349. https://www.currentscience.ac.in/Volumes/121/03/0348.pdf.

Directorate of Plant Protection, Quarantine & Storage (DPPQS); https://ppqs.gov.in/. (Accessed on 20.11.2023).

Douglas, A. E., 2006, Phloem-sap feeding by animals: problems and solutions. J Exp. Bot., 57(4): 747–754.

Dudwal, R., Jakhar, B. L., Pathan, A. R. K., Jan, I., Kakralya, B. L., Dhaka, S. R., ... & Choudhary, S. K. (2022). Dissipation kinetics, risk assessment, and waiting period of spiromesifen on chili fruits using gas chromatography–electron capture detector. Biomedical Chromatography, 37(4), e5577.

Ensley, S. M. (2018). Pyrethrins and pyrethroids. In Veterinary toxicology (pp. 515-520). Academic Press.

Ernest Hodgson, Ronald J. Kuhr. (2020). Safer Insecticides - Development and Use, Marcel dekker, INC.

Farnsworth, W. R. (1997). Treated filter paper bioassay for detection of insecticide resistance in Haematobia irritans exigua de Meijere (Diptera: Muscidae). Australian Journal of Entomology, 36(1), 69-73.www.bayercropscience.com.

Federation of Indian Chambers of Commerce and Industry (FICCI), Study on Sub - Standard, Spurious/Counterfeit Insecticides in India (2015) - Report. Available from: https://croplife.org/wp-content/uploads/2015/10/Study-on-sub-standard-spurious-counterfeit-insecticides-in-India.pdf.

Finney, D. J., & Stevens, W. L. (1948). A table for the calculation of working probits and weights in probit analysis. Biometrika, 35(1/2), 191-201.

Forgash, A. J. (1984). History, evolution, and consequences of insecticide resistance. Pesticide biochemistry and physiology, 22(2), 178-186.

Fujiwara Y, Takahashi T, Yoshioka T, and Nakasuji F. 2002. Changes in egg size of the diamondback moth Plutella xylostella (Lepidoptera: Yponomeutidae) treated with fenvalerate at sublethal doses and viability of the eggs. Appl Entomol and Zool 37:103-109.

Gallardo, E., & Barroso, M. (Eds.). (2022). Pesticide Toxicology. Humana Press.

Grinberg Nelu, Rodriguez Sonia. (2019). Ewing's Analytical Instrumentation Handbook, Fourth Edition", CRC Press.

Guest, M., Kriek, N., & Flemming, A. J. (2020). Studies of an insecticidal inhibitor of acetyl-CoA carboxylase in the nematode C. elegans. Pesticide Biochemistry and Physiology, 169, 104604.

Guidance document on analytical quality control and method validation procedures for pesticide residues analysts in food and feed. Document No. SANTE/11312/2021 V2.

Hajek, A. E., 2004, Natural enemies: an introduction to biological control. Cambridge University Press, Cambridge, UK.

Hamilton, D., Ambrus, A., Dieterle, R., Felsot, A., Harris, C., Petersen, B. ... & Bhula, R. (2004). Pesticide residues in food-acute dietary exposure. Pest Management Science: formerly Pesticide Science, 60(4), 311-339.

Hardin, M. R., Benrey, B., Coli, M., Lamp, W. O., Roderick, G. K. and Barbosa, P., 1995, Arthropod pest resurgence: an overview of potential mechanisms. Crop Protection, 14, 3- 18.

Harischandra Naik R, Pavankumar K, Pallavi M S, Bheemanna M, Udaykumar Nidoni R. "Simultaneous determination of 34 chemical pesticides in red chili using gas chromatography-tandem mass spectrometer (2021). Separation science plus.

Heinrichs, E. A., and G. B. Aquino, S. Chelliah, S. L. Valencia, and W. H. Reissig., 1982, Resurgence of Nilaparvata lugens (Stål) populations as influenced by methods and timingof insecticide application in lowland rice. Environ. Entomol. 11:78-84.

Henderson, C. F., and Tilton, E. W., 1955, Tests with acaricides against wheat mites. J. Econ.

Holland, P. T. (1996). Pesticides report 36. Glossary of terms relating to pesticides (IUPAC Recommendations 1996), Pure and Applied Chemistry.

Horacio Heinzen, Leo M.L. Nollet, Amadeo R. Fernández-Alba. (2017) Multiresidue Methods for the Analysis of Pesticide Residues in Food", Routledge.

http://cibrc.nic.in/.

Indian Standard insecticide - Cartap Hydrochloride G specification, IS 14184: 1994, Availablefrom:https://standardsbis.bsbedge.com/BIS_SearchStandard.aspx?Standard_Number=14184&id=0. (Accessed on 15.11.2023).

Indian Standard Methods of Test for Insecticides and Their Formulations, IS 6940: 1982, https://standardsbis.bsbedge.com/BIS_SearchStandard.aspx?Standard_Number=6940&id=0. (Accessed on 20.11.2023).

Indian Standard of Acetamiprid, Soluble powder Specification, IS 16328: 2017, Available from:https://standardsbis.bsbedge.com/BIS_SearchStandard.aspx?Standard_Number=16328&id=0.(Accessed on 20.11.2023).

Indian Standard of Chlorpyrifos, Emulsifiable concentrate Specification, IS 8944: 2005, Available from:https://standardsbis.bsbedge.com/BIS_SearchStandard.aspx?Standard_Number=8944&id=0. (Accessed on 20.11.2023).

Indian Standard of Fenpropathrin Technical Specification, IS 15161: 2002, Available from:https://standardsbis.bsbedge.com/BIS_SearchStandard.aspx?Standard_Number=IS+15161&id=0. (Accessed on 20.11.2023).

Indian Standard of Fenvalerate, Emulsifiable concentrate Specification, IS 11997: 1987, Availablefrom:https://standardsbis.bsbedge.com/BIS_SearchStandard.aspx?Standard_Number=11997&id=0. (Accessed on 20.11.2023).

Indian Standard of Glyphosate Technical Specification, IS 12502: 1988, Available from: https://standardsbis.bsbedge.com/BIS_SearchStandard.aspx?Standard_Number=11997&id=0. (Accessed on 18.11.2023).

Indian Standard of Imidacloprid, Technical Specification, IS 15443: 2004, Available from:https://standardsbis.bsbedge.com/BIS_SearchStandard.aspx?Standard_Number=15443&id=0. (Accessed on 18.11.2023).

Indian Standard specification for Ethion Emulsifiable concentrates, IS 10319: 1982, Availablefrom: https://standardsbis.bsbedge.com/BIS_SearchStandard.aspx?Standard_Number=IS+10319&id=0 (Accessed on 20.11.2023).

International Encyclopedia of Statistical Science (2011).

International Rice Research Institute, 1969. Annual Report for 1968. Los Banos, Phllippines, p: 402.

International Rice Research Institute, 1971. Annual Report for 1968. Los Banos, Phllippines, p: 402.

Jack Cazes. (2019). Analytical Instrumentation Handbook, CRC Press, 2019.

Jayaraj, S. and Regupathy, A., 1987, Studies on the resurgence of sucking pests of crops in Tamil Nadu. In: S. Jayaraj (ed.) - Proc. Nation. Sympo. on resurgence of sucking pests. Tamil Nadu Agric, Univ., Coimbatore, India, pp. 225-240.

Jayaswal, A p. and Singh, O. P., 1987, Resurgence of whitefly on cotton and its management- Proc. Nat. Symp. (Ed.) S. Jayaraj, Tamil Nasu Agric. Univ., Coimbatore pp: 78-84.

Jeschke, P. (2016). Propesticides and their use as agrochemicals. Pest Management Science, 72(2), 210-225.

Jeschke, P. (2024). Recent developments in fluorine-containing pesticides. Pest Management Science, 80(7), 3065-3087.

John Kenkel. (2019). Analytical Chemistry for Technicians, CRC Press.

Johnson, D. E., & Anderson, G. L. (2014). Green Chemistry and Toxicology. Green Chemistry and Engineering: A Pathway to Sustainability, 325-353.

Jose L. Tadeo, Jose L. Tadeo. (2019). Analysis of Pesticides in Food and Environmental Samples, CRC Press.

Karam, Nilima. (2020). Investigations on the Bio- Efficacy of a Ryanodine Receptor Agonist in Controlling Two Lepidopteran Borer Pests and Its Impact on Non Target Organisms, Bidhan Chandra Krishi Viswavidyalaya University (India).

Kato, K., Kiyonaka, S., Sawaguchi, Y., Tohnishi, M., Masaki, T., Yasokawa, N., Mizuno, Y., Mori, E., Inoue, K., Hamachi, I., Takeshima, H., and 1424 34 Mori, Y. (2009) Molecular characterization of flubendiamide sensitivity in the lepidopterous ryanodine receptor Ca^{2+} release channel. Biochemistry, 48, 10342.

Kobayashi, T. 1961. The effect of insecticide applications to the rice stem borer on the leaffolder population. Sprc .Rep. Pest. Minist. Agric. Forest., 2: 1-126.

Lakshman Chandra Patel. (2023). Applied Entomology - Insect Ecology and Integrated Pest Management", CRC Press.

Leo M.L. Nollet, Hamir S. Rathore. (2019). Handbook of Pesticides - Methods of Pesticide Residues Analysis", CRC Press.

Lewis, K.A., Tzilivakis, J., Warner, D. and Green, A. (2016) An international database for pesticide risk assessments and management. Human and Ecological Risk Assessment: An International Journal, 22(4), 1050-1064.

Li, R., Pan, X., Wang, Q., Tao, Y., Chen, Z., Jiang, D. ... & Zheng, Y. (2019). Development of S-fluxametamide for bioactivity improvement and risk reduction: systemic evaluation of the novel insecticide fluxametamide at the enantiomeric level. Environmental science & technology, 53(23), 13657-13665.

Manual on the development and use of FAO and WHO specifications for chemical pesticides. (2022). Food and Agriculture Organization of the United Nations (FAO).

Martin, T., Ochou, O. G., Vaissayre, M., & Fournier, D. (2003). Organophosphorus insecticides synergize pyrethroids in the resistant strain of cotton bollworm, Helicoverpa armigera (Hübner) (Lepidoptera: Noctuidae) from West Africa. Journal of economic entomology, 96(2), 468-474.

Mary Louise Flint, Robert van den Bosch. (1981). Chapter 4 A History of Pest Control", Springer Science and Business Media LLC.

McCann, S.F., Annis, G.D., Shapiro, R., Piotrowski, D.W., Lahm, G.P., Long, J.K., Lee, K.C., Hughes, M.M., Myers, B.J., Griswold, S.M., Reeves, B.M., March, R.W., Sharpe, P.L., Lowder, P., Barnette, W.E., Wing, K.D., 2001. The discovery of indoxacarb: oxadiazines as a new class of pyrazoline-type insecticides. Pest Manage. Sci. 57, 153–164.

Metcalf, R. L., 1986, The ecology of insecticides and the chemical control of insects. In: Ecological Theory and integrated pest management practice (Ed. Marcos kogan), pp. 251- 298.

Mohan, S and Jayaraj, S. and Rangarajan, A. V., 1987, Influence of fenvalerate in inducing resurgence of sorghum aphid, Rhopalosiphum maidis- Proc. Nat. Symp. (Ed.) S. Jayaraj, Tamil Nadu Agric. Univ., Coimbatore pp: 40-38.

Morse JG and Zareh N. 1991. Pesticide-induced hormoligosis of citrus thrips (Thysanoptera: Thripidae) fecundity. J Econ Entomol 84:1169-1174.

Natarajan, K., Sundaramurthy, V. T. and Chidambaram, P., 1987, Pyrethroid induced aphid resurgence in the cotton eco system- Proc. Nat. Symp. (Ed.) S. Jayaraj, Tamil Nadu Agric. Univ., Coimbatore pp: 155-159.

Patil, B. V., 1987, Studies on the resurgence of spider mites on cotton- Proc. Nat. Symp. (Ed. S. Jayaraj, Tamil Nadu Agric. Univ., Coimbatore pp: 181-183.

Pavlidi, N., Vontas, J., & Van Leeuwen, T. (2018). The role of glutathione S-transferases (GSTs) in insecticide resistance in crop pests and disease vectors. Current opinion in insect science, 27, 97-102.

Pedigo, L. P. and Rice M. E., 2006, Entomology and Pest Management. 5th Ed. Pearson, Prentice Hall, Upper Saddle River, NJ.

Peña-Fernández, A., Evans, M. D., & Cooke, M. S. (Eds.). (2021). Toxicology for the health and pharmaceutical sciences. CRC Press.

Ranjan Kaushik, Ankit Kumar, Rekha Phogat, Rakesh Gehlot, Neha Rani. (2024). Chapter 21 Principles of Food Analysis and Food Laws", Springer Science and Business Media LLC.

Rehman, H., Aziz, A. T., Saggu, S., Abbas, Z. K., Anand Mohan, Ansari, A. A., 2014, Systematic review on pyrethroid toxicity with special reference to deltamethrin. J. Entomol. Zool. Stud., 2(6): 60-70.

Ripper, W. E., 1956, Effect of pesticides on balance of arthropod populations. Annu. Rev. Entomol., 1: 403-438.

Safety Evaluation of Pesticide Residues in Food. (1997). Chemistry International - Newsmagazine for IUPAC.

Selby, T. P., Lahm, G. P., & Stevenson, T. M. (2016). A retrospective look at anthranilic diamide insecticides: discovery and lead optimization to chlorantraniliprole and cyantraniliprole. Pest management science, 73(4), 658-665.

Singh, J. P., Lather, B. P. S. and Banergee, S. K., 1987, Effect of date of sowing on thepopulation of cotton leafhopper, Amrasca biguttula biguttula (Ihida) - Proc. Nat. Symp. (Ed.) S. Jayaraj, Tamil Nadu Agric. Univ., Coimbatore pp: 202-204.

Spalthoff, C., Salgado, V. L., Balu, N., David, M. D., Hehlert, P., Huang, H. ... & Göpfert, M. C. (2023). The novel pyridazine pyrazolecarboxamide insecticide dimpropyridaz inhibits chordotonal organ function upstream of TRPV channels. Pest Management Science, 79(5), 1635-1649.

Specification for Malathion Emulsifiable concentrates IS 2567: 1978, Available from: https://standardsbis.bsbedge.com/BIS_SearchStandard.aspx?Standard_Number=2567&id=0. (Accessed on 18.11.2023).

Standard of Fipronil Suspension concentrate Specification, IS 16145: 2013, Available fromhttps://standardsbis.bsbedge.com/BIS_SearchStandard.aspx?Standard_Number=16145&id=0. (Accessed on 11.11.2023).

Stevens, W. L. (1948). A table for the calculation of working probits and weights in probit analysis", Biometrika.

Subbarami reddy, Reddy, A. S. and Venkateswara rao., 1987, Effect of synthetic pyrethroids on Leucinodes orbonalis Guen. And Myzus persicae Sulzer on Brinjal-Proc. Nat. Symp. (Ed.) S. Jayaraj, Tamil Nadu Agric. Univ., Coimbatore pp: 44-49.

Sun, R., Liu, C., Zhang, H., & Wang, Q. (2019). Benzoylurea chitin synthesis inhibitors. Journal of agricultural and food chemistry, 63(31), 6847-6865.

Sutapa, Chowdhury. "Investigation on the Chemistry and Transformation of Clothianidin – a New Neonicotinoid Insecticide.", Bidhan Chandra Krishi Viswavidyalaya University (India), 2020.

Takagi, K., Hamaguchi, H., Nishimatsu, T., & Konno, T. (2007). Discovery of metaflumizone, a novel semicarbazone insecticide. Veterinary parasitology, 150(3), 177-181.

Takagi, K., Ohtani, T., Nishida, T., Hamaguchi, H., Nishimatsu, T., Kanaoka, A., 1996. Hydrazinecarboxymide derivatives, a process for production thereof and uses thereof. US Patent 5,543,573

Takagi, K., Wada, Y., Yamaguchi, R., 2005. Ant controllers and method for application thereof. European Patent EP 1191847B1.

Testud, F. (2014). Insecticides néonicotinoïdes. EMC-Pathologie professionnelle et de l'environnement. EMC Toxicol Pathol. doi, 10, S1877-7856.

Thammali Hemadri, Hanchinal, S.G., Prabhuraj, A., Pavankumar K., Somashekhar Gaddanakeri, Shrikant, Meghana V Kodler, Archana, P., Sangeeta and Prakash A Jituri (2024). Pesticide formulation testing: importance and Protocols. Indian Entomologist. 5(1): 74-80, 2024.

Training manual: Training programme on estimation of pesticide residues in cereals and vegetables. Edited by Dr. Vandana Tripathy, PC-cell-All India network project on pesticide residues and other contaminants, ICAR-IARI, New delhi-110012.

Tsipi, D., Botitsi, H., & Economou, A. (Eds.). (2015). Mass spectrometry for the analysis of pesticide residues and their metabolites. John Wiley & Sons.

Tuzimski, T., & Sherma, J. (Eds.). (2015). High performance liquid chromatography in pesticide residue analysis. Crc Press.

Verma, A. 1981. Pest complex in ratoons in the north Indian sub-tropical conditions and how to tackle it. Proc. Natn. Semi. Ratoon Mgmt., Lucknow p.92-95.

Wang, Y., Chen, C., Qian, Y., Zhao, X., Wang, Q., & Kong, X. (2015). Toxicity of mixtures of λ-cyhalothrin, imidacloprid and cadmium on the earthworm Eisenia fetida by combination index (CI)-isobologram method. Ecotoxicology and Environmental Safety, 111, 242-247.

Woods, T. S. (2003). Pesticide formulations. Encyclopedia of Agrochemicals, Wiely.

www.syngenta-us.com.

Yamaguchi, R., Nishimatsu, T., Takagi, K., 2005. Ectoparasitic insec pest controllers for animals and their usage. US Patent 6903237B2.

Yan, Q., Lu, X., Zhang, Z., Jin, Q., Gao, R., Li, L., & Wang, H. (2023). Synthesis, Bioactivity and Molecular Docking of Nereistoxin Derivatives Containing Phosphonate. Molecules, 28(12), 4846.

Yang, T. (2013). Cytochrome P450s and Their Roles in Insecticide Resistance in the Mosquito, Culex quinquefasciatus (Doctoral dissertation, Auburn University).

Yu Y, Shen G, Zhu H, and Lu Y. 2010. Imidacloprid-induced hormesis on the fecundity and juvenile hormone levels of the green peach aphid Myzus persicae (Sulzer). Pestic Biochem Phys 98:38-242.